CRITI

"I found John Spencer's "The Eternal Law" to be enormously refreshing; for here we have someone willing to speak out forcefully in favour of Platonic ideals lying at the roots of modern science."

Sir Roger Penrose (OM FRS)
Emeritus Rouse Ball Professor of Mathematics,
Mathematical Institute, University of Oxford,
author of *The Road to Reality*

"This book gives a spirited defence of the continuing importance of Platonic philosophy. The author shows how Pythagorean and Platonic ideas influenced the thinking of the creators of modern science, from Kepler in the seventeenth century to the founders of quantum mechanics in the twentieth."

Dr. Stephen M. Barr
Professor of Particle Physics, University of Delaware,
author of *Modern Physics and Ancient Faith*

"The new physics has revealed a congruence of the mental and the physical and of the spiritual and the rational, which is so difficult to accept that it amounts to a metamorphosis of our consciousness. In this process John Spencer's book is a much needed guide that explains challenging concepts of physics and philosophy in a competent, inspired and easily understandable way. "The Eternal Law" will be of immense value to teachers in both the sciences and the humanities, and to everybody who wants to live with an enlightened understanding of the world."

Dr. Lothar Schäfer
Distinguished Professor of Physical Chemistry (emeritus), University of Arkansas, Fayetteville, author of
In Search of Divine Reality: Science as a Source of Inspiration

"This is an exciting and thought-provoking book from a young scholar whose base is in the philosophical foundations of quantum physics, but who has gone on as well to master the whole philosophical tradition, particularly that of Platonism, to bring us a wide-ranging study of the parameters of reality. All of us can learn from it."

Dr. John Dillon
Regius Professor of Greek (Emeritus), Trinity College Dublin, author of *The Middle Platonists: 80 B.C. to A.D. 220*

"Spencer takes the reader on a fascinating journey through the terrain where quantum physics, mysticism and philosophy meet. Profound challenges are addressed in highly accessible ways, and Spencer guides the reader through the complexities of his topic with remarkable clarity. This is a work for the interested general reader who wants to understand how ancient philosophical ideas resonate with the latest discoveries in science. Spencer's vision enlivens the human quest to find meaning and purpose in our world. A highly accessible, informative and profoundly stimulating book."

Dr. Brian "Les" Lancaster
Professor of Transpersonal Psychology at Liverpool John Moores University (retired), former Chair of the Transpersonal Psychology Section of the British Psychological Society, President of the International Transpersonal Association, author of *Approaches to Consciousness: the Marriage of Science and Mysticism*

"I can honestly say "I love this book". It appealed to both the scientist and the philosopher in me, and should be compulsory reading for anyone who thinks they know anything."

Dr. Kathryn Hopkins
Ministry of Justice, UK

"You need not agree with everything Spencer says in order to benefit from this readable, lively and thought-provoking book."

Dr. John C. Taylor
Professor of Mathematical Physics, Cambridge University, author of *Hidden Unity in Nature's Laws*

"John Spencer discusses complex problems with an engaging, easy-to-read clarity that has significantly improved the clarity of my own thinking, about philosophy generally and about philosophy of science in particular. Spencer's incisive metaphysical reasoning has also illuminated for me why I always feel gaslighted in exactly the same way by the seemingly antithetical agendas of positivist materialism and postmodern relativism. Spencer shows how the expositors of these wildly incompatible ways of thinking have all drunk the same philosophical Kool-Aid—called "anti-realism". So it must be their anti-realism that makes me feel crazy because *in reality*—as Spencer demonstrates—anti-realism is logically incoherent in a way that does violence to the human spirit. Spencer's prescription for this cultural illness? Unembarrassed rigorous metaphysical thinking. Sounds difficult, but the way Spencer does it, metaphysical thinking has the refreshing clarity of common sense (and elucidates the philosophical foundations and implications of quantum physics too!)."

Dr. Elio Frattaroli
Psychiatrist, Psychoanalyst, and author of *Healing the Soul in the Age of the Brain: Becoming Conscious in an Unconscious World*

"The poison of anti-realism has been dripped into the ear of modern humanity for too long, bringing with it a curious fever of body and lethargy of mind: 'The Eternal Law' is an antidote of the finest potency. The author exposes the fatal contradiction at the heart of anti-realism with such ruthless clarity that I strongly suspect anyone trying to defend it after this will only succeed in further demonstrating its absurdity."

Tim Addey
Chairman of the Prometheus Trust,
author of *The Seven Myths of the Soul*

"Reading his work is, at first, like listening-in on conversations with the major thinkers of philosophy and Quantum physics discussing the significance of this new and dramatic turn in science and then coming to realize that you are being lead to the surprising realization that, indeed, every recourse to systems of philosophy rather than the Platonic system will lack the scope and precision necessary to understand the depth of this profound Quantum turn in science."

Dr. Pierre Grimes
Professor of Philosophy, Golden West College (retired), Korean Buddhist Dharma Successor, co-author of *Philosophical Midwifery: A New Paradigm for Understanding Human Problems With Its Validation*

John Spencer's THE ETERNAL LAW offers a spirited challenge to the claims and widespread influence of anti-realism, insisting on key connections between modern physics and ancient philosophy, particularly Platonism, while reminding us that the theoretical and metaphysical foundations of physics have always aimed at unity, simplicity and beauty. Accessible to non-specialists, this richly interdisciplinary book remains highly relevant to philosophers, classicists, physicists, and historians of science."

Dr. John E. MacKinnon
Department of Philosophy, Saint Mary's University

"An important contribution to the philosophical and scientific exploration of the concept of reality."

Dr. Mario Beauregard
Department of Psychology and Neuroscience Research Center, Université de Montréal, author of *Brain Wars*

"The Eternal Law is comprehensible to a general reader who is not well read in philosophy or in the subtle nuances of cutting edge theoretical physics. As an experimental scientist, I found it to be a real eye-opener regarding the misunderstandings and misinterpretations of modern physics that persist among various academic disciplines."

Dr. Nicholas P. Blanchard
Research Engineer, Institut Lumière Matière
(UCBL-CNRS), Université de Lyon

This robust defence of ancient thought, objective truth and metaphysical realism also provides along the way a most valuable education in philosophy, philosophy of science and the history of science. Whether or not they agree with the author's conclusions, specialists and non-specialists alike will be rewarded by engaging with this important and wonderfully interdisciplinary book.

Dr. Stephen D Snobelen
History of Science and Technology Programme, University of King's College

"Whatever else *The Eternal Law* may bring to readers, it will make them think, ponder and contemplate the source of *gnosis*. Spencer's book is situated at the intersection of science, ancient philosophy and spirituality. It serves as an introduction to these three domains as well as a discourse on them. Spencer conjoins Platonism and Neo-Platonism with quantum physics in order to demonstrate that many of the great scientific luminaries throughout history and up through Einstein, Bohr, Gödel, Heisenberg, Schrodinger and others were deeply indebted to Plato's *One Truth* as the explanatory hierarchy underlying all being.

Spencer has a remarkable command of three vast disciplines which he delivers to the reader with stunning clarity. His numerous examples are at once elegantly simple and informative, often humorous. The gentle jibes at contemporary analytical philosophers double as a critique of philosophy as it is practiced and taught today. His analysis of the realism/anti-realism debate is thorough and subtle.

Of lesser, but nonetheless significant importance, I could mention his overview of Western philosophy in general along with interdisciplinary connections to arts, literature and Eastern thought. We hear echoes of these interconnections in the recent voices of Krishnamurti, the teacher/philosopher, and David Bohm, the physicist.

As a linguist, I should note a very important attribute of Spencer's work. The language of his text is formidable: He does not banter words about but reduces them to their very *semes* thus avoiding ambiguity surrounding definitions, terms, laws and theories...Spencer's book inspires us to extend the boundaries, or rather to go beyond the frontiers of our current knowledge in an effort to ameliorate the human condition. For this alone, he is to be lauded and thanked. I am grateful for this journey he has taken me on and I encourage readers to follow the same path."

Dr. James W. Brown
Retired Professor of French Studies, Dalhousie University

The following evaluations are based on John's original PhD thesis, which has formed the basis of The Eternal Law.

"Essential reading for anyone engaged with modern science or late antique philosophy - and also anyone who wishes a better personal understanding of reality."

Professor Stephen R. L. Clark
Department of Philosophy University of Liverpool

"Reading this work helped me to understand physics better."

Dr. Dominic Dickson
Department of Physics University of Liverpool

"Highly distinctive...original...intellectually committed...should do much to shake and dislodge these complacently orthodox habits of mind [in academic philosophy]."

Professor Christopher Norris
Cardiff School of English, Communication and Philosophy, Cardiff University

"Will be of direct interest to scientists who want to engage with fundamental issues, as well as to philosophers...original and incisive, and notable for its clear presentation of difficult issues."

Dr. Peter Rowlands
Department of Physics University of Liverpool

"Highly provocative...original...sure to stir up very lively debate."

Dr. David Leech
(Formerly at the) Ian Ramsey Centre for Science and Religion, University of Oxford

The Eternal Law

Ancient Greek Philosophy
Modern Physics
and Ultimate Reality

JOHN H SPENCER

Param Media
Publishing

PARAM MEDIA PUBLISHING
Vancouver, BC
© 2012 by Param Media Publishing

All rights reserved. Published 2012
Printed in Canada

ISBN: 978-0-9868769-0-5

Cover design by Param Media

TheEternalLaw.com
ParamMedia.com

For my wife, Ryoko, without whom this book would not have come into existence.

ACKNOWLEDGMENTS

If everything in the universe is intrinsically interconnected, then logically I should either thank no one and nothing, or everyone and everything. But logic and practicality are two different things, and I feel such deep gratitude toward so many people that it is impractical to try to name everyone. At least I can say that the following very partial list of people have had the most direct and significant impact, and I have been extremely fortunate to have received their continuing support: Jim Brown, Pierre Grimes, Stephen Clark, Peter Rowlands, Tim Addey, David Leech, Juan Tellez, and Andrea Blackie.

CONTENTS

PREFACE	1
ONE: A Panoramic Prelude	6
The most beautiful experience	6
Science, religion, and fragmented thinking	9
Visionary thinking	12
Who are we?	15
An ordered universe	17
Trans-logical trust	19
An unreal earth?	21
Scientia	25
Unified knowledge	27
TWO: Clarifying Key Concepts	30
An ancient hierarchy	30
Realism and antirealism (or, where is Mexico?)	31
Quirky quantum theory	35
What is Platonism, and what on earth does it have to do with physics?	42
Metaphysics, mysticism, and mind journeys	52
THREE: The Curtain Falls for Analytic Philosophy	59
Is philosophy dead (again)?	60
A brief (but revealing) history	61
Staying out of touch (five false assumptions)	66
Analyzing analysis	76
Prying philosophers	78
FOUR: Metaphysics and Physics	82
Beauty and symmetry	82
Distinct but inseparable	85
Three categories of metaphysics	90
Uncertain uncertainty	95
Coming full circle	96
FIVE: Faith in Physics	98
What is faith?	98
Rationally inconsistent physicists	103
Faith-filled physicists	109
Fanatical physicists	117
SIX: Expanding Realism	120
Outlining the struggle	120
What is reality?	121
More fun challenges	124

We are all realists ... 130
As good as it gets? ... 138

SEVEN: Imploding Antirealism ... 140
What, really, is antirealism? ... 140
Apparent antirealists ... 144
Diehard antirealists ... 152
Amorphous antirealism (or, an anti-antirealist antirealist closet realist) ... 156
Where are my keys? ... 161

EIGHT: The Laws of Nature and the Nature of its Laws ... 163
Aristotelian abstractionism: clever isn't good enough ... 163
Ducks and rabbits (and the Mona Lisa) ... 171
The big bang computer (or, nature is one smart cookie) ... 174
However you slice it (it's still one pie) ... 179

NINE: Unity and Ultimate Reality ... 185
Disunity and disengagement from reality ... 185
Systematic unity ... 188
Soul in science and science in soul ... 192
Breathing eternity ... 201
New frontiers ... 205

APPENDIX A ... 207

APPENDIX B ... 219

APPENDIX C ... 230

APPENDIX D ... 232

NOTES ... 236

WORKS CITED ... 289

INDEX ... 317

EDITOR'S NOTE

It is generally considered to be the convention to recommence footnote numbering at the beginning of each new chapter. However, we have found it to be easier to navigate between the text and the footnotes by simply numbering all of the footnotes consecutively.

PREFACE

You may be shocked to discover that many professors have been teaching their students that there is no such thing as objective truth or reality. If they are right, this spells the end of not only religion, but also the end of science and rationality. The very notions of objective truth and reality are often misrepresented to be relics of a naïve bygone era, despite the fact that everybody, rich or poor, and regardless of social or cultural background, believes that at least some things are objectively true or real. Is it even possible for anyone to believe that it is really true that there is no such thing as truth?

There are many subtle distinctions and clarifications that are needed in order to get a clearer picture of what exactly is meant by the notions of 'truth' and 'reality', but the fact is that we all believe that there is truth when the issue becomes personal. When the bank demands your mortgage payment, will it do you any good to tell them that there is no such thing as reality and that mathematics is nothing more than a subjective fiction? Would anyone want to say that there is no objective truth if they were on trial for a murder they did not actually commit, or would they want the jury to recognize the *truth* of their innocence? All of this probably seems so obvious that you have to wonder why we need to bother discussing it. I share such sentiments myself. Why, indeed, do so many professors teach their students, the future of our civilization, that there is no such thing as objective truth?

A very short and partial answer to this question is that we want to free ourselves from fundamentalism. Fundamentalists, whether they be religious, ideological, or any other type, say that there *is* real truth but that they, and *only* they, have access to it, and that if you fail to follow them you will be punished in one way or another. Thus, if we can deny that there is objective truth, we can deny power to fundamentalists. But if there is no objective truth, then all knowledge claims are reduced to nothing more than mere opinion. In fact, the claim that there is no objective truth would be just as true (and, consequently, just as false) as any fundamentalist belief, and, in that case, there is no rational defense against fundamentalism. If there were no objective truth, even the claim that you need oxygen would not necessarily be true, so we should be able to deprive you of it indefinitely without any harm coming to you.

The view that there is no objective truth easily leads to the view that there is no objective reality. If there is no objective reality, you should be able to create any reality that you want, and nothing would have any existence until you thought of it or experienced it. So, do you believe that your car exists even when you do not look at it? If you answer 'yes', you are a *realist*. If you answer 'no', you are an *antirealist*, which is to be a denier of objective reality and truth. Regardless of your answer, I am quite sure that if you go outside tomorrow morning and discover that your car has been stolen, you are not likely to tell your insurance company that your car just disappeared because you stopped looking at it. In this sense, at least, we are all realists.

The question of whether or not there really is objective truth or reality is at the heart of our personal and collective belief systems. Our beliefs, in turn, directly shape our daily behavior, which in turn shapes our society. These questions and issues are of profound importance for each and every one of us, and so it is in our own best interests to begin to understand how antirealism has been spreading, and especially how we ourselves have probably been caught by it at one time or another.

It is quite easy for various professors in the Arts and Humanities to endorse antirealism to one degree or another, since there do not seem to be any immediate and obvious negative repercussions of denying objective reality or truth. I can write the words 'reality is an illusion' or 'all truth is a fabrication' and nothing bad seems to happen. But if an engineer gets a calculation wrong, a bridge or a building may collapse. In the sciences, at least, one may expect to find a serious pursuit of objective truth. How shocking, then, to discover that the greatest support for antirealism has actually been coming from the foundational science of all the sciences. Modern physics, the common story now goes, endorses antirealism.

The irony is that pioneering physicists such as Albert Einstein could not have made such great scientific achievements without the belief that there is objective reality and that we are capable of discovering certain aspects of it. But it is not merely that Einstein needed to hold this belief in order to make his scientific discoveries, but rather that his belief in objective reality must actually be true. The sciences have often been seen as the protector against whimsical fashions, opinions, and prejudice, but physics is now being used to deny that there is any objective truth or reality at all, which is to destroy the very foundation of all of science. There have been various contemporary philosophers, sociologists, and other theorists who have sought to diminish the power of the sciences, and if the sciences can be used to dethrone themselves, then such people will happily propagate this idea. Some scientists do falsely believe that they hold the only key to unlock the mysteries of reality, but such fundamentalism cannot be successfully overcome with antirealism.

So, we find ourselves in a terrible muddle. On one side, there are those who claim that there is no truth about anything, and thereby wrongly assume that this belief brings them genuine freedom (or even mystical power). By believing that there is no objective truth, they assume that they can never be wrong, and so they are free to do or be anything they like. They seem to forget, however, that they are not free to jump from a thirty-story building without falling. On the other side, there are those who believe that there is objective truth and reality, but these people often do not want to face the fact that this belief necessarily has metaphysical and mystical implications. After all, we cannot physically locate objective truth, and if it has no physicality, it must be nonphysical, which is immediately to admit some sort of mystical position.

It is here that the essence of this book begins to shine, where some of the vital connections between ancient Greek philosophy and modern physics are revealed as we explore what it means for something to be real and yet not physical. Think about the fact that science is only possible if the universe behaves in predictable ways. The universe will only behave in predictable ways, however, if there is some sort of order, rather than total random chaos. If everything were total random chaos, we could not make reasonable, accurate predictions, which would make science impossible. Since science is possible, there must be order in the universe. But we then need to ask what this order is and how it is possible. At the very least, we know that we find order in the mathematical laws of physics, where such constant laws enable us to make extraordinarily accurate predictions about seemingly unrelated phenomena, such as falling apples and the rising and falling of the tides.

The most significant scientific progress occurs when we discover an underlying mathematical law that unites or supersedes previously discovered laws, explaining all that the previous laws explained and even more. If we follow this reasoning process, where underlying mathematical laws become more abstract, more unifying, and therefore simpler, we must logically arrive at one ultimately unifying, simple law, from which all others are derived. This ultimate mathematical law must be real, nonphysical, and unchanging, which is to say that it is eternal. This eternal law, in turn, requires further explanation, leading us through higher metaphysical notions, such as absolute beauty, all the way to ultimate reality, which is pure, absolute One.

Many of the arguments I offer in support of the reality of this eternal mathematical law have their roots in ancient philosophy, specifically in the tradition known as Platonism. This philosophical tradition has encountered much prejudice in many contemporary philosophy departments despite the fact that so many of the most significant founders of modern science throughout history have either implicitly or explicitly endorsed its essential aspects. Roger Penrose, a distinguished mathematician, theoretical physicist, and former collaborator of Stephen Hawking, is a preeminent example of a contemporary scientific pioneer who sees the importance of Platonism. 'It is undoubtedly the case', writes Penrose, 'that the more deeply we probe Nature's secrets, the more profoundly we are driven into Plato's world of mathematical ideals as we seek our understanding'.[1]

Platonism presupposes objective reality, and it demands the highest levels of abstract rational thought and intuition, which are all quite unfashionable notions in many academic circles. The word 'philosophy' literally means 'the love of wisdom' but, unfortunately, many contemporary philosophers have been trying to convince us that there is no such thing as truth at all and that the search for wisdom is nothing but old-fashioned nonsense. Such misguided teachings are not at all helpful in the face of the genuine problems we face.

Contrary to such misguided academic fashions, the aim of the Platonic tradition is not only to help us understand the universe and develop the sciences, but also to help us understand ourselves and live better lives. It aims for a complete and totally integrated understanding of all aspects of reality, so far as we are able to achieve. Given such powerful and noble goals, why is there so much resistance to embracing this tradition in our universities, precisely where it is most urgently needed? The irony, of course, is that the very concept of academia finds its origins in Plato's school, the Academy.

Our universities are at the beginning stages of a major crisis, which is going to necessitate considerable societal changes. There are a growing number of people within the university structures who are aware of these problems to one degree or another, but not many people have any idea what to do about them. Universities have traditionally been considered the vanguards of knowledge, but they are losing this status. Remember that there was once great resistance to translating the Bible into the common tongue, precisely in order for the Church to maintain control of its followers; if the people could not read the Word of God themselves, they would have to rely upon the priests to do it for them. There is a similar transformation happening right now, where the Internet and new technologies are making knowledge accessible in ways virtually unimaginable by previous generations. If whatever you want to know is likely to be found online, what function do the universities serve? They are not preparing students for a life of contemplation, and they tend not to be very good at preparing them for specific careers either, especially those courses in the Arts and Humanities. There are exceptions, of course, but just ask your local barista what university degree they graduated with.

I am not denigrating the importance of getting an Arts degree, or the value of working as a barista, but you do not need to study for four years and spend a hundred thousand dollars to learn how to serve coffee. A degree in the Arts should ideally foster genuine creativity, which is a vital aspect of any visionary pursuit, whether you are a painter, scientist, or entrepreneur. The problem is not that an Arts degree has no practical or intrinsic value; rather, it is that our overly profit-driven, materialistic culture is too often unable to recognize it. Indeed, many universities seem to be more concerned with acquiring greater and greater amounts of money, often at the expense of the quality of education they offer.

Profit is good and necessary. There is nothing wrong with making money. But if maximizing immediate, short-term profit is our main goal, we are likely to adversely affect our potential for even greater profit in the future, because we are not going to have a full understanding of all the integrated aspects of our business. To be rich tomorrow requires intelligent planning today. But obviously, being rich is not sufficient, either. We can become as wealthy as we can possibly imagine, but if we poison our water and air in the process, then we are behaving quite stupidly. It would be like having a fifty million dollar mansion but being unable to live in it because the air within its walls makes us sick. What is far more important, and what is mostly ignored in our culture today, is how badly we poison not only our physical environment and our own bodies, but especially our own minds. We would never put sand and rocks into the gas tank of our car, but many of us, from the richest to the poorest, nonchalantly fill our bodies with unhealthy foods, and we also uncritically accept false and harmful beliefs, which then harm us right at our core.[2] It often seems easier to give more care and attention to our material possessions than to ourselves.

It is becoming increasingly important for us to take responsibility for our own continual education, which is one of the reasons I have written this book particularly with non-specialists in mind. That does not mean that the ideas are easy—we are talking about philosophy, physics, and ultimate reality, after all—but it does mean that the writing style makes this book accessible for those who do not have a specialized background in such topics. I have also made reference to an extensive number of philosophers, physicists, and other theorists, from antiquity to the present, providing a rich panoramic view of the main issues. Those who specialize in one or more areas of research that I have drawn from will also find this book to be highly relevant to important questions and topics in their respective disciplines.

By acquiring deeper understanding and knowledge of the topics discussed in this book, you will have greater potential for attaining higher forms of power. There are many different manifestations of power, but the deepest emanates from absolute beauty, simplicity, and unity, the same metaphysical notions that are at the foundation of modern physics. To reach our highest potential, we must seek, embrace, and implement these genuine aspects of ultimate reality.

While further developing specialized areas of research, we especially need to find ways to integrate, synthesize, and creatively transcend traditional divisions, whether in the arts, science, or business. From Plato to Einstein to contemporary business visionaries, the greatest minds throughout history have known that an appropriate balance between rationality and intuition is necessary in order to achieve this cohesive understanding. As this book shows, such understanding is accessible to us all.

ONE: A Panoramic Prelude

The most beautiful experience

Every physical aspect of the entire universe must obey the one eternal mathematical law. For some people, this claim will appear to be an obvious truth. For others, it will sound like sheer nonsense. The truth is, it is neither nonsense nor obvious. Whatever our beliefs about the nature of reality, whether we are atheistic, agnostic, religious, or otherwise spiritually inclined, we cannot physically circumvent the one eternal mathematical law. It does not matter whether you are a scientific genius, a world champion athlete, a movie star, a lawyer, a teacher, or a pizza cook; the law does not discriminate in unfolding its power. The same law applies equally to all of us. It also applies equally to everything else, from amoebas to galaxies to black holes and to every single aspect of the universe that has ever existed, exists now, or ever could exist.

All of the laws of physics are partial reflections of the one eternal mathematical law, which is a kind of super-law, the foundation of all the mathematical laws of the universe. These ideal laws are actually more fundamental and more real than the physical world. A thoroughly rational investigation of this truth will reveal that the one eternal mathematical law depends upon higher metaphysical principles, such as symmetry and beauty. Unfortunately, many of us would have been taught in grade school, and even in university, that physics is simply about measuring things, which seems to have nothing to do with philosophy, let alone anything ancient or beautiful. But, as we shall see, this parochial view of physics distorts and misrepresents some of the most fundamental aspects of the entire discipline.

There is another way of misrepresenting physics, which has become especially popular in recent years. If you have been following any of the relevant New Age literature or films, you may have come to believe that physics has nothing to do with measuring anything, but instead that it has proven Eastern mysticism to be true. According to this viewpoint, quantum physics is supposed to be magical, giving you the power to create any reality you want just by wishing it to be so, which is an unfortunate misrepresentation and trivialization of the powerful ancient traditions of the East. It is not that there is nothing mysterious about modern physics, or that there are no interesting connections to be found between this discipline and the Eastern traditions; it is that we have no idea what we mean when we say that we 'create reality'. If anyone really believes that they can create any reality they want without any limitations whatsoever, then they are welcome to snap their fingers and instantly end world poverty, or transform into a dinosaur.

Perhaps due to the understandable fear of slipping into irrationality and undermining scientific progress, many philosophers and physicists have turned away from what they believe to be *mysterious*. If we must believe that we create reality and that rational thought is unnecessary or even a deception, then we would be wise to reject mysticism. But genuine mysticism is neither irrational nor does it entail that we create reality. In fact, quite the opposite is true. One of the formal meanings of mysticism is the belief that we can discover mathematical proportions everywhere,[3] which has been fundamental to the inception and continual development of physics. Genuine mysticism requires total rationality, and it demands that we try to discover reality as it really is and then transform ourselves, or allow ourselves to be transformed, accordingly. Modern physics does have deeply mystical and metaphysical implications, but such implications have nothing to do with irrationality or the belief that we can create any reality we want just by wishing it instantly into physical existence.

It is generally considered more respectable in academic circles to talk about science and religion, as opposed to science and mysticism. But whichever view one may hold, we would still have to ask what we mean by such terms as 'God', 'reality', 'religion', 'mysticism', and 'science', and by attempting to answer such questions we are immediately entering into philosophical territory. Can we even coherently use the word 'science'? After all, there is no such thing as *the* scientific method. There are different technical methodologies for the varying sciences, and even within the same scientific discipline there are different conceptions of and approaches to scientific research. Those who work at the theoretical foundations of physics will be thinking about and approaching their research much differently than an experimental biologist aiming for some kind of practical application. Thus, it makes more sense to speak of 'the sciences' than to speak of 'science'. Yet it is a common and convenient practice to use the word 'science' (I usually do so as well), so long as we understand that there are different sciences, in a similar way that there are different religions. This does not mean that there is no underlying unity in the sciences, or in reality. That underlying unity, however, requires a very profound philosophical investigation.

If you believe that philosophy has nothing to do with science, then what you are saying is that rational thinking is not necessary (or even possible) for scientists. You would also have to admit that you are in disagreement with many of the greatest pioneering theoretical physicists throughout history. Nevertheless, there are many important examples of basic philosophical questions that are left unanswered and usually unasked by physicists. Such questions are so basic that their answers are often taken for granted, even though physicists have little or no idea how to provide proof for these assumptions. These kinds of questions remain ignored, allowing us to retain the illusion that we know what we are talking about.

Consider the following seemingly simple questions. What is a 'law of physics'? Are the laws of physics real? If they are real, then why can't you touch them or perceive them like other physical objects? If you say that they are not real, then why can't you make up your own law of physics in any way you please, as if you were writing a novel? When attempting to answer these challenging questions, we will soon find ourselves entering into the realm of ancient philosophy. Any philosopher or physicist who says that such questions are meaningless will be admitting ignorance about some of the most important questions underpinning the whole discipline of physics. Even the claim that philosophy is irrelevant to physics is itself based upon philosophical assumptions. Nevertheless, many academics still claim that the laws of physics are nothing but 'useful' fictions and that philosophy is totally irrelevant to physics. Those who hold such views are usually known as *positivists*. Interestingly, the common academic myth is that quantum theory endorses positivism even though, as we shall discover, the leading pioneers of quantum theory actually rejected it.

Once we realize that philosophical understanding is a vital part of physics, we have opened up the way to making arguments about the nature of ultimate reality based upon our knowledge of physics. In other words, we can in fact make theological or atheistical arguments based upon physics. It is true in one sense that purely observational aspects of physics cannot tell us anything about ultimate reality, but when we consider the underlying assumptions that make physics possible, then we can, in fact, make rational claims about ultimate reality while drawing support from physics.

In any case, it is not even possible to make a purely observational statement in physics, or about anything at all, without having underlying (or unconsciously held) philosophical assumptions about the nature of reality. Even asking the question, 'What is physics?' is to enter immediately into the realm of philosophy. In other words, physics cannot escape philosophy. When we are engaged only in the 'practical' physics that ignores the underlying philosophical assumptions of the entire discipline, we have to pretend that we know what we are talking about and ignore the deeper unknown. But without those brave pioneering theoretical physicists who were forced to enter into the unknown mystical and metaphysical depths underpinning modern physics, we would not have our laptops, mobile phones, or any of the high-tech wizardry of today.

Defining and fully clarifying the terms 'metaphysics' and 'mysticism' is a task that could itself take up an entire book, or even several books. I offer more detailed explanations in the next chapter and in relevant places throughout the book, but here we need only grasp the following clarifications. *Metaphysics* is a technical term in philosophy that generally refers to our attempt to offer a comprehensive rational account of the most fundamental aspects of all reality, while *mysticism* is concerned with spiritual experiences and what is beyond the limits of logic and science, including the assumption that the physical universe is in some way mathematical. There is much more to explore concerning these topics, but for now let us simply recognize that we cannot claim to be scientifically sophisticated while at the same time ignoring the foundational metaphysical beliefs of so many of the most important theoretical pioneers of physics. Similarly, we cannot pretend to be saving science from the clutches of the mysterious when many of the greatest pioneering physicists throughout history have been openly mystical.

Clearly, however, we should not jump to the conclusion that just because a brilliant physicist endorses some form of mysticism that, therefore, every mystical claim made by any so-called guru is true. We must combine extensive study with independent arguments and our own experience to discover the truth for ourselves.[4] Nevertheless, those who dismiss metaphysics and mysticism must admit that they are rejecting the work of some of the most brilliant, creative pioneering physicists in history.

The unknown can be extraordinarily frightening, but our psychological fear, while understandable, must not deter us from our goal of seeking what is true. So long as we remain mesmerized by the illusion of comfort within the bounds of what we believe we know, we will never be able to fathom the depths of mind explored by great scientific pioneers like Einstein, who writes that 'the most beautiful experience we can have is the mysterious. It is the fundamental emotion which stands at the cradle of true art and true science. Whoever does not know it and can no longer wonder, no longer marvel, is as good as dead, and his eyes are dimmed'.[5] Einstein is here telling us that 'the mysterious' is the most beautiful experience we can have and that 'true art' exists. He is also echoing the famous words of Socrates: 'the unexamined life is not worth living'.[6] How prevalent are such views in our culture today?

Science, religion, and fragmented thinking

Much of the content of this book will be relevant to academic research in various fields, but I have especially turned my attention toward the intellectually and spiritually curious non-specialists. Quite understandably, there is a growing interest in topics related to science and spirituality, but if we are going to make claims that our preferred spiritual system, religion, or alternative healing methodology has been proven by the 'magic' of quantum physics, then we had better become much more serious about understanding what we are actually talking about.

Despite having a PhD specializing in the philosophy of quantum physics, I have no hesitation in admitting my extensive ignorance and limitations in this field. It is true that the more we know and understand, the more we come to know and understand how little we know and understand. It is only a shallow, insecure mind that pretends to know everything. But none among us is immune to ignorance, and so it makes sense to challenge each other and examine ourselves constantly, as Socrates so long ago advised. Nevertheless, most of us have very busy schedules full of all sorts of practical and recreational activities, so why should we bother thinking about whether or not things exist when we are not looking at them? Why should we care if there really is an eternal mathematical law, or if beauty could possibly have any relation to mathematics and physics?

As will become clear, we all need to be concerned with such questions to some degree. In fact, our implicit assumptions about such topics have a direct, though often unconscious, impact on how we live our daily lives. If you believe that the universe has no order and that it is made up of purely random chaos with no meaning or purpose to anyone's existence, then you are likely to experience life quite differently than someone who holds the opposite view. One thing is for certain: if the universe were not ordered, science would be impossible. But as soon as we realize that science presupposes an ordered universe, there is an immediate fear (or joy) that this fact might have religious implications.

Since some of the most contested, sensitive, and potentially volatile issues in society today center around religion, often manifesting in the misguided form of the 'science vs. religion' debate, it is in our best interests to become as intellectually informed on this topic as possible. Although it is not my main goal to discuss this topic itself, an important benefit of understanding the arguments and information presented in this book is that you will be able to better defend your views, regardless of whether you are spiritual, or an atheist, or a spiritual atheist. Regardless of whether or not you happen to like or agree with the views I present, you will nevertheless be better prepared to engage in rational dialogue with those who oppose you, thereby facilitating the possibility of learning from one another.

At present, many of the discussions (and shouting matches) surrounding the relations between science and religion are full of historical misrepresentations, philosophical errors, and scientific misunderstandings. We all need to slow down and take a long, hard look at our own assumptions. Not everything we believe to be true is actually true, and at least some of what our opponents believe to be true is probably true, so let us learn from one another. I am not, however, advocating a forced tolerance or unnecessary compromise. Sometimes we are wrong, and sometimes our opponents are wrong, but the truth is often a bit different than what either of us believes. We should argue as strongly as possible for our views, and then alter them if we find we are mistaken. Our goal is not so much to defend our own views, but rather to discover truth.

If we want to discover truth, we have to be prepared to question our own assumptions and abandon them when we realize that they are false. For example, if you are an atheist, you will need to admit that many of the most important pioneering theoretical physicists (previously known as 'natural philosophers') in the last several centuries have believed in God or a supreme unifying power. You will also have to acknowledge those metaphysical beliefs that both science and religion share. If you are religious, however, you are going to have to relinquish those particular beliefs that are no longer amenable to contemporary knowledge, and which may never have made any sense in the first place. In other words, the claim that 'the universe is rational' is part of both religious and scientific metaphysical foundations, but Paul's decree in the First Book of Corinthians that 'women should remain silent in the churches'[7] has nothing to do with physics and has no place in society, regardless of whatever interpretative story we may be able to spin from it.[8]

Our society has been dizzied into oblivion by all the truth-spinning that occurs around us every day. It is easy to blame mainstream media, but we usually behave no better in our private lives. We spin stories all the time to justify our own weaknesses and irrational behavior. We, too, ignore important truths about ourselves and avoid challenging conversations about the most meaningful topics. Psychiatrist Sigmund Freud believed that 'groups have never thirsted after truth', and that 'they demand illusions, and cannot do without them'. He goes on to say that groups 'constantly give what is unreal precedence over what is real; they are almost as strongly influenced by what is untrue as by what is true. They have an evident tendency not to distinguish between the two'.[9] If Freud is correct, what does that say about our hope for humanity? After all, if we deny that there is objective truth, or that there is any difference between what is real and what is unreal, how can we possibly make any rational decisions about how to handle our personal, societal, or global problems?

While it is not my goal to try to convince 'groups' to pursue the real over the unreal, it is an unavoidable requirement of scientific and technological development for researchers to at least implicitly be seeking or utilizing certain objectively true aspects of reality. When investors provide money to a high-tech startup developing new software, for example, they are not expecting the company to hire people who have no idea how to turn on a computer; they are expecting them to hire people who are knowledgeable and competent in their respective relevant fields, which is to say that they have the best understanding of certain objective facts and are most likely to have the potential to discover new objective facts that can then be used to develop the new software (and make money for the investors). Two major caveats to take into account here are that:

1. we may believe we have discovered an objective fact when we are actually mistaken; and,
2. the totality of all known objective scientific facts does not constitute the limit of reality.

The totality of all reality will forever remain beyond the potential for complete and final scientific elucidation, which is precisely what makes it possible for our scientific understanding and knowledge to increase. However, the continual increase in specialized areas of scientific knowledge has driven us into more fragmented thinking patterns, at the cost of a holistic understanding of the unity underlying all of this complexity.

Moreover, the latest research in the study of the human brain tells us that continual multitasking, a habit that is common for many of us today, can cause us to 'constantly seek new information', even though we 'have difficulties in concentrating on its content'.[10] There is little value in amassing information if we do not understand it or know what to do with it. To whatever degree multitasking may be partly to blame, it is certainly true that many of us are lacking the ability for intensely focused concentration on one train of thought for an extended period of time. But if we are really serious about confronting the multifaceted problems that we face today, from environmental disasters to cultural and religious clashes, we are going to have to learn (or unlearn) how to think creatively, deeply, and beautifully.

The complexity of our globally interconnected problems is continually increasing, and if we want any hope of being able to deal with these problems, we need to change our thinking and behavior, both personally and collectively, in ways that more closely reflect reality. Genuine change requires clear direction and unwavering determination to see the intended changes put into practice, but in order to achieve sufficient clarity, we need the right insights or intuitions coupled with creative rational thinking.

Visionary thinking

To think deeply is very different than employing superficially clever logical arguments. Just as a mechanized assembly line will produce identical products, so too will mechanized thinking patterns produce identical or very similar results. Such predictability leaves us vulnerable and unable to adapt to new circumstances, as we cannot expect genuine creativity to flow from someone who is only capable of repeating or doing what was said or done before. It is not that we need to reject the past, but rather that we cannot remain chained to those past views (or present views) that are irrelevant or false. We must have the insight to recognize what was good and true in the past, the creative capacity for novel development, and the strength to sustain us during the arduous process of implementation.

To return to the factory analogy: after having a creative insight for a wholly new type of product, we may very well need to build a new type of assembly line in order to produce it. In other words, there is an unavoidable interplay between standardization and the creative force that explodes beyond the known and the predictable. By necessity, our daily life is mostly taken up by performing repetitive or habitual tasks, so much so that we can easily forget the importance of breaking such patterns. There must be pattern-breaking and pattern-recreating, an endless, tension-filled interplay between these apparently opposing yet complimentary forces.

One problem is that those who are stuck in a common pattern easily overlook the importance of the creative aspect, the power of pattern-breaking. We become rooted in habits that bore us to death, and yet we are too scared to jump outside the parameters of what we know and into the unpredictable unknown. After attaining a higher state of mind, we also need to be able to understand and integrate our new insights in ways that are relevant or connected to our former beliefs, which are still necessary for our survival. A classic example can be found in Robert Pirsig's *Zen and the Art of Motorcycle Maintenance*. Pirsig had travelled far into the depths of mind but could not find a way to integrate his profound experiences or communicate them meaningfully to other people. Once he found a way to solidify his insights and capture them in a groundbreaking novel, he was hailed as a genius. Before this publication, he had been subjected to electroconvulsive 'therapy'.

We must respect the potential dangers of the unknown, but we must also persevere and break the chains of common consciousness, allowing us the opportunity to explore unfamiliar territory. Any visionary, whether in science, art, or business, knows this to be true. Bruce Lee, the famous martial artist and movie star, understood this point. His philosophical approach to fighting resonated deeply with the teachings of J. Krishnamurti, the former 'world teacher' of the theosophy movement, and he 'found in the teachings of Krishnamurti the foundation of Jeet Kune Do'.[11] Both Lee and Krishnamurti recognized that we cannot remain chained in set patterns, be they the kata (forms) of the martial artist or the ideologies and habits of thought of the scholar (or of anybody else).

We must dedicate a tremendous amount of time and energy to repetitive training in order to perform any task well, while being prepared to drop the patterns that we have learned so that we may attain genuine creativity and spontaneity. Lee writes that 'the secondhand artist blindly following his sensei or sifu accepts his pattern. As a result, his action and, more importantly, his thinking become mechanical. His responses become automatic, according to set patterns, making him narrow and limited...You can see clearly only when style does not interfere'.[12] We need to learn patterns in order to develop our capacity for spontaneous creative expression, but there is no method for being creative. However, there are methods that can help us break through the internal blocks that stop our natural creativity from flowing more freely.

Genuinely creative, penetrating thinking also requires us to turn our attention to what is good and beautiful. We must be able to sit comfortably with our internal silence in order to open ourselves to receiving higher insights, and we must also be courageously aggressive in order to clear new intellectual frontiers. Such deep thinking requires great effort, energy, patience, persistence, intuition, direct insight, understanding, logical clarity, and the desire to find the truth to the best of our ability. Truth, in the most abstract sense, is beautiful, and so by seeking pure beauty, we are seeking pure truth. By reflecting clearly about our more superficial desires of lesser forms of beauty, we allow ourselves the opportunity to discover our hidden, innate longing for pure beauty.

Our educational system, however, from grade school to university, is usually not very good at fostering the sort of experiential understanding required to develop a significant appreciation of the role of beauty in the sciences, or even in our daily lives. As a result, many of us may need to learn a new way of experiencing and thinking about the world, especially since so much of our popular culture denigrates thinking and rationality, misrepresents the sciences, distorts beauty, trivializes morality, and even denies that there is any such thing as truth.

We need to develop a deeper understanding and appreciation of rationality and science, as well as truth, beauty, and the good. In this way, we can discover many novel possibilities to enrich our lives and attain clearer direction for how to achieve our hidden or suppressed potential. Einstein, for example, was not simply born a genius. He had to work extremely hard to develop his depth of mind, passion, and creative intelligence, and he was especially critical of the educational system he had endured along the way. In contrast to the overemphasis on specialization found in this system, he believed that students must aim to 'acquire a vivid sense of the beautiful and of the morally good'.[13] This noble goal is something that he himself lived by. He writes that 'the ideals which have lighted my way, and time after time have given me new courage to face life cheerfully, have been Kindness, Beauty, and Truth'.[14]

We may more readily accept the idea that beauty plays an important role in art and in various aspects of our practical lives, such as when choosing a romantic partner, painting a picture, planting a flower garden, or browsing for new clothes, but the claim that beauty could have anything to do with physics may seem to be quite far-fetched. Penrose, however, would disagree. He writes that 'Beauty and Truth are intertwined, the beauty of a physical theory acting as a guide to its correctness in relation to the Physical World'.[15] In other words, pursuing beauty in mathematics helps us to discover the most accurate scientific theories and laws about the physical world. But the claim that beauty is foundational to physics is not the only controversial claim I set out to defend. In fact, much of what will be explored in this book may initially seem odd or just plain false when held up against our supposedly sophisticated modern scientific assumptions. Our assumptions about the nature and history of the sciences, however, are often mistaken.

Who are we?

Atheists tend to adhere to the belief that 'faith' is nothing but unscientific nonsense. Richard Dawkins, former Professor for Public Understanding of Science at the University of Oxford and well-known champion of atheism, has even said that faith seems to qualify 'as a kind of mental illness'.[16] It is true that *blind faith* is a symptom of underlying fear and ignorance, but blind faith in a political party, capitalism, communism, a scientific theory, or your partner's absolute fidelity are all equally problematic. The stubborn refusal to question our assumptions about anything in our lives is a form of fundamentalism that requires blind faith. But, obviously, not all religious followers are blind believers, and not all scientists are atheists. In fact, the founders of modern science, the chief ones being Copernicus, Kepler, Galileo, and Newton, were decidedly not atheists, and their spiritual understanding was actually the foundation of their scientific endeavors.

Many modern physicists, we may be surprised to learn, have held similar views. Max Planck, for example, was the originator of quantum theory, the 'new physics' that superseded classical or Newtonian physics by postulating that energy comes in discrete, discontinuous packets or 'quanta'. In 1932, Planck stated that 'anybody who has been seriously engaged in scientific work of any kind realizes that over the entrance to the gates of the temple of science are written the words: Ye *must have faith*. It is the quality which the scientists cannot dispense with'.[17] According to Dawkins, therefore, the founder of quantum physics must have suffered from mental illness. If this is the case, shouldn't more scientists wish to have the same kind of 'mental illness' with which Planck was supposedly afflicted?

Planck, however, is far from being alone in his understanding of the connections between spirituality and the sciences. Erwin Schrödinger, another important pioneering physicist in quantum theory, agreed with psychiatrist Carl Jung that 'all science...is a function of the soul, in which all knowledge is rooted'.[18] Schrödinger would probably have preferred the word 'mind' or 'consciousness', but he does not shy away from Jung's use of the word 'soul'. In either case, they are both referring to the same thing—the very subject of awareness, the nonphysical locus of all experience, the self or being or consciousness that 'I am'.

My position transcends the polarity of two equally misguided viewpoints. At one extreme, we have those who believe they are being thoroughly scientific, when they end up reducing humans to nothing more than rather clumsy machinery with no conscious awareness. At the other extreme are those who claim that the soul is real, despite seldom having any idea of how to go about rationally explaining their position. The soul is beyond direct scientific scrutiny, but defending this view is anything but straightforward or easy. Nevertheless, by being relentlessly rational and while remaining in harmony with the sciences we can conclude that the immaterial 'self', 'I', or 'soul' is real.

In contrast to the standard medical model that reduces you—the 'I' or 'experiencing self' that you are—to nothing more than chemicals sloshing about in the brain, psychiatrist Elio Frattaroli has recently taken the bold step of seeking 'to reclaim the soul itself as the most important element of our existence as individuals, as a society, and as a culture'.[19] Psychiatrists typically aim to treat what they refer to as a chemically imbalanced brain while, as Frattaroli puts it, 'ignoring the experiencing self'.[20] He argues that 'the idea that chemical imbalances cause mental illness is no more scientific today than it was [in Freud's day] in 1897'.[21]

Whether you believe that we are self-aware souls or purely physical things no different in principle than rocks, you can be sure that the sciences are an intimate part of our process of self-discovery. Schrödinger made a similar point when he claimed that the only task of science that really counts is to help us answer the 'one great philosophical question that embraces all others, the one that [third-century Platonic philosopher] Plotinus expressed by his brief [question]...*who are we?*'[22]

All of our scientific endeavors are directly or indirectly leading us back toward a fuller understanding of ourselves, which is why Schrödinger believed that the most important task of science was to help us answer the question of who we really are. The problem is that the sciences can never offer us a complete picture of reality, nor can they take us directly to what it is that we are, to our essential nature. As Planck realized, 'science cannot solve the ultimate mystery of nature. And that is because, in the last analysis, we ourselves are part of nature and therefore part of the mystery that we are trying to solve'.[23]

Although our growing body of scientific knowledge is vital for increasing our potential for self-understanding and exploration, such knowledge is no guarantee that we will attain enlightened self-awareness. There is also no guarantee that we will not use this knowledge for self-destruction. Similarly, however, a profound spiritual experience is not in itself sufficient to give us our technological wonders. Science and spirituality are deeply interrelated, but they often require very different modes of expression and tend to have unique functions.

Spirituality is often misconstrued as being illogical when it makes claims that are beyond strict logical scrutiny. However, any logical system is forever limited by its own starting assumptions and, therefore, cannot be of much help once we begin to seek something deeper than those assumptions. We can never allow ourselves to forgo the importance of logical reasoning and scientific methodology but, equally, we must not allow ourselves to be fooled into believing that the limits of logic and science are the limits of reality. Scientific knowledge is capable of growing precisely because our current knowledge is always limited. Just as there is no logical starting point with which to begin logic, so too is there no scientific method with which to begin science. Both logic and science are ultimately dependent upon insight, intuition, or direct knowing or understanding, coupled with a tremendous amount of hard work. We need to have faith in logic and use logic to understand faith.

An ordered universe

We also need faith that the universe is ordered, and that it will remain ordered. None of our scientific endeavors would be possible at all if the universe were not ordered, a vital point that tends to be overlooked by atheists afraid of facing its possible religious implications. Wolfgang Pauli, another key quantum pioneer, also believed that there is a 'cosmic order independent of our choice and distinct from the world of phenomena'.[24] What he means is that there is some kind of cosmic order that exists independently of the physical universe, and which nevertheless guides the unfolding of all physical reality. When we understand the implications of his words, we have to admit that he does not sound so different from a medieval theologian describing the infinite intelligence and power of God. Einstein, too, was devoted to striving to 'comprehend a portion, be it ever so tiny, of the Reason that manifests itself in nature'.[25] Pauli and Einstein are saying that the cosmic order or universal reason, what the ancient philosophers would have called the *logos*, permeates the universe. In other words, there must be an inherent rationality to the universe that never changes, but always is just what it is.

It can be very difficult for us to grasp the notion of something simply being what it always is, eternally, without ever changing. It may even sound as if such a thing would be very boring indeed; after all, we constantly seek change and are easily bored by familiarity. However, this common way of thinking is not applicable here. Whatever is absolutely incapable of change is actually more powerful and more fundamental than that which can change. In other words, something perfect could never change, lest it become imperfect.

Although I will explain this difficult concept in more detail in later chapters, a common example can help us to grasp the main point. If an architect were to suddenly change the blueprints of a house, it would require a lot of effort on the part of the builders in order to accommodate these changes. Imagine a crazy architect who kept randomly changing the plans for a house on a daily or even hourly basis. The builders would soon quit because they could never keep up with the changes and complete the construction of the house. In other words, the blueprints, the foundational plans, must remain stable, without change, or else the physical building will never come into existence. Of course, relatively minor revisions in the architectural plans may be required after construction has begun; nevertheless, such changes must be minimal and the plans must remain relatively stable.

We must also ask how the logos (the 'cosmic order' or 'Reason') itself is possible, and we need to know how we can even know anything about the logos in the first place.[26] To ask questions about how we can know things, or about the very nature of knowledge itself, is to engage in what philosophers call *epistemology*. When I ask how we can have *epistemic* access to something, be it an idea or an object, I am asking how I can know it, and how I can know that I know it. In this case, we need to ask how we can have epistemic access to the logos, or how we can know it. As we shall discover, the answer inevitably leads us into mystical territory, which is one reason many academics would like to stay away from such questions altogether.

Most physicists simply assume that they can know, and do know, all sorts of things without ever thinking about justifying such knowledge claims. This is quite interesting, especially since physicists expect theoretic or experimental proof to be provided for any claim made about the physical universe. Yet most physicists seldom ask themselves what it means to say that we actually have *proved* something in the first place.

To assume that deeper, underlying metaphysical questions have nothing to do with physics is as misinformed as a taxi driver believing that drilling for oil and producing steel have nothing to do with the job of transporting people around city streets in a motor vehicle. A taxi driver may not need to know such things, but someone certainly does. If nobody knew how to drill for oil or produce steel (or design electric or solar powered batteries), we would not have any taxicabs. Similarly, an experimental physicist may be unconcerned with Plato's story of the creation of the universe in his *Timaeus*, but Werner Heisenberg, another key founder of quantum theory, was profoundly influenced by this dialogue.

In order to help sort out the persistent confusion embedded in the heart of modern physics, we must be prepared to explore questions that may at first seem not to be relevant to science, in the same way that questions about how to increase our capacity to produce steel may not at first seem to be related to driving a taxicab. We must be ready to navigate what may for many of us be unfamiliar philosophical terrain. We must even be willing to accept the possibility that Schrödinger was correct when he declared that the final explanation for our shared material world 'is mystical and metaphysical'.[27]

Although the logos is a nonphysical principle, it is similar in its function in the creation of the physical universe as the architectural plans are in the construction of a house. One essential aspect of the logos is expressed in the physical universe through the eternal mathematical law, similar to what Einstein would have called the abstract 'lawful structure'.[28] This law must be beyond and prior to all physical phenomena, while being responsible for ordering all such phenomena. It is precisely because the universe is rational and ordered that we are able to do science in the first place.

Imagine if there were no cosmic order, no logos or reason-principle guiding the unfolding of the physical universe. If this were the case, we could never make any scientific predictions, because we would never know how the universe would behave from one instant to the next. My book may fall to the floor when I drop it right now, but it may not the next time I let it go. It may float away, or turn into a pink elephant. If there were no inherent rationality, no cosmic order that remained absolutely stable and without change, then we could never rule out the genuine possibility of my book instantly turning into a floating pink elephant.

If we are to understand why these Nobel Prize-winning and knighted theoretical physicists held such views about beauty, cosmic order, reason, and faith, we are necessarily going to be led into the sort of metaphysical and mystical journey that this book undertakes.

Trans-logical trust

The interconnections between the sciences, metaphysics, and mysticism should not be so surprising to us. Even the claim that the universe exists or does not exist is, in the end, purely metaphysical. For example, as soon as we ask ourselves how we know that the universe exists, we will immediately end up in philosophical hot water. Most people would say that they know the universe exists because their senses allow them to observe or experience it in some way, but relying solely upon our senses to tell us anything certain about reality will inevitably lead us to the skeptical position. If our senses have ever deceived us, even once, then we have reason to doubt them at any other time. Since we have all been deceived by our senses at some point, such as when seeing a stick appear to bend when submerged in water, we are justified in doubting our senses at any other point.

But this situation requires further reflection, since I only know that my senses have deceived me about the bent stick once I have removed the stick from the water and realized that it is still straight. In other words, my senses themselves have informed me that my senses had previously deceived me, leaving me in a very puzzling predicament: I need to trust my senses to tell me not to trust my senses. The only way out of the paradox of the senses is to recognize that I am not merely relying upon my senses. I must go beyond the limits of my senses and use metaphysical reasoning, while having rational faith in the laws of physics. Sticks, after all, do not bend by placing them in a pond (at least not as much as they appear to). Similarly, we know that the earth moves even though our senses indicate that it is stationary. But my knowledge of the unbending stick and the moving earth does not arise purely from my senses. Consequently, adherence to strict *empiricism*, the view that all knowledge comes only through our senses, is false.

If we can justifiably doubt our senses, we seem to have justification for doubting that there even is a universe at all. We cannot even appeal to scientific evidence to defend our view that there really is a universe, because we can justifiably doubt any and all scientific data, which we only perceive through our senses. But despite any reasonable doubt we may be able to muster, we all (or at least most of us) still take for granted that the universe really exists. Most of us believe, and all of us certainly act as if we believe, that we are all part of the same physical environment, even though we all experience the world in more or less different ways. I experience the world a bit differently than you do, and a great deal differently than birds do. Nevertheless, you and I and the birds all share the same physical environment. When we try to logically prove this fact, however, we realize that we have taken on a nearly impossible task. When we first start to think deeply about the issue, we realize that it appears to be logically impossible to prove with absolute certainty that anything exists beyond our own experiences.

But we must not be fooled by our superficial cleverness. We must dig deeper and reflect further, for what we know is often beyond what we can logically prove, a claim validated every day by the success of the sciences. We are able to achieve tremendous practical results by believing that the laws of physics, within their own domains, apply equally to a variety of phenomena. However, we forget, or do not even realize, that we cannot logically prove for certain that the laws of physics as they are now will remain the same in the distant future, or even in the next second. We also cannot logically prove that they apply to all the phenomena that we have not yet tested. Nevertheless, most scientists believe that the same laws of physics that were present at the Big Bang are still applicable now, billions of years later, and will continue to be applicable until the end of time (if there *is* an end of time).

Even if the Big Bang theory is totally false, it is still the case that the universe is obeying the laws of physics, and scientists, whether they are atheists, theists, or believe in the tooth fairy, must have rational faith in the reality of these laws. The sciences depend upon logical thought, while at the same time necessarily reaching beyond the constraints of logic. The realization of the distinctions between logic and science leaves us in the seemingly awkward position of admitting that essential aspects of scientific inquiry are, strictly speaking, not logical. However, this does not mean that science is illogical. It means, rather, that certain aspects of scientific inquiry are beyond logic. They are *trans-logical*.

An unreal earth?

Partly because it is so difficult, if not impossible, to prove logically and with absolute certainty the reality of the laws of physics, or even the reality of a material world that exists beyond our own consciousness, some theorists have postulated that perhaps there really is no material world that exists independently of us. The physical world may merely be an illusion, and we can therefore will to happen whatever we desire. This view, that there is no mind-independent or objective reality, no reality outside of our own consciousness, or beyond our particular knowledge limits, is referred to as *antirealism*. The opposing view, known as *realism* (not to be confused with 'political realism'), claims that reality is mind-independent. In other words, reality embraces and yet is beyond our limited perceptions and thoughts. More simply stated, the realist will say that even though I cannot see my wife when she in the next room, she still continues to exist. The antirealist will say that if I cannot see her, then she does *not* exist.

If I really believe that there is no reality outside of my own consciousness, then you do not really exist except when I happen to want to bring you into existence in my own mind. But if you only exist in my mind, then I can do to you whatever I wish. If you are nothing more than one of my dream characters, then there is no real harm done if I rip out your eyeballs or set you on fire. You cannot really suffer, because you do not really exist.

Although it may seem as if a psychopath would be the best example of an antirealist, many highly intelligent and more or less moral people claim to be antirealists. In fact, if you have ever said something like, 'Well, that is *your* truth, not *my* truth', then you too have slipped into antirealism. There is no such thing as 'your truth' or 'my truth'. There is only truth. Obviously, we may have very different interpretations of the truth, some more accurate than others, but truth is still truth. You may believe that you are perfectly safe as you walk down the road listening to music through your earphones—you may believe all you want in *your* truth of perfect safety. Unfortunately, the *actual* truth is that you are just about to be hit by a bus that you cannot see or hear heading straight for you.

We need to understand antirealism at its core because, in one way or another, it has subtly permeated various areas of our society, so much so that we are usually not even aware of it. Even some physicists are falling prey to antirealism due to a lack of philosophical acumen and historical perspective (not to mention a basic sense of rationality). We can be sure that something is wrong when out of 534 submitted questionnaires 3% of physicists believe the earth is not real and 2% are not sure it is real.[29] In other words, about 26 of these physicists either deny or doubt the reality of the earth. There are intricate and important philosophical subtleties to be discussed in relation to what it means for something to be 'real' but, in all seriousness, is it *really* possible to deny that the earth is *real* and still do physics—or even to be classified as sane?

Even the once-cherished eternally unchanging laws of physics have come under fire from someone who really should know better. Theoretical physicist Lee Smolin puts the cart before the horse by making physics depend upon biology, claiming that the laws of physics are not constant but are themselves susceptible to Darwinian-style natural selection. But if the laws can change, then either they change according to a higher order that itself cannot change, or they are changing purely randomly. If the laws of physics are changing in a purely random fashion, there is no reason to believe that they will hold from one moment to the next and there is every reason to believe that they will, in fact, constantly change in any unimaginably haphazard way from moment to moment, which would be the death of physics.

In one way or another, antirealism is responsible for the denial of the reality of the laws of physics and the denial of the reality of the earth. There seem to be far too many people determined to cajole us into believing that, as philosopher Nancy Cartwright puts it, the laws of physics lie.[30] However, she would have a very difficult time trying to explain why she does not simply float away from the surface of the earth. Contrary to Cartwright, my position is that physics and philosophy, when properly pursued, are possible pathways to the Divine. Newton was not wrong on this point.

Despite their overwhelmingly realist commitments, however, a few of the quantum pioneers did make some comments that, at first glance, do appear to imply a qualified antirealism. Consequently, the kind of mysticism that has become most commonly associated with quantum physics is usually nothing more than some version of antirealism. There are profound mystical implications to be found in modern physics, but they do not support antirealism. The whole of reality, in other words, is not limited by my particular thoughts, intentions, or actions. With appropriate metaphysical reasoning, however, we will be able to set aside antirealism and discover for ourselves the rational mysticism that is harmonious with and ultimately underpins the sciences.

My metaphysical position in this book is a result, first and foremost, of my own mystical experiences. To admit such a thing in most philosophy publications would not usually be good for one's academic career (although this would not usually be an issue due to the unlikelihood of work of this nature being accepted for publication in the first place). Anything personal, let alone mystical, is usually anathema to academia. But everyone (or nearly everyone) at some moment in their life has had or will have some kind of mystical experience, whether or not they recognize it as such. Denying the possibility of mystical experience just because you have not had, or are unable to recognize, such an experience for yourself would be like a man trying to denial the possibility of giving birth.

Mystical experiences, however, are not enough. Not nearly enough. We must also be relentless in our reasoning. Merely claiming to have had a mystical experience and then denying the importance of rational thought is asking not to be taken seriously whatsoever. In any case, not having a mystical experience is no grounds for denying the possibility of having one, while having a mystical experience is no guarantee that we will properly understand it or know what to do with it. While we need to be completely committed to rational understanding, those who pursue reason to its absolute limits will be taking themselves to the edge of direct mystical understanding. Those who think that none of this has anything to do with physics will be quite surprised to learn otherwise. Mystical experiences have been a guiding light for many great theoretical physicists, and metaphysics is inextricable from physics.

Despite the importance of metaphysics in interpreting, understanding, and guiding physics, quantum theory has ended up becoming something of a metaphysical mess. Although it is immensely powerful, quantum theory is not the final arbitrator of reality, and most interpretations of it tend not to make much sense. The 'many-worlds' interpretation of quantum theory, for example, ends up implying that at every moment I am creating and living in infinite universes (or something to this effect). This interpretation seems to logically entail that there are infinite 'me's' in infinite universes, where each 'me' at every moment in each universe is creating further infinite 'me's' in further infinite universes.[31] Whether you find this view intriguing or just totally bizarre, I am willing to bet that your angry spouse will not be appeased when you try to explain that, although you had an affair in *this* universe, you did not have an affair in any of the other infinite universes that you also inhabit.

Many physicists have said that they do not really understand quantum theory, and we may already be starting to appreciate why they would feel this way. But the problem is not so much that we cannot understand quantum theory specifically. A close analysis of classical physics or the neo-classical theory of relativity leaves us with just as many deeply puzzling mysteries and unanswered questions. The more fundamental problem is that we too easily forget that quantum theory is not the end, or limit, of physics. When we assume that any scientific theory can encapsulate every aspect of reality, we are setting ourselves up for great disappointment.

Ideally, philosophers should have been able to provide significant assistance to physicists as they developed quantum theory and faced the various ensuing conceptual challenges. But the general lack of genuine communication between philosophers and physicists has led many of the former to hold outdated views about physical reality and many of the latter to mistake their metaphysical mess for incontrovertible empirical evidence.

The majority of the influential philosophers of the last century were quite out of touch with the philosophical ways of thinking that engaged several of the most important pioneering physicists. For example, Heisenberg openly expressed the view that the ancient philosophical tradition(s) of Platonism and Pythagoreanism are foundational to modern physics. One reason for his belief was that these traditions advocated the idea that mathematics was foundational to the study of nature. Aristotle, on the other hand, who was Plato's student for many years, supposedly denied the foundational importance of mathematics. If we had followed Aristotle instead of Plato, modern physics would never have been possible and we would not have achieved the technological progress that we have.

While many of the founders of quantum theory held philosophical views that were explicitly or implicitly akin to the views of the Platonic tradition, which at once embraces mysticism, the sciences, and the outermost reaches of rational thought, the majority of contemporary philosophers have thoroughly rejected it. Many philosophers have also dismissed all of ancient philosophy as being dead or useless. However, many of those same philosophers of the last century who most strongly rejected the history of philosophy are now deceased, and their writings have themselves passed into the history of philosophy. Although this rejection of ancient philosophy was a colossal error, I am not at all saying that we should humbly bow to tradition.[32] Plato certainly would have abhorred such a view. Nevertheless, much of the Platonic tradition remains, and will remain, vital for both scientific and spiritual inquiry.

The ancient metaphysical views propagated by philosophers in the Platonic tradition pointed to eternal truths underlying the constant fluctuations and changes of the physical world, truths that unveil the unity between the sciences and spirituality. We, too, can rediscover these eternal truths, the same ones pointed to in various spiritual traditions. Numerous popular books, especially those written since the publication of Fritjof Capra's *The Tao of Physics*, have tried to show similarities or parallels between quantum theory and Eastern mysticism.[33] But most of these authors have ignored the Platonic tradition, despite the fact that Platonism played a vital role in the beginnings of modern science and continues to provide the metaphysical foundations for modern physics. We have become strangers to our own cultural heritage.

In this book, we will undertake a detailed examination of how ancient *Western* philosophy, specifically Platonism, not only parallels quantum theory but actually *underpins* all of physics.[34] Cross-cultural comparative studies are very important, as they often reveal profound metaphysical similarities while also providing the potential for mutual enhancement through the examination of their differences. But we must not allow ourselves to believe that the grass is always greener on our neighbor's side of the fence. The grass is not necessarily greener in our own backyard, either. What we often forget is that the grass in our own yard and in our neighbor's yard draws its nutrients from the same underlying earth. Even the distinctions between the East and the West become blurry as we trace our way back through history.[35] In the end, what will matter most is not whether the philosophical ideas are Eastern or Western or which tradition thought of which ideas first, but whether or not the ideas are actually *true*. As Newton wrote almost 350 years ago, 'Plato is my friend and Aristotle is my friend but my greater friend is truth'.[36]

A foundational truth underlying all of physics is the one eternal mathematical law. As a modern astronomy textbook puts it when discussing the birth of the universe, the four fundamental forces of physics 'are said to have been *unified* at this early time—there was, in effect, only one force of nature'.[37] The eternal mathematical law is outside of space and time—it is independent of the physical universe, as Pauli had suggested when referring to the cosmic order. The fact that the foundation of all physical reality is itself nonphysical is part of the genuine mystical implications of modern physics, and such implications are not dependent upon the strangeness of quantum theory in particular.

Regardless of how much we may know about physics or spirituality, we should exercise caution when claiming that our spiritual or metaphysical beliefs have been proven by quantum physics. Scientific theories and interpretations change, so if you have staked your worldview on an interpretation of quantum theory (or on any scientific theory) that may be replaced or revised in the future, you will find yourself trapped in an existential corner. We are avoiding such a potential pitfall here. The main arguments that I am putting forth are not dependent upon any particular interpretation of quantum theory, nor are they dependent upon any possible future developments in physics. In fact, quite the opposite is true. Any future physics will also be dependent upon the basic metaphysical assumptions of Platonism that are being examined here. Nevertheless, as contemporary physical chemist Lothar Schäfer has shown, it is still true that quantum physics has made the implications of spiritual reality underlying the physical universe much harder to deny, and so it is very important to undertake further exploration of specific quantum phenomena and to analyze their metaphysical implications.[38]

Scientia

As some of the arguments in this book have obvious religious implications, we must here make clear the distinctions between primary and secondary religious beliefs. *Primary* religious beliefs are those that claim that reality is ultimately one, many, or nothing, or good, evil, or neutral, or rational and intelligible or random and unintelligible, and so forth. *Secondary* beliefs have to do with the conventions of each particular religion, which encourage adherents to live a certain way of life within society. Atheists and theists who are committed to scientific and rational inquiry should come to realize that they often share, whether consciously or not, primary beliefs about the ultimate nature of reality. If atheists want to say that we should reject religious ideas because they are out of sync with modern science, then they have walked straight into their own trap. In actuality, primary religious beliefs, such as unity and cosmic order, are far more in sync with the metaphysical foundations of modern physics than materialism, empiricism, positivism, or antirealism.

The polymath philosopher William Whewell introduced the term *scientist* 'at the 1833 meeting of the British Association for the Advancement of Science'.[39] The word 'science' comes from the Latin root *scientia*, meaning 'knowledge',[40] symbolizing the general feeling in our culture that a scientist is 'one who knows'. However, it is not always clear what exactly it is that scientists are supposed to know. Is their knowledge supposed to be limited to telling us how to make gadgets, such as power drills or nuclear missiles, or are they also able to tell us something about reality beyond technological applications? Should we humbly accept every scientist's claims about religion, such as the view expressed by physicist Frank Tipler that theology needs to be absorbed into physics?[41] What about other scientists who tell us that science has proven theism to be nonsense, or that science has proven traditional theism to be true? What if they tell us that there is no reality whatsoever beyond our own minds, or that we have infinite selves living in infinite universes? The fact is that any scientist who makes any claim about reality is necessarily going to have to rely upon philosophy, so we can judge their views as we would any other philosophical position.

We must be willing to challenge some of our deeply held convictions, including assumptions about the nature of reality that we may not even be aware we have. The implications of such understanding reach far and wide into virtually every aspect of society in general, and into our daily lives in particular. For example, some people may find psychological comfort in reflecting on the necessity of physically obeying mathematical laws, especially one eternal law, while others may strongly reject it. After all, we want to feel free. We do not want to believe that our bodies are nothing more than an expression of a mathematical equation. Some of us, however, may believe ourselves to be totally indifferent to this idea, since whether or not my claim is true, it does not seem likely that it will affect our daily lives in any significant way. We will still go to work and buy groceries, attend weddings and funerals, rejoice and cry, go jogging and take a shower, and otherwise continue making choices and living our lives.

Reality, however, is unconcerned with our personal beliefs or our supposed indifference. We may believe we can fly, but after stepping off the rooftop of a high-rise building we soon realize that we are mistaken. Gravity does not care whether or not we believe it is real or whether or not we understand anything about the way it works. Gravity acts on us all with equally proportionate force. But if we all must obey the eternal mathematical law, of which the law of gravity is only a partial reflection, have we thereby eradicated free will? If precise and invariable mathematical laws dictate all of my actions, how can I be free to make my own choices? The fact is, I *can* choose, and I *do* have free will. The conscious entity that I am, what used to be referred to as 'soul', is *not* reducible to physicality. Although we are here oversimplifying a very complicated issue, the basic idea is that I am free to act in any way I may desire within the constraints of the laws of physics. In other words, we may be experiencing anguish and confusion, which entails various neurological reactions, but it still remains our free choice whether or not to jump from a building. Once we have committed to and followed through with this decision, however, our body is bound by the law of gravity.

Our bodies are biological organisms, and as such they are subject to the laws of physics. But we are not *only* biological. We are also social, psychological, and spiritual creatures. Biology also depends upon chemistry, which depends upon physics, which in turn depends upon intangible laws and metaphysical principles. We are material and something more than material. We are complicated creatures.

Unified knowledge

Although I am offering an original perspective and approach, many of the arguments and topics discussed in this book find their roots in ancient philosophy. I also rely upon research in contemporary philosophy, physics, sociology, classics, and the history of science. The interdisciplinarity of this book reveals the unity underlying the more obvious differences between these various disciplines. It can open the way for scholars to gain a better understanding of these different branches of knowledge and of how their own particular research relates to many other fields. For example, classicists will be better able to understand how their discipline has been, and will continue to be, important for the development of modern physics, while contemporary philosophers will be able to better appreciate how and why so many founders of quantum theory openly embraced a metaphysical and mystical understanding of physics while rejecting mainstream philosophy. Most importantly, we will be able to discover some of the key interconnections between ancient philosophy and modern physics.

The essence of this book, however, is not reducible to any of these particular disciplines, and it would be impossible to address every point of debate of every topic I discuss here. No amount of detail is ever truly satisfactory for the expert, and this lack of complacency is part of the driving power of scholarship. But we cannot forget that words seldom, if ever, adequately capture our insights, no matter how much detail we may provide. In any case, if physicists remain locked in technical jargon and mathematical equations, scholars from the arts and humanities are not likely to take much interest in their work. Similarly, an overly detailed analysis of a dispute between opposing contemporary classicists regarding an interpretation of one passage in a 2000-year-old text is not likely to hold the attention of the physicist or sociologist. Not to mention that most contemporary philosophy is largely unintelligible to all outsiders (and to many insiders as well).

Too many professional philosophers have been unwilling to entertain the possibility that there really are mystical implications in quantum theory, or that ancient philosophy is at all relevant to modern physics (or to anything else). This obstinate refusal to acknowledge what many physicists themselves have openly declared has served only to sideline philosophers even further, both inside and outside of academia. This unfortunate fact has, to a significant extent, been brought about by philosophers themselves for various reasons, including their insistence upon using unnecessarily recondite jargon that excludes non-specialists, and by refusing to take the 'love of wisdom', the true meaning of philosophy, seriously. Biologists and lawyers have their own specialized terminology too, but the general populace knows that they produce practical (though not always benign) results, whereas many people, including academics in other disciplines as well as university administrators, often have little understanding of what philosophers actually do. Since they seem not to produce any easily recognizable, tangible results, many people are left wondering what the point of philosophy really is.

Some philosophy departments (as well as other academic departments) have recently been closed or threatened with closure across North America and the United Kingdom. While it is important for academic philosophers to continue to try to keep saving deaprtments from closure or unnecessary reduction of faculty, we also need to continue developing philosophical organizations outside of the university structure. Genuine philosophy includes logical analysis, but it also asks the deepest possible questions while endeavoring to bring about meaningful transformation in its participants. Such deep philosophical inquiry cannot remain solely at the mercy of those who may not really understand or appreciate it.

As we come to understand the vital role of philosophy for physics, we can develop a greater appreciation of the importance of philosophy for quantum physics. Although the development of quantum physics has given us tremendous technological benefits, there are also potentially dangerous consequences. Peter Gibbins, a specialist in the philosophy of quantum mechanics, writes that 'one must of course admit that through nuclear weaponry, the transistor, and now the microchip, our system of communication, indeed our technoculture as a whole, has come to be based on and threatened by an ultimately quantum-mechanical technology'.⁴² But the potential danger is not limited to technological applications. Many of our current troubles in society can be traced back to various philosophical misunderstandings of the implications and presuppostions of quantum physics.

Our metaphysical journeying can greatly benefit from reviving and further developing ancient philosophy, especially the Platonic tradition, because we are dealing with many of the same fundamental questions that engaged the minds of so many of the greatest philosophers in this tradition for more than two and a half millennia. Even the famous business visionary of Apple Inc., Steve Jobs, apparently believed that the Macintosh computer has always existed.⁴³ When we try to explain that this kind of foundational belief was one of the key ingredients for Apple's tremendous success, we soon discover that a mystical metaphysics is required.

Speaking about Jobs, his biographer Walter Isaacson states that 'his whole life is a combination of mystical enlightenment thinking with hardcore rational thought'.⁴⁴ Those of us who facilely dismiss such views can hardly deny the technological and financial power of Apple. An important part of understanding the mind of Jobs, at least the part of his mind that could tap into these higher levels of reality and thereby envision and guide the development of some of the most powerful technological products available to the public, requires understanding the sorts of ideas and arguments presented here.

So often we hear people extolling the importance of change, but when we are offered the possibility of genuine transformation, many of us turn and run. This reaction is understandable; deep, genuine change is truly frightening, and who among us has never run away from or avoided some deeply meaningful challenge? But we must be brave and not be deterred from such transformation. After all, it is part of our very nature to seek higher understanding and greater possibilities. By recognizing more clearly the underlying unity of ultimate reality, we are more likely to achieve our highest potential, whether in science, business, art, or our personal lives.

TWO: Clarifying Key Concepts

In this chapter you will find brief but sufficient clarification of several key concepts, such as realism, antirealism, quantum theory, Platonism, metaphysics, and mysticism. The basic aim of this chapter is to help level the playing field, so to speak, by ensuring the necessary introductory understanding of the main topics.

An ancient hierarchy

Many of the key ideas that we will be investigating will simultaneously require greater generality and more explicit specificity, as a thorough analysis of any topic needs to integrate both of these approaches. We will need to revisit many of these ideas as the process unfolds, but revisiting does not just mean repeating. On the contrary, revisiting allows us the opportunity to gain greater depth of understanding and develop a more sophisticated appreciation, especially for ideas or concepts that so many of us take for granted. The notion of 'reality', for example, is extraordinarily difficult to clarify. It is so difficult that many of us don't even bother thinking about it. If we do, we tend to console ourselves with comforting definitions and explanations that take the edge off our fear of the unknown. This intellectual subterfuge does not eliminate the unknown; it just makes us less prepared to face it.

Professional philosophers tend to display the highest skill in this game of self-consolation. Artful application of their logical training can easily fool us (and themselves) into believing that they actually have achieved the direct experience and understanding of the topic they are addressing. We may be able to offer a definition of the word 'war', for example, without having any personal experience of fighting in or living through a real war. Similarly, talking about 'mystical experience', or apparently defining it away as nothing more than a neurochemical glitch, does not mean that we actually know what we are talking about.

Consider the mystery of how something intangible, such as a law of physics, has more reality than something physical like a chair. Not only is a law of physics more real than a chair, it is also more real than the universe itself. How is that possible? First, whatever is responsible for the universe coming into existence and continuing to behave the way it does must necessarily be at least as real as the universe. Second, whatever cannot change and yet is real must be more real than something that exists but always changes. After all, if something always changes, even in the infinitesimally smallest way, it has become something else, so at each moment we would have to say that it no longer exists in the way in which it formerly existed. Conversely, whatever does not change will always remain, and that which always remains what it is without any change whatsoever must necessarily be more real than something that no longer exists as it previously was. In other words: something that never changes will always be just what it is, whereas something that constantly changes will never remain exactly what it is for any length of time. Therefore, something that never changes would be eternal and thus more real than something that could, at best, only be said to exist as it is for a brief moment before becoming something else.

The one eternal mathematical law is an important example of something that always remains what it is, and yet its reality is dependent upon absolute beauty. What follows is that absolute beauty is actually more real than the chair I am presently sitting on. I realize how strange this may sound; but sounding strange does not make something false. This does not mean that all strange sounding things are necessarily true; it just means that being strange does not necessarily make something untrue. In fact, unless symmetry and beauty were real metaphysical principles, the eternal mathematical law would not be possible. Symmetry and beauty, in turn, are dependent upon absolute good, which has historically been called God, Goddess, Tao, Emptiness, the Divine, and the One, among other names. While many of us might mistakenly believe that such ideas have nothing do with physics, Heisenberg recognized that 'the search for the "one," for the ultimate source of all understanding, has doubtless played a similar role in the origin of both religion and science'.[45]

As we can see from this very simple, introductory outline, there is a hierarchical metaphysical structure extending from the One down to beauty and symmetry, then further down to the eternal mathematical law, finally reaching all the way to the physical world. Of course, we will need to unpack, expand, and refine this hierarchical structure, which, as specialists in ancient philosophy will immediately recognize, has its roots in Platonic metaphysics. For many of us, however, this way of reflecting on the nature of reality will be quite new.

The notion of hierarchies has received very bad press from within the insular walls of academia—a place replete with hierarchical power. But hierarchies are everywhere, and without them we could not function. I speak English better than Akihiro, so I would have a higher place in any hierarchy relevant to English speaking ability. Akihiro speaks Japanese better than I do, so he would have a higher place in any hierarchy related to Japanese speaking ability. This obvious point is totally lost on many academics who feel hieracrhically superior in denouncing all hierarchies as evil.

Realism and antirealism (or, where is Mexico?)

First of all, we need to set aside the common use of the terms 'realism' and 'idealism'. Realism is usually taken to mean being 'realistic', to face the cold, hard facts of everyday life, such as getting a job, paying bills, and so forth. Being idealistic, on the other hand, often has the negative connotation of not being realistic, of being a dreamer unwilling or unable to face reality. Here we will use these terms in a much more technical way, and this section provides the first step in such clarification.

As we will see later in this chapter, one form of idealism is actually realist, but another form is antirealist. I know that this sort of terminological confusion can be frustrating, but it is vital to bring clarity here in order to deepen our understanding. Such clarity also helps us to avoid unnecessary disagreements that are simply a result of using different definitions for the same term, or using different terms that have the same meaning or function.

At the most basic level, *realism* is the view that we discover truth, whereas *antirealism* says that we merely invent truth or that truth is non-existent. For example, the realist will believe that '2+2=4' is an objective fact in reality, whereas the antirealist will believe that it is nothing but a human construct, a mere fiction. Although this distinction may seem clear at first reading, there are numerous intricate difficulties lurking just below the surface, which we will investigate in detail. For example, there has been a continuing confusion of terms whereby realism has been equated with materialism and determinism. Since quantum theory is generally considered to have overthrown materialism and determinism, then it would follow, according to these assumptions, that realism would also be false, and so antirealism must be true. We shall soon discover why this conclusion is false.

Despite subtle variations which will not concern us here,[46] all materialists seem to adhere to the following basic definition given by theologian and philosopher Keith Ward: *materialism* says 'that the only things that exist are material things in space'.[47] Ward is not a materialist, so let us see if a materialist philosopher can define the concept any better. Daniel Dennett claims that 'there is only one sort of stuff, namely **matter** – the physical stuff of physics, chemistry, and psychology – and the mind is somehow nothing but a physical phenomenon'.[48] It is true that defining 'matter' is not easy, but Dennett does not help, for all he is saying is that matter is physical stuff, and physical stuff is matter.[49] This sort of circular reasoning is not something Dennett would tolerate from someone arguing for the reality of God, so we have to wonder why he allows it for materialists such as himself. It is not that I think I can offer a better definition of materialism than Dennett; rather, I do not pretend that the issue is so simple.

Determinism, in its most basic form, assumes that any particular outcome or effect could not have been otherwise. In other words, every action in the universe is the only action that could have occurred, because every action is completely determined by prior actions. However, there are many different uses of the terms 'realism' and 'antirealism', as well as ambiguities concerning the notions of 'materialism' and 'determinism'. When such confusion has been clarified and we have taken seriously and critically evaluated the relevant philosophical views of the founders of quantum theory coupled with appropriate metaphysical reasoning, we will then be in a much better position to understand how modern physics presupposes and implies the truth of essential aspects of Platonism.[50] Two important examples are that physics (both classical and quantum) presupposes the notion of a unified, hierarchically ordered universe, and that what is most real or fundamental in physics is the ideal.

An explicit return to a refined realism in physics is essential, although it is more accurate to say that we need to recognize—or remember—that realist assumptions have always remained at the foundation of physics, despite the apparent bizarreness of quantum theory. For those physicists and philosophers who have become accustomed to assuming that realism equals materialism, an important terminological and conceptual shift is required. An essential point to keep in mind from the outset is that the theoretical and metaphysical *foundations* of physics have never been congenial to materialism and have always aimed at unification, simplicity, and the abstract. Antirealists believe that the abstract laws of physics are mere fictions that are somehow constructed from the phenomena, even though all phenomena, by antirealist criteria, must also be fictitious, which complicates the issue, to say the least.

Realists, however, believe that we discover these laws and that they are actually real. Since these laws themselves have no physicality yet underpin all of physical reality, realism entails that materialism is false, because we have to admit that something can be real and yet not be physical. As a result, materialists cannot be realists. Conversely, many of those who have rejected materialism and thereby felt compelled to reject realism as well should realize that they are actually realists.

But saying that materialism is false does not mean that immaterialism is therefore true. *Immaterialism*, which is conceptually associated with antirealism, denies that there are any physical objects in reality. It essentially says that all of reality is pure mind, or that all so-called material objects are nothing more than a collection of nonphysical qualities.[51] Platonic realism, as will become clear, transcends both materialism and immaterialism. I discuss realism in detail in Chapter Six, and antirealism in Chapter Seven, so here I will clarify only the most essential realist assumption, which the antirealist denies.

The realist believes in 'verification-transcendent truths'. 'In other words', as philosopher Christopher Norris notes, 'there exist many features of reality that lie beyond our knowledge or present-best powers of understanding, but which nonetheless obtain quite apart from what we happen to think or believe'.[52] A verification-transcendent truth refers to some truth that transcends our ability to verify it. For example, on July 6, 1590 at 10 a.m., William Shakespeare either sat on a chair or did not. There may be no possible way for us to verify whether or not he sat on a chair at that exact moment in time, but realists will say that it is the case that he either did or did not do so, despite our inability to know for certain one way or another. This may seem to be nothing more than pure common sense, but it is something that many antirealists do their best to deny. (There may be some ambiguity surrounding the precise definitions of 'to sit' and 'chair', but these potential issues are not relevant to my point here.)

While there is much that is wrong with antirealism, the common form of realism, sometimes referred to as 'naïve realism', has its own inherent difficulties. For example, naïve realism usually just means materialism. Such realists tend to believe that reality is exactly as it is represented to them through their senses. In this case, we would have to say that other animals (and even other humans) do not see reality clearly because their perceptual systems are different than our own.

Another relevant concern has to do with the meaning of the verb 'to exist'. Existence inherently involves change, and so, strictly speaking, it is not correct to say that the laws of physics actually exist. Since the laws of physics are verification-transcendent truths that are objective, eternal, and unchanging, they cannot actually exist because they do not ever change. A law of physics, after all, is not one thing at this moment and then something completely different at another moment. A truth that is unchanging, such as a law of physics, is what it is and remains what it is, and yet is still real. Something that always is just what it is cannot properly be said to exist if the notion of existence inherently implies change. Consequently, we need to describe a law of physics as being real despite not actually existing, and the way to make this distinction in the English language is to say that something 'has being'. To say that something has being is to denote its timeless and unchanging nature, but such phrasing is awkward in normal conversation. However, this idea was a common assumption of Platonism apparently capable of clearer exposition in ancient Greek than in modern English.[53]

If at this point you still find it difficult to accept the reality of timeless, unchanging truth, you can think of realism as entailing *belief*-independent truths. For example, I may believe that Japan is a country on Mars rather than on Earth, but my belief would be false. Or I may believe that time travel is impossible, but it may very well turn out that this belief is also untrue. In other words, I can genuinely, objectively be mistaken. I will also make the more difficult argument that reality is *mind*-independent, but this view will require important subtle clarification as well. Nevertheless, belief-independent reality is enough to vindicate basic realism.

The opposing view, antirealism, was apparently first given its name by philosopher Michael Dummett in 1978, decades after the inception of quantum theory. Thus, in the same way that many philosophers feel justified in calling quantum theory antirealist even though the founders of quantum physics did not use the term themselves, we are even more justified in calling it Platonic, since some of these physicists did explicitly endorse Platonic realism, while implicitly doing so as well. In general, antirealists believe that the limit of what we know is the limit of reality. We can talk about many different types of possible antirealists, from those who deny that there is any reality whatsoever to those who claim that reality is limited to what each individual personally happens to believe. Whichever type of antirealism one may espouse, however, for the antirealist, ontology is subservient to epistemology.

Ontology refers to the study of existence itself and to what there is in reality, while epistemology, as has already been mentioned, refers to the nature and limits of knowledge. To say that ontology is subservient to epistemology, as we have already learned, is to say something to the effect that the limit of our knowledge is the limit of reality. This may seem a bit crazy, for how could reality be limited by my particular knowledge? How can our knowledge, the knowledge held by us tiny specks of biological dust floating on a lonely, spinning ball in the vast, infinite sea of (mostly) darkness, define the limit of reality? If my knowledge really is the limit of reality, and I have no knowledge about Mexico then it could not exist in reality. That would be bad news for the many millions of people who currently reside there. Although antirealism may seem not to make much sense, nevertheless, antirealists do raise some worthwhile objections against realism, especially in relation to quantum theory. But I will not be seeking a compromise;[54] the antirealist denial of objective reality is unequivocally false.

Quirky quantum theory

Although my arguments apply to the foundations of any past, present, or future physics, I have chosen to focus mostly on quantum theory for five main reasons:

1. it is one of the most powerful scientific theories in history;
2. it is usually assumed that the chief architects of the Copenhagen interpretation of quantum physics were positivists and antirealists, the opposite of what I argue;
3. its founders were profoundly metaphysical by necessity of the results of empirical and theoretical research;
4. due to several unnecessary assumptions about the goals and permitted methodology in mainstream academic philosophy, far too many philosophers have ignored or misrepresented the metaphysical reflections of these pioneering physicists, to the detriment of both philosophy and fundamental thinking in physics; and,
5. many people have a naïve, misconstrued, or completely false view of quantum physics.

Physical chemist Jim Baggott is a good example of a very intelligent scientist who has been mistakenly proclaiming that quantum theory is antirealist. He writes that quantum theory is the 'most important fundamental theory of science that is truly anti-realist both in its mathematical formalism and in its orthodox interpretation'.[55] An important goal of this book is to rectify this error and show how the rejection of naïve realism is really a rejection of materialism and not a rejection of nonphysical reality, and certainly not a rejection of truth. The three main physicists involved with the development of quantum theory who were accused of being antirealists are Heisenberg, Pauli, and Niels Bohr. A thorough examination of their views, however, shows that they were in fact assuming the truth of Platonic realism. To whatever degree they may have veered from this view, they were incorrect.[56]

Quantum theory is extraordinarily difficult, but as Baggott acknowledges, 'students are likely to blame themselves for failing to understand quantum theory'. 'This is a great pity', he continues, 'because this non-understandability can, in fact, be traced to the anti-realism of the Copenhagen interpretation. The theory is, quite simply, not *meant* to be understood'.[57] However, the real problem is that it is not possible to understand it from a materialistic perspective. It is even less possible to understand it through the distorted antirealist lens. The difficulty of the metaphysical implications of quantum theory has led some physicists, such as Richard Feynman, to say that nature is absurd,[58] while Dawkins has gone so far as to imply that the universe is 'too queer' for us to understand.[59]

But reality is neither queer nor absurd. Reality is quite happy being just what it is, not caring about our personal struggles to understand it. Galileo believed that nature cares 'nothing whether her reasons and methods of operating be or be not understandable by men'.[60] Proclus, the fifth century Platonic philosopher, made a similar point: 'deliberation is the mark of thought's encounter with difficulties: this is why Nature produces and knowledge says what it says without deliberation'.[61] In less gnomic terms, 'Nature' cannot be in contradiction with itself. But even if nature simultaneously could be what it is and what it is not, it would simply say what it says, or do what it does, without deliberation, not caring whether or not it violated our cherished beliefs. Quantum theory may be a bit quirky, but if nature appears to be contradictory to our reason, then *our* inability to reason correctly is the cause of the apparent contradiction.

I can sympathize to some degree with the views of Feynman and Dawkins due to the numerous conceptual difficulties found in modern physics. Dawkins also makes the plausible (but not original) assertion that our biological equipment and our perceptual and cognitive limitations may in principle prevent us from ever understanding the universe completely. But this worry is misplaced. Just because we cannot understand *everything* about the universe in completely exhaustive detail, it does not follow that we cannot understand *some* things. The universe in its absolute entirety may forever remain beyond the conceptual and experiential grasp of any sentient creature that has arisen, or will arise, from within the cosmos, but it does not follow that the confusion resulting from our lack of total comprehension entails that the universe or reality itself is absurd or queer. Apparently the universe is not too queer for us to understand basic aspects of our own evolution, which requires an understanding of geology, biology, chemistry, and astronomy, all of which are dependent to varying degrees on quantum theory. If anything is queer, it is the view that the universe itself is queer.

While the universe is not queer, Gibbins correctly points out that quantum mechanics is '*deeply* mysterious…because it subverts the classical picture of the world, of which classical mechanics and electromagnetic theory are refinements'.[62] The classical picture was supposed to be one of mechanistic materialism, presumably stemming from Newton, which is in fact false.[63] Newton was not a mechanist, but was primarily a theological thinker who, as theoretical physicist Peter Rowlands states, believed that 'the ultimate causes of things were abstract rather than mechanical'.[64]

While Newton himself was not a materialist, the common view is that Newtonian (or classical) physics necessarily implies the truth of materialism. This view, however, is false. The mathematical laws discovered in classical physics are themselves not physical, and so, unless we want to say that all of the laws of physics are fictions (which would in turn make all of science a fiction), these laws must be real and yet nonphysical, which necessarily defeats materialism. This basic argument applies equally to classical, quantum, and any future physics, and I discuss in greater detail the reality of the laws of physics in Chapter Eight.

But we must here differentiate between a mechanistic view and a materialistic view. One could be a mechanist and not be a materialist, or one could be a materialist and not be a mechanist. A *mechanist* believes that any two physical phenomena can interact only when there is some sort of direct physical contact between them. Of course, these physical interactions are governed by forces, which are themselves govenered by mathematical laws, and these mathematical laws are themselves not physical. Hence, mechanism inherently denies materialism.

Indeed, the only sort of materialism that we could possibly accept is the ancient Epicurean view that there are only atoms moving randomly in a void. In such a universe, there would be no mathematical laws, no cosmic order, nothing underpinning and dictating the movements of the atoms. Therefore, we would be able to say that there is nothing in reality that is nonphysical, and that only physical things are real. (Of course, that still does not explain what the 'void' is, because if the void is physical, then it too would be made of randomly moving particles, and if it is nonphysical, then materialism is defeated.)

In any case, if the entire universe were without order or mathematical laws, then science would be impossible because we could never make any predictions. Thus, if we want to allow for the possibility of science (and for a universe even to exist in the first place), then we must allow for cosmic order that is not reducible to physicality, which is to defeat materialism.

However, it is possible to be a mechanist about all physical phenomena, in the sense that physical things can only be directly affected by other physical things (including imperceivable physical forces), and yet also recognize that underpinning all such ordered physical interactions are abstract, nonphysical, unchanging causes. This classical view, which Newton seemed to hold, does not adequately explain how nonphysical laws could interact with physical things.

Of course, it is also understandable why a mechanistic view of classical physics developed in the first place, since Newton rejected the idea that two physical bodies could in anyway interact without some intermediary physical force or contact.[65] In other words, he denied the apparent implications of his own theory of gravity, which was to deny 'action at a distance', which means that he probably would not have liked the notion of nonlocality that we find in quantum physics. Nevertheless, nonlocality is itself an expression that physicists really do not know how to explain. Simply saying that some form of interaction is 'nonlocal' does not at all explain what that means.

All we need to be able to understand here is that nothing can interact with anything unless they are in some sort of physical or nonphysical relationship. The irony, of course, is that atheists such as Dawkins champion the worldview of materialism, which is completely out of touch with the assumptions and explanations of quantum physics. Moreover, such atheists refuse to acknowledge that many of the founders of quantum physics not only rejected materialism but also were openly mystical. Consequently, these atheists are refusing to look at the evidence because it contravenes their most cherished beliefs about reality.

This anti-mechanical view was inherent but often hidden in classical physics, but now it is more openly revealed by quantum physics. Such a view has always tended not to be popular amongst those physicists who are more concerned with the technical application of the principles discovered by the pioneering theoreticians, but for the past four hundred years it has been the most fruitful in fundamental physics. It has, in fact, always been at the foundation of physics. Plato, for example, recognized the importance of mathematical reasoning for solving certain problems instead of relying on mechanical models.[66]

Philosopher and historian of science Giorgio de Santillana notes that Archimedes also thought that 'mechanical gadgets had nothing to do with true science which built abstract and rigorous mathematical models impossible to translate into mere clockworks'.[67] In other words, Archimedes, like modern day theoretical physicists twenty-two centuries later, was more concerned with abstract mathematics that transcend the physical realm of mechanical processes. Rowlands is also correct that although 'the abstract aspects of the Newtonian and Maxwellian theories were long resisted on account of their intrinsically abstract nature, quantum mechanics has forced modern physicists into the same abstract positions'.[68] The point to note here, therefore, is that whatever differences exist between ancient, classical, and quantum physics, they all share the same fundamental metaphysical assumption that the physical universe obeys abstract mathematical laws.

Despite the mystery (or because of it), philosophers and physicists (and lots of people who have seen the film *What the Bleep Do We Know?*) use the phrase 'quantum physics' quite casually, as if they and their listeners or readers all know exactly what this means. However, Gibbins notes that despite its extraordinary predictive power, 'philosophers and physicists are in total disagreement about what, again if anything, quantum theory tells us about the way the quantum world is'.[69] Yet, as Heisenberg reminds us, 'it is in quantum theory that the most fundamental changes with respect to the concept of reality have taken place'.[70] We can cite some obvious aspects of quantum theory (which I do below), which will be sufficient for our purposes here, but it seems impossible to pin it down more concretely. This lack of concreteness would seem to be apropos since the theory itself contravenes materialism.

Planck was the first to understand that the allowed energy levels of an electron are discontinuous and determinate—'you never have half a photon'[71]—and the amount permitted is proportionate to Planck's constant (6.62×10^{-34} Joule-seconds) multiplied by the frequency. But what is probably the most significant aspect of quantum theory, setting it apart from classical assumptions, is the claim that energy is discontinuous. When electrons move into higher or lower states of energy, they can only do so when the energy level reaches a certain point. For simplicity, we can say that there can be energy levels 1, 2, or 3, for example, but not 1.5 or any other fraction in between. In other words, energy comes in discrete packets and is not continuous, which is what prompted the phrase 'quantum jump'. Beyond this, it is difficult to say for sure what else comprises quantum theory exclusively. However, most physicists would not be too averse to accepting the basic ten-point outline that I offer below.

Today we refer to the old and new quantum theories. The old quantum theory includes Planck's explanation of black-body radiation in 1900, Einstein's explanation of the photoelectric effect in 1905, Bohr's model of the atom (optical line spectra) in 1913, and Louis de Broglie's explanation of the particle/wave character of the electron in 1923, which seems to be an essential point in the transition to the so-called 'new quantum theory'. This 'new' phase was inaugurated by Heisenberg's matrix mechanics and a year later by

Schrödinger's wave mechanics, two approaches that seemed to be in contradiction until Schrödinger proved that they were equivalent.[72] Gibbins observes that the physicists usually refer to quantum theory as consisting of a 'user-friendly blend' of Heisenberg's and Schrödinger's approaches, but what the 'contemporary mathematician calls quantum mechanics is an abstraction from both due to John von Neumann'.[73] Despite all the confusion surrounding the meaning of quantum theory, it has at least ten prominent features, which could be contested or added to, but such simplification is sufficient for our purposes here:

1. discretely quantized energy;
2. Schrödinger's equations (time-independent and time-dependent);
3. Heisenberg's matrix mechanics;
4. the Heisenberg uncertainty principle;
5. wave/particle duality;
6. principle of complementarity;
7. nonlocality;
8. entanglement;[74]
9. Pauli exclusion principle; and,
10. irreducibly statistical laws.

Physicists can argue about whether or not other aspects should be included in quantum theory, but these points contribute to the strange non-classical implications when we search for a physical description, and so this list, however incomplete or contestable, will suffice for our purposes. I will only discuss these aspects of quantum theory to the degree that they are directly relevant to our purposes in this book, so it will not be important to know anything more about them other than what I discuss later on.

There are also different interpretations of quantum theory, such as the Copenhagen, hidden-variables, and many-worlds, but my focus will be on the former, because it is the standard view and because it is supposedly positivist and antirealist. The Copenhagen interpretation was mostly, but not exclusively, the result of the views of Bohr, Heisenberg, and Pauli, which, despite their internal disagreements, was 'founded on the dual wave-particle properties of quantum entities'.[75] However, there is still no one 'single consistent interpretation of the theory',[76] and the founders themselves never used the phrase 'Copenhagen interpretation'.[77] Einstein disagreed with the Copenhagenists about their view that quantum theory was complete and consistent[78] (and a recent experiment is being hailed by some physicists as vindicating his belief[79]). However, all experiments need to be interpreted, and each interpretation rests upon prior metaphysical commitments and requires at least some applied metaphysical reasoning (which I explain further in Chapter Four).

Our philosophical interpretation of quantum mechanics is not merely a curious afterthought of so-called 'real' science. Our interpretation has a direct effect upon how we understand physics, which is directly related to any possibility of further developments in the discipline. It can also impede potential research by limiting which projects receive funding. For example, if I hold the view that a particular interpretation, such as hidden-variables, is wrong from the outset, then I will not support spending money on research that is based upon this interpretation. It is essential to understand, therefore, that such interpretations are also philosophically based and not purely a result of indisputable scientific fact.[80]

Our interpretations, and our metaphysical assumptions that underpin our whole approach to and understanding of physics, have a direct pragmatic effect on our research.[81] In showing how quantum theory is realist (though not materialist), there will be important philosophical and scientific implications. For example, by seeing that a rejection of hidden-variables alternatives is very much rooted in philosophical assumptions, we can open the way to such empirical and theoretical research once we critique and move past such assumptions. An objection that a hidden-variables approach will not provide any more empirical accuracy than what is already given by the standard view is nullified (or at least attenuated) by the fact that string theory, one of the currently dominant investigative routes, has not yet been experimentally confirmed, even after so many years.[82] It may, however, prove fruitful in the future—and we may also discover hidden variables.[83]

Bohr, however, does seem to have believed (in some moments) that there was no mind- or measurement-independent quantum world, and that therefore no quantum particle has any intrinsic properties until measured. He seems to have been led to this belief due to the fact that all relevant physical descriptions depict the quantum system *plus* the macro measuring apparatus as a whole—they cannot be separated and, therefore, there would not be any independent reality to be claimed about the quantum world. All of these points have earned him the reputation of being an antirealist and, admittedly, at first glance such a reputation seems justified. But despite important subtleties that will be discussed later, in effect Bohr was not really saying anything more than that any object, such as a tree, cannot have independent reality because the tree requires oxygen, soil, sunlight, and water, as well as physical space to grow. There is no such thing as a tree existing in complete isolation from the rest of the universe, and so all the properties of a tree are interconnected with the properties of all the elements it requires for its existence, which does not in any way mean that trees do not exist until we measure them.

Bohr also provided the qualitative counterpart to the uncertainty relations (discussed further in Chapter Four) by proposing his principle of complementarity, which he also wanted to extend to other disciplines.[84] The essence of *complementarity* is that all the apparently mutually contradictory notions in quantum theory, such as wave/particle duality or the uncertainty relations, are two sides of the same coin, each of which could not exist on its own. The principle of complementarity is a good example of an explicit metaphysical principle used in physics.

There are already numerous popular books that explain various aspects of quantum physics for non-specialists, so there is no need to repeat such explanations here. For now, the only other thing we need to be aware of is that the Copenhagen interpretation is essentially philosophical. It aims to set limits on what can be known, not just in physics but also in the other sciences and in philosophy as well. Nevertheless, it is, as Gibbins says, difficult to state exactly what the Copenhagen interpretation is other than the fact that it is really a 'philosophy of physics'. In fact, it is so 'deep' that 'it is extraordinarily unclear as to what the Copenhagenist philosophy of physics asserts'.[85]

What is Platonism, and what on earth does it have to do with physics?

I am sure that most readers know something about Plato, an Athenian citizen born in the fifth century BCE who was a student of Socrates. There has been so much written about Plato, including a wealth of information to be found online, that there is little point in discussing his life and works in detail here. In this book, I am not really concerned with Plato the man, nor with his predecessors or successors as such. Platonism is something beyond the particular idiosyncrasies of the men and women who have contributed to and participated (consciously or not) in this tradition. In fact, 'Aristotle tells us that Plato "followed the Italians (i.e., the Pythagoreans) in most things", while Plotinus tells us that Plato was not the first to say the things that we today widely identify as elements of "Platonism," but he said them best'.[86] Platonism is, obviously, deeply connected to the writings of Plato, but the Platonic tradition is far beyond the abilities of any one person to encapsulate in words, or in any form of expression.

If it is a challenge to clarify the essence of quantum theory, then finding agreement between philosophers and classicists as to what unambiguously and definitively constitutes Platonism seems to be impossible—we are still arguing about it even after twenty-four centuries. Not only are there a variety of aspects of Platonism, but there have also been various interpretations and manifestations of Platonism throughout history.[87] Identifying the essence of Platonism is as important as it is difficult and controversial, but Lloyd Gerson, a specialist in ancient philosophy, has already made significant progress in this respect.[88] For example, he shows how two of the most fundamental ingredients of any reasonable understanding of Platonism are that the universe has a systematic unity, and that this systematic unity is also an explanatory hierarchy.[89]

As a very brief explanation of these two fundamental aspects of Platonism, consider the fact that if the universe were not in a unity, then nothing could exist, since anything that exists must in some way exist as a self-contained unity that is in some way related to other self-contained unities, which entails a whole network of unified unities. For example, my chair is a self-contained unity. It consists of many different parts that are held together in a unity, and this self-contained unity is distinct from other self-contained unities, such as my desk. But at a deeper level, the desk and the chair are part of different unities, such as the unifying set of all my belongings, while also being part of the totality of my household, which is also part of the totality of the entire cosmos. But if unity were impossible, nothing could ever exist because all things that exist always exist in some sort of unity. All such unities must fit within the universal explanatory hierarchy, which states that the *simple* comes before the *complex*, and that the *intelligible* comes before the *sensible*. In other words, a law of physics, which is simple compared to the complexity of the cosmos, is metaphysically prior to the cosmos, and the intelligible (pure consciousness or mind) is metaphysically prior to the sensible aspects of reality, such as my desk and chair.

Any philosophical, scientific, or spiritual system that embraces these fundamental metaphysical principles could be classified as Platonic, whether it comes from the East or the West, or from another planetary system. On the other hand, and what may be even more difficult to understand, is that there is a sense in which there is no such thing as Platonism at all. If the philosophers in this tradition are correct that ultimate or absolute truth is eternal, then it is the same for everybody at any point in time, regardless of different cultures, personal opinions, and so forth.

Truth is truth, even though we have varying degrees of ability to attain it, understand it, speak intelligibly about it, and express it in our own unique way. But if truth is truth, anyone who glimpses it or understands it to whatever possible degree must inevitably (if they are to report accurately what they have attained) provide a very similar perspective. The great mystics of the various spiritual and religious traditions make it clear that they are all pointing to the same fundamental truth about the ultimate nature of reality. The multiplicity of rituals and expressions can deceive us into believing that there is no underlying unity, but when one achieves the same sort of mystical insight for oneself, such unity becomes obvious. As we shall see, Heisenberg himself went through a similar process.

Consider the fact that any personal relationship, such as the one shared between a married couple, requires a certain essential unity in order for that relationship to function optimally. If my main goal in life is to see how many pints of beer I can drink each weekend, whereas my partner is striving to attain a PhD in nuclear physics, then we are likely to have a lot of tension because we are lacking a foundational unity in our relationship. My partner is pursuing one way of life while I am pursuing another. We can only be mutually enriched by our respective differences so long as we are bound together by an underlying unity, whatever that may be. In a similar way, the laws of physics indicate the underlying unity of all the diverse phenomena in the physical universe.

It seems easier to focus on differences rather than unity, for we are surrounded by endlessly different objects and creatures in the physical world, but it is vital for us to understand underlying unity if we are to have any hope of really understanding each other or the universe we inhabit. Each and every thing in the whole universe is in some way different and in some way similar to everything else, and it is from this fundamental abstract understanding that we can best approach our problems, be they those of our personal relationships or those found in physics.

Besides referring to Gerson and various other contemporary scholars, the three main Platonists I draw from are Plato, Plotinus, and Proclus.[90] Plotinus (204-270 CE) is widely considered to be one of the greatest 'pagan' philosophers, but recent scholarship has also elevated Proclus (C. 411-485) to the role of 'spokesman of mature Neo-Platonism'.[91] Kevin Corrigan, a specialist in ancient philosophy, notes that Proclus is 'perhaps the greatest systematizer of all time…[his book *The Elements of Theology*] condenses the whole of Neoplatonic metaphysics into 211 propositions (each deduced from its predecessors as in Euclidean geometry)'.[92] This book is one of the most challenging philosophical/spiritual texts ever written and is essential contemplative material for anyone seriously interested in integrating rationality and mysticism. In any case, my position is that those philosophers who we now refer to as Neoplatonists were among the best exponents, expanders, and developers of the Platonic tradition.

It is generally accepted that *Neoplatonism* refers to a designated historical period that began with Plotinus and ended in 529 CE, a few decades after the death of Proclus, when Emperor Justinian suppressed the philosophers. Although these philosophers simply considered themselves to be Platonists, they also had subtle, and sometimes quite significant, disagreements, yet they were all united by their love and pursuit of wisdom and truth.[93] Indeed, Ralph Waldo Emerson's well-known statement about Plato is not strictly true. He writes that 'Plato is philosophy, and philosophy, Plato – at once the glory and the shame of mankind, since neither Saxon nor Roman have availed to add any idea to his categories'.[94] Many philosophers have, in fact, developed the Platonic tradition to a great extent in original and inspiring ways.

The incredible depth and power of the Platonic tradition has not stopped various scholars from electing to use it as a target of abuse. Richard Rorty, an influential contemporary philosopher whose antirealist (and pretty much anti-everything) views we will examine in Chapter Seven, was no friend of Platonism. 'The trouble with Platonic notions', he writes 'is not that they are "wrong" but that there is not a great deal to be said about them—specifically, there is no way to "naturalize" them or otherwise connect them to the rest of inquiry, or culture, or life'.[95] All that Rorty's criticism amounts to saying is that Platonic 'notions', such as mathematical laws, symmetry, or absolute justice and beauty, cannot be explained according to a materialistic metaphysics. Rorty also claimed that 'there are few believers in Platonic Ideas today',[96] suggesting that he must not have known anything about the founders of quantum physics.

Nineteenth century philosopher Alfred Benn provides a good example of the sort of prejudice that still exists in one form or another in many humanities departments today. Even though Benn admitted the importance of Platonism for the foundations of modern physics,[97] he also held the misconstrued belief that Plotinus essentially destroyed everything good about Plato, Aristotle, and the Stoics. According to Benn, Neoplatonism 'is dead, and every attempt made to galvanise it into new life has proved a disastrous failure'. He continues by saying that 'the world of culture…will not read Plotinus or his successors' and that Neoplatonism is not only 'false' but also 'out of relation to every accepted belief.'[98]

This sort of view still resonates in much of academic philosophy, as evidenced by Rorty's misguided remarks, even though those scholars who specialize in this tradition generally regard Plotinus as one of the most important philosophers in history. The problem is that such specialists are a small minority in academia, although their numbers appear to be growing. It is also a serious problem that most philosophers and scientists seem not to know that, for example, Kepler, one of the most important founders of modern science, was thoroughly immersed in Neoplatonism. If Benn was correct, then even a great physicist such as Schrödinger could not be considered part of 'the world of culture' because he read (and quoted) Plotinus. Platonism is anything but dead.[99]

I could simply stipulate that the sort of realism I am defending in relation to quantum theory relies precisely upon the same sort of essential or fundamental aspects of Platonism as outlined by Gerson, without actually worrying about any historical continuity. After all, my arguments stand or fall on their own. Nevertheless, I am able to show that the two key foundational assumptions of Platonism also provide the metaphysical foundation of modern physics. Platonism is a huge tree with many branches, but they are unified at the base by the key metaphysical principles that have been identified by Gerson. The following diagram should be helpful in making our way through the various aspects of Platonism.

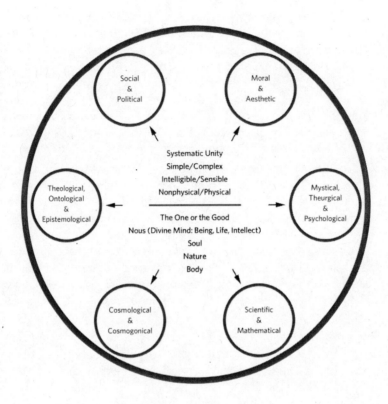

I have developed the central points of the diagram based mainly on the work of Gerson and Lucas Siorvanes, another specialist in ancient philosophy,[100] while the diagrammatical representation and the concomitant system of classification are my own. The top central principles in the above diagram (i.e., Systematic Unity; Simple/Complex; Intelligible/Sensible; and Nonphysical/Physical) provide the metaphysical foundation for all aspects of Platonism, while the bottom central outline provides the hierarchical divisions from the absolute simplicity, unity, and goodness of the One all the way down to physical bodies in motion. I am not going to discuss the many diverse aspects of Platonism that are represented in the circles, for we need to focus instead on what is of most importance for our immediate purposes, such as offering an initial clarification of mathematical and scientific Platonism.

Mathematical Platonism generally refers to the claim that mathematical objects, laws, relations, or structures exist eternally in nonphysical reality.[101] As mathematician Bharath Sriraman adds, 'according to this view, mathematicians do not invent or create mathematics—they discover mathematics'.[102] To put it another way, if you interpret mathematics as being ideal and divorced from sense experience and physical phenomena, and if you believe that valid mathematical structures apply to or dictate the behavior of physical reality, then you are a mathematical Platonist.[103] However, if 'you pretend that physics needs no other basis than experience and must be built directly on perception, that mathematics has to content itself with the secondary and subsidiary role of a mere auxiliary, you are an Aristotelian'.[104] Even if the physical world *just is* mathematical, the principles or deeper laws behind the mathematical laws of physics, such as order, symmetry, and unity, are not reducible to mathematics or anything physical.

Scientific Platonism adds to the claims of mathematical Platonism that there really are physical entities in the universe that obey abstract mathematical relations and metaphysical principles. It also adds that our theories can approximate a true representation of the actual dynamic interplay between the physical and nonphysical aspects of reality, while also allowing for the creative freedom to discover new ways of expressing these theories.

As my diagram implies, one could maintain a belief in scientific and mathematical Platonism while rejecting, say, political and social Platonism. Einstein, for example, probably would have rejected *theurgy*, or working on the Gods,[105] even though he believed in moral and aesthetic Platonic assumptions, as well as scientific and mathematical Platonism. However, when one truly begins to understand Platonism more deeply, it becomes apparent that most if not all of these diverse aspects are inextricably interconnected. As Siorvanes notes, 'metaphysics, the cosmos and the human condition are connected. What constitutional government (*politeia*) is to the state, a sense of justice is to the soul, and order is to the cosmos'.[106] Consequently, contemporary theorists who talk about integral systems are, to varying degrees, participating in this ancient philosophical tradition of Platonism.[107]

Einstein, too, never (so far as I know) said explicitly that he was a Platonist, but he did hold key beliefs that necessarily made him a Platonic realist. In complete opposition to the anti-historical and anti-metaphysical fashions of academic philosophy, Einstein knew the power of ancient thought: 'in a certain sense, therefore, I hold it true that pure thought can grasp reality, as the ancients dreamed'.[108] In this sense, Einstein was a rationalist, as opposed to an empiricist. *Empiricists* believe that we can only acquire knowledge through sense experience, whereas *rationalists* believe that we can also attain knowledge *a priori*, which basically means that we can achieve knowledge prior to or without physical experience by using rational thought.

The rationalist need not deny that we can also attain knowledge empirically, for it is not reasonable to believe that we can use only pure thought without ever relying to some degree upon our senses. However, it is even less reasonable to believe that we can *only* attain knowledge through our senses without ever thinking rationally about what our senses are actually presenting to us.

Platonism is, in this sense, rationalist, but also recognizes the importance of our senses, even though our senses alone can never provide us with the actual truth of higher metaphysical aspects of reality. I need my senses to help me measure the rate of a falling object, but I need my reasoning and intuitive mind to discover the higher level of reality of the mathematical law responsible for how objects fall in the first place. Since the mathematical laws of physics are nonphysical, I can never experience them with my senses and so, according to empiricism, they must be unreal. Empiricism, therefore, would make science unreal.

Although we may now better understand why Einstein believed that we could grasp reality with pure thought, he was not always an adherent of Platonism. John Norton, a specialist in the history and philosophy of science, has shown how Einstein's initial rejection of Platonism in his earlier years 'delayed the completion of general relativity by three years', which meant that he 'nearly lost priority in discovery of its gravitational field equations'.[109] Our philosophical positions really do have tangible effects, and it is fortunate that Einstein realized his philosophical error and returned to a Platonic position.

That we should return to the Platonic tradition to help us bring conceptual clarity to contemporary physics should not be surprising—after all, it was a return to Platonism that inaugurated 'modern exact science' in the first place. As Heisenberg reminds us, Galileo paid no attention to the authority of Aristotle but instead followed the teachings of Plato and Pythagoras and tried to 'find mathematical forms corresponding to the facts obtained by experiment'. In this way, Galileo was able to arrive at his law of falling bodies and initiate the 'beginning of modern exact science'.[110] The two key points here are that the Platonic tradition provided the main inspiration for modern science, and that nonphysical, abstract mathematical laws are real principles that we can discover. Paul Shorey, a specialist in ancient philosophy, also showed how Plato's desire for the soul to gaze upward actually referred to 'the study and contemplation of abstract mathematical relations and principles in their application to solids in motion'.[111] There can hardly be a better directive for contemporary theoretical physicists.

The chief founders of modern science—Copernicus, Kepler, Galileo, and Newton—were directly or indirectly influenced by Platonism, or at least held fast to guiding principles that were in essential harmony with or identical to Platonic realism. There has been much debate, for example, about whether or not Copernicus was a (Neo)Platonist, but such debates tend to miss the point. Whether or not he read Plato or Plotinus, or associated with people who did, is not nearly as significant as the fact that he believed that the universe followed real mathematical laws, which makes him a Platonic realist in the sense relevant here.[112] The mathematical laws of physics have their source in the one eternal mathematical law, which in turn is only possible because of the higher metaphysical principles, such as symmetry, beauty, and simplicity. Santillana writes that contrary to those stuck in 'rigid Aristotelian categories of substance, quality', and so forth, 'we have the creative minds of science—Copernicus, Kepler, Galileo—starting their search from what is to them an obvious assumption, viz., that the world is "a perfect work of art," and that hence the clues to unraveling its complexities must be sought in beauty, simplicity and symmetry.'[113]

The search for symmetry, beauty, simplicity, and unity is inextricable from the foundations of physics. Unless we are able to understand this point, it will be very difficult to comprehend the guiding beliefs of so many theoretical pioneers of physics. Mathematician and philosopher Alfred North Whitehead also notes that 'the Platonic world of ideas is the refined, revised form of the Pythagorean doctrine that number lies at the base of the real world.' Thus, 'when Einstein and his followers proclaim that physical facts, such as gravitation, are to be construed as exhibitions of local peculiarities of spatiotemporal properties, they are following the pure Pythagorean tradition'.[114] In a similar way as Heisenberg, Whitehead recognized the vital importance of Platonism for modern physics: 'Plato and Pythagoras stand nearer to modern physical science than does Aristotle.... The popularity of Aristotelian Logic retarded the advance of physical science throughout the Middle Ages'.[115]

Galileo and Newton also owe a huge debt to Platonism. Regarding Galileo, philosopher Edwin Burtt writes that 'the Neo-Platonic background of mathematical and astronomical development of the times had strongly penetrated the mind of the Italian scientist [Galileo], as in the case of so many lesser figures'.[116] Mitch Stokes, philosopher and former engineer, helps to clarify Newton's Platonism. He notes that Plato and the Pythagoreans believed that 'God orders the universe according to perfect mathematical laws, laws that are intelligible to human beings', which 'became natural philosophy's [science's] underlying assumption'. Stokes then adds that 'although Newton wholeheartedly accepted' this Platonic view, it was even more 'fundamental than his natural philosophy—it explained it, made sense of it, made it possible'.[117] As Newton himself wrote, 'this most elegant system of the sun, planets, and comets could not have arisen without the design and dominion of an intelligent and powerful being'. Moreover, 'if the fixed stars are the centers of similar systems, they will all be constructed according to a similar design and subject to the dominion of *One*'.[118]

Newton was a mathematical and scientific Platonist, since he believed that there really were material objects and forces and that the laws of physics were written in the language of mathematics. Newton represents, in this respect, what the sciences as a whole really are in practice. We need both our senses and the reasoning and intuitive aspects of our mind, just as we need both empirical research and deep theory. The problem is that, with so much focus being placed on practical research, we can too easily forget the importance of the metaphysical thinking required by such pioneers. Newton believed he could work his way back to God by studying the empirical world, leading to eternal laws of physics written in mathematics, which ultimately leads us back to the One.[119]

Kepler, whose work was important for Newton's later developments, was more upfront about his Platonism. Siorvanes writes that Kepler 'was saturated in Proclus' mathematics and philosophy'.[120] It is quite significant that Kepler quotes Proclus on the Title Page of three out of the five Books that comprise *The Harmony of the World*, one of Kepler's most important works. In Book IV he even quotes Proclus in full for five straight pages,[121] and he hopes that relying upon Proclus will remove from him 'the blame for rejecting Aristotle'.[122] In a recent scientific textbook, Peter Bergethon also writes that Kepler was 'one of the pioneering biophysicists'. Bergethon notes that although we may not accept Kepler's astrological rules today, his assumption that 'biological behaviour can be understood via an understanding of the physical rules governing our universe is the fundamental assumption of this book and modern biophysical chemistry'.[123] This 'fundamental assumption' is also one of the essential aspects of Platonism.

Alexandre Koyré, who was a specialist in the history and philosophy of science, has been criticized for attempting to reveal the Platonic foundations of science. For example, in his review of Koyré's *Metaphysics and Measurement: Essays in Scientific Revolution*, philosopher L. Laudan exemplifies the sort of misunderstanding of Platonism (and the foundations of physics) that I have already mentioned. Laudan claims that Koyré has not clarified sufficiently the metaphysical principles of which Platonism consists. He also believes that the Renaissance resurrection of Neoplatonism was a 'violent departure from, and corruption of' Plato's own philosophy, and that Kant, Newton, and every mathematical physicist, on Koyré's account, would 'emerge as a disguised Platonist'.[124] Laudan may be partially correct that Koyré did not sufficiently clarify the metaphysical principles of Platonism, but we are remedying that problem here.

Laudan's assertion that Renaissance philosophers violated Plato, however, reveals a significant misunderstanding of the entire Platonic tradition, and faulting Koyré for arguing that all mathematical physicists must be Platonists is thoroughly misplaced. It is *exactly* my point that all physicists, including self-professed positivist physicists such as Hawking, necessarily hold beliefs that make them at least mathematical Platonists. However, it does not follow that Plato was the sole 'architect of science', single-handedly guiding all the great mathematicians and astronomers of his day.[125] But we can no longer deny the importance of the Platonic tradition for modern science, which has been permitted to occur in academia for far too long. While my opponent will object that I am looking for Platonism everywhere, the fact is that it is actually to be found everywhere.

Despite the debate surrounding Plato's views of experimental science,[126] he would not have discouraged our modern scientific enterprise. To the degree that it is aiming for the good, he also would have reminded us to keep seeking the underlying unity of the diverse fields of research. It would be misguided to argue that Plato was anti-scientific because he did not examine the internal organs of animals or conduct other physical experiments. Anyone who even implicitly holds such a view would not only be misrepresenting the sciences but would also be excluding many of the most significant scientific pioneers in history. Surely we will not object to calling Einstein a scientist even though he was not an experimentalist. Empirical research is, obviously, indispensable for physics, but so, too, is the role of the theoretician seeking the underlying mathematical laws—and metaphysics underpins and permeates both approaches. I am not saying that Platonism is the only philosophical tradition that has been important to the development of modern physics (or the sciences in general), but I am saying that Platonic realism provides the unifying, metaphysical foundation for physics (and thus for the sciences in general).

Platonism has often been presumed to be a form of idealism[127] when, in fact, it is realist. But this issue is not so simple. Philosopher Thomas Whittaker makes the following crucial distinction between ancient and modern idealism: *ancient idealism* assigns greater reality to the unchanging (such as a law of physics), whereas *modern idealism* claims that there is no definable reality beyond the appearances of things in consciousness,[128] which is a form of antirealism. But is Platonism really realist? Yes, it is *extreme* realism, the only sort that is capable of accounting for the foundation of all reality, including nonphysical objective truths, the physical universe, and sentient creatures. As Gerson notes, 'Platonism is firmly committed to the existence of an intelligible, that is, immaterial or incorporeal realm, that is ontologically prior to the sensible realm. Thus, Platonism is a form of explanatory realism, in principle similar to theories that posit neutrinos or the unconscious to explain certain phenomena'.[129]

Philosopher Roberto Poli recognizes the extraordinarily difficult problem of the relationship between the physical world (what he calls the 'real world') and the ideal world. He writes that 'until a way to coordinate forms of reality and ideality is found, it will be impossible to understand the unreasonable effectiveness of mathematics, as well as the ability of values to become dimensions of reality'.[130] Eugene Wigner, a physicist and mathematician who made important contributions to quantum theory, wrote a paper on the 'unreasonable effectiveness of mathematics', where he explored a special kind of miracle: 'the miracle of the appropriateness of the language of mathematics for the formulation of the laws of physics is a wonderful gift which we neither understand nor deserve'.[131] Through Platonism, however, we can hope to understand to a greater degree the role of mathematics in physics, as well as how physics is underpinned by higher metaphysical principles that are inherently full of value. Poli is correct about the importance of values, not in the sense that they need to *become* dimensions of reality but in the sense that we need to *discover* them in reality.

Given that quantum theory is so poorly understood at its conceptual foundations, providing a metaphysical framework within which one can begin to grasp its various apparent paradoxes is clearly an important step. Antirealism in academia, and in society in general, is often derived from misinformed and misunderstood assumptions about the supposed antirealism of quantum theory, and so by showing explicitly how quantum theory necessarily presupposes realism we will help put an end to such false and harmful beliefs. If Platonic realism is true, then the pre-eminent unifying principle known as the One, which Platonists have identified as the ultimate nature of reality, must be supreme good. The metaphysics that the Platonists put forward entails that the universe is intelligible, beautiful, and good, and that humans need to be transformed in order to follow these ultimately real moral/spiritual aspects of reality. As Gerson aptly puts it: 'one might say that the first principle of Platonic ethics is that one must 'become like god.'"[132]

Metaphysics, mysticism, and mind journeys

The terms 'mystical' or 'mysterious' may be used to conceal the fact that we do not really know what we are talking about, and 'leaving something in a mystery' does not at all help us to develop a more profound understanding or explanation of it. But we need not be concerned about eliminating mystery through rational comprehension, as new knowledge inevitably leads to more questions than it answers. The term 'mysticism' is sometimes misused to represent the anthropocentric view that human minds *create* reality, which is part of the antirealist agenda. Our beliefs, actions, and perceptions have a direct impact on how we react to and interpret our surroundings, but that obvious fact in no way entails that we can instantly create any reality we want. It also does not mean that we create electrons just by measuring them or looking at them. After all, we cannot look at or measure something that is not there. Those who disagree should be happy to admit that they can physically observe nonexistent and unreal entities.

However, the issue is a bit more complicated. As Schäfer writes, 'quantum entities observed act like particles; not observed, like waves'.[133] Thus, to say that observation or measurement creates quantum entities is really to say that this process is necessary in order to physically actualize a wave state. Waves, however, which are 'non-material' and 'mass-less',[134] are already real; otherwise we would not be able to measure them or actualize them. But when we physically actualize the wave state as a localized particle through the measurement process, we are not literally creating electrons out of nothing because wave states are already real, even though they are non-material (or at least not material in any obvious sense). Despite the mysterious nature of the process of physically actualizing particles, it does not mean that we can instantly create anything at all; after all, I cannot instantly physically actualize a dinosaur in my living room.

In any case, by arguing that we cannot look at or measure something that is not there, we have just employed a very simplified form of metaphysical reasoning. We have used reason to show how it makes no sense to say that we can physically measure or look at something that does not physically exist. Philosophers do not usually carry out experiments (certainly not the sort of physical experiments that take place in a physics laboratory), but instead they rely upon reasoned arguments to tell us something about the nature of reality, which is to engage in metaphysics.[135]

Even if we say that metaphysics cannot tell us anything about reality, we have just used reason, rather than experiments, to tell us something about the way things actually are in reality. Thus, arguing that we cannot say anything about reality is actually telling us something about the nature of reality, which is to do metaphysics. Jennifer Trusted, who specializes in the history and philosophy of science, notes that 'all science presupposes some metaphysical system of beliefs, and mystical beliefs have been an important part of most systems'.[136] In fact, there was a time when metaphysics—not science—was thought to be the highest form of knowledge.[137]

The depth of our knowledge increases to the degree that we understand abstract principles and laws, which themselves do not change. For example, knowing how to turn on a car engine does not necessarily mean that I am a competent mechanic. I know that if I turn the key in the ignition my car engine will start, but if the engine does not start then I cannot fix it because my knowledge of car engines is limited. My knowledge of car engines increases as I gain technical competence, but such knowledge significantly deepens as we further understand the relevant scientific principles and laws that make possible our technical knowledge in the first place.

To the degree that we know the highest or most fundamental simple, abstract, unchanging, and unifying laws, we possess greater knowledge. Sometimes this greater knowledge may not help us practically, such as if we were to find ourselves stranded in the forest incapable of building a decent fire despite knowing how to build a nuclear reactor. Nevertheless, our depth of knowledge depends upon our understanding of the deeper underlying principles and laws, which eventually requires at least some degree of metaphysical understanding. Plato uses the term 'knowledge' more precisely than we tend to today, and he was correct that genuine knowledge, or the deepest knowledge, is only possible once we have apprehended that which really is, that which is unchanging and yet is responsible for all changes.

In Chapter Four I clarify a helpful tripartite distinction between pure, applied, and presupposed metaphysics, so here we need only understand that metaphysics embraces both epistemology and ontology. In other words, metaphysics deals with the most abstract questions concerning knowledge, existence, and the higher nonphysical aspects of reality, with its main tools being rational arguments, relevant physical evidence, and intuition.

Metaphysics had begun to fall into disrepute during the Age of Enlightenment in the eighteenth century, especially with the widespread turn against the idea of 'natural kinds', the Aristotelian category of final cause, and other metaphysical entities considered unnecessary for explaining the supposed constant conjunctions of events observed in the physical universe.[138] Auguste Comte, arguably the founder of positivism and one of the founders of sociology, misguidedly attacked metaphysics: 'theology and metaphysics, said Comte, were earlier stages in human development, and must be put behind us, like childish things'.[139]

But metaphysics cannot be vanquished, since it underpins every thought and utterance that has ever been made and ever could be made. Trusted correctly notes that, 'if thrown out of the house, metaphysics has a tendency to re-enter through the back door'.[140] Burtt makes a similar point: 'there is no escape from metaphysics, that is from the final implications of any proposition or set of propositions. The only way to avoid becoming a metaphysician is to say nothing'.[141] To clarify: just because one chooses to speak, it does not follow that one is therefore a metaphysician. But making any sort of relatively coherent utterance whatsoever does necessarily entail that one is holding explicit or implicit metaphysical presuppositions. Philosopher A. E. Taylor also observed that 'every great metaphysical conception has exercised its influence on the general history of science, and, in return, every important movement in science has affected the development of Metaphysics'.[142]

Taylor's point is correct, except that the most fundamental metaphysical ideas, such as symmetry, order, truth, beauty, and unity, must always be presupposed and cannot be eliminated by *any* development in science. For example, we can argue about what the concept of order actually means, and we must inquire into what it is, how it is, and how it operates throughout the universe, but no science can ever say that there is no order whatsoever. Order is presupposed by the very possibility of science and by any coherent attempt to say anything at all about anything. It is in this sense that fundamental metaphysical ideas cannot be overturned by any future science.

Philosopher Errol Harris was a strong critic of contemporary analytic philosophy, which I discuss in the next chapter, and he also excelled at showing the importance of metaphysics. However, he argues that his holistic metaphysical position is a 'consequence of the current scientific paradigm',[143] which is an assertion from which I wish to distance myself.[144] Platonism *is* implied by physics but, more significantly, I am arguing that Platonism is *presupposed* by physics and that physics is only possible *because* Platonic realism is true. To put it another way, I am not saying that Platonism is true because it is a consequence of modern science; rather, I am saying that classical, quantum, and any future physics could only be possible if Platonism is true. Harris claimed that his holistic metaphysics would transcend the standard dichotomies of realism versus idealism, and empiricism versus rationalism. The metaphysical view I am offering in this book does exactly what Harris had hoped.

Arthur Eddington, an astrophysicist who provided confirmation of the theory of relativity in 1919, writes that 'the mere questioning of the reality of the physical world implies some higher censorship than the scientific method can supply'.[145] Even inquiring into the meaning of 'science' requires metaphysics. As Whitehead notes, 'if science is not to degenerate into a medley of ad hoc hypotheses, it must become philosophical and must enter upon a thorough criticism of its own foundations'.[146] However, 1,500 years earlier we find Proclus making the correct observation that 'no science demonstrates its own first principles or presents a reason for them; rather each holds them as self-evident, that is, more evident than their consequences'.[147] But if science is not prepared to give some account of its first principles, which requires a metaphysical explanation, then the danger of dogma sets in. Consequently, philosophers and scientists should work together as much as possible, providing mutual enrichment for their respective disciplines.

We might very well be able to convince many people that metaphysics is essential to physics, but mysticism seems to cross the unspoken line of academic respectability. Nevertheless, as mentioned in Chapter One, there have been several popular books, many written by physicists, which have claimed something along the lines of physics proving Eastern mysticism. Usually the position offered is antirealist, whereby we are apparently able to create any reality we want just by thinking about it or perceiving it. This naïve view of mysticism is not what I am talking about here. The original Greek word *mustikos* referred in a more general sense to something mysterious, secret, or hidden.[148] Bernard McGinn, a specialist in Western mysticism, admits that referring to Plato as a mystic is a controversial issue, but that he nevertheless has no hesitation in doing so.[149] But this claim should not be controversial at all. A thorough reading of Plato's dialogues makes it unambiguously clear that he prized direct inspiration from the gods above (but *not* in opposition to) rational analysis, which is always essential for testing and clarification. This view of the interplay between intuition and rationality also forms the cornerstone of scientific genius, where the logic inherent in some intuitive knowledge may not become clear until confirmed experimentally.

There are at least eight (and probably more) basic ways of categorizing mysticism, none of which have to do with being anti-rational. Mysticism can be characterized as:

1. a part or element of religion;
2. a process or way of life;
3. an attempt to express a direct consciousness of the presence of God;[150]
4. referring to some hidden or secret teaching, doctrine, or revelation;
5. an 'individual's direct experience of a relationship to a fundamental Reality'[151] [to expand: a direct experiential understanding or knowing of deeper levels of the nature of reality, ultimately involving a greater or lesser degree of intangible contact or union with God (or the ultimate nature of reality) and direct realization of the nature of the self that is having the experience];
6. a vital part or element of the most fundamental creative processes, whether in the sciences or the arts, or in any other field or endeavor;
7. the foundation of rationality and logic; and,
8. 'seeking everywhere for evidence of mathematical proportion'.[152]

For someone who has had a direct experience of ultimate reality, it seems as though nothing can really be said about this experience, for anything we say is in some way moving us away from the essence of the experience. There is also a way in which appropriate silence can offer powerful personal insight or guidance for others. But if we want to attain a fully integrated life where our mystical insight is related to science, business, and the arts, then there *is* a lot to say.

An excellent example that brings together several of the above meanings of mysticism is provided by Fanchon Fröhlich in her Biographical Notes of her late husband, Herbert Fröhlich, who was a highly acclaimed physicist at the University of Liverpool. His scientific contributions are extensive, including 'providing the first successful explanation of superconductivity as the result of an electron-phonon interaction', and being 'a pioneer in introducing quantum field theory into solid-state physics'.[153] However, the foundation of his scientific genius was direct mystical experience. F. Fröhlich, herself a philosopher and an artist, relates how her husband believed 'that there is an impersonal, non-individualistic path or Tao embedded both in the world and in the mind, and that at some deep level of insight they coalesce'. He also regarded the role of abstract mathematics 'as a source of wonder and mystery', and he often said that 'in the creative process of thinking his mind goes out from his human frame and *becomes* the physical particle and field situation, feeling directly how they tend to behave'. He would then use mathematical techniques 'both to capture this unknown situation and as an anchor so the mind can return to his own brain or everyday personality'.[154]

Fröhlich's mind journeys into the subatomic realm were a vital part of his scientific endeavors, but it is this very sort of experience that is almost ubiquitously ignored because it does not fit our psychologically comforting model of merely collecting empirical data and providing straightforward logical analysis. How many times do we need to be reminded that so many pioneering theoretical physicists have had such deeply mystical inclinations permeating their whole approach to physics? It seems that no matter how often such accounts are offered, many scholars are still incredulous, or they try to explain them away as malfunctions of the brain. Yet the fact remains that Fröhlich was an incredibly successful scientist, which cannot be explained away as mere errors in the functioning of his brain. Even if his mind did not actually go 'out from his human frame', the experience that he underwent needs to be explained, and not just explained *away* with a dismissive wave of the hand. Whatever the actual nature of his experience was, it played an important role in his scientific results.

Understandably, however, deep metaphysical reflections and mystical experiences do not usually get discussed in scientific journals. But even if 99% of all scientists never have such experiences, it is still a fact that some do, and these experiences and metaphysical ways of thinking have shaped or informed their understanding of, and approach to, science. This fact is enough to provide scientific and logical justification for further inquiry into these domains. By ignoring such facts, we are left assuming that every aspect of scientific methodology can be reduced to nothing more than postulating and experimentally testing a hypothesis, a misleading image to which many scientists cling as well. If your hypothesis is that mysticism and metaphysics are nonsense, or not relevant to science, you now have evidence that proves your hypothesis to be false.[155] As good scientists, we cannot ignore (or superficially explain away) the evidence.

This mystical aspect of pioneering foundational physics is not at odds with empirical evidence or logical analysis, for we must always aim for logical coherence and rely upon empirical data so far as possible, but mysticism does underpin both logic and our data. The creative and intuitive aspects of the scientific enterprise cannot be ignored without forsaking genuinely novel scientific advancement. This one insight alone is enough to necessitate appropriate changes in how various relevant funding bodies operate and how we teach the sciences.

It is unfortunate that so many philosophers feel the need to scoff at mystical claims. I can only assume that the philosophers Julian Baggini and Peter Fosl, who depict mysticism as being unintelligible, unreliable, and inconsistent, are aiming at some other meaning of mysticism than the ones given above.[156]

Physicist Victor Stenger is also intent on removing any notion of Platonism and other such apparent nonsense from physics. He is 'telling people things that many do not want to hear: that according to our best knowledge, the world of matter is all that exists'.[157] Stenger, who is a good example of a philosophically confused physicist with an atheistic axe to grind, has just eliminated mathematics (among many other things). He is also far from the mindset of Einstein and other great visionary pioneers.

Some highly influential philosophers in the analytic tradition, which I discuss in the next chapter, had actually rejected the sorts of metaphysical questions at the core of the most important debates amongst the pioneers of quantum theory. They dismissed anyone who suggested that Platonism might be relevant to modern physics. This sort of prejudice, while not as strong as it was a few decades ago, is still prevalent, and it is part of the task of this book to bring to light and correct such misguided beliefs. One cannot have a penetrating understanding of the history of science and philosophy, while also maintaining that the Platonic tradition has had or still has little or nothing to do with modern science. I am not saying that Platonism is the only tradition that has been important for the development of modern science, but I am saying that it has played a very important role, one that has been ignored for too long, especially by most analytic philosophers.

THREE: The Curtain Falls for Analytic Philosophy

What exactly is analytic philosophy? Philosopher John Heil notes the difficulty in answering this question: 'as in the case of chicken-sexing, it is relatively easy to identify analytic philosophy and philosophers, though difficult to say with any precision what the criteria are'.[158] Rorty tells us that analytic philosophers generally believe that the 'linguistic turn' (which basically entails ignoring our own experiences, feelings, and states of mind and focusing solely on language) coupled with symbolic logic can 'turn philosophy into a scientific discipline'.[159] Another philosopher, Tom Sorrell, adds that 'analytic philosophy is not uniform, but it usually aspires to a very high degree of clarity and precision of formulation and argument, and it often seeks to be informed by, and consistent with, current natural science'.[160] The irony is that while pretending to be emulating the sciences, such philosophers have generally ignored both the metaphysical and mystical thinking of many of the greatest pioneering theoretical physicists throughout history.

This chapter will offer a deeper understanding of the roots of many of the misconceptions regarding the relations between philosophy and physics. If you have a background in academic philosophy, then much of this chapter will probably be somewhat revealing. If you have no such background, you may at first wonder what all the fuss is about. Who really cares about seemingly arcane disputes between academic philosophers? I can assure you that these issues have had, and will continue to have, extraordinary consequences for society. In a similar way, the powerful conversations between Bohr and Heisenberg ended up having major consequences in society through the development of quantum physics.

After a brief but revealing historical background of analytic philosophy, we will be in a better position to appreciate five false assumptions embedded in this tradition that are antithetical to the metaphysical underpinnings of modern physics. These five false assumptions are:

1. the history of philosophy is irrelevant;
2. metaphysics (the deep kind) should be eradicated;
3. the linguistic turn provides the only proper methodology for philosophy;
4. the relation between nonphysical consciousness and matter is not a serious philosophical question;[161] and,
5. there are logically and physically separate atomic facts.

I also discuss the limits of analysis itself, which opens the way for more confidently confronting many perplexing questions and paradoxes embedded in modern physics. This chapter shows how quantum physics has gone in the opposite philosophical direction as contemporary analytic philosophy,[162] the branch of philosophy to which the dominant group of academic philosophers belong.

Is philosophy dead (again)?

The tension between philosophy and physics is not as obvious as that felt between science and religion, but it is actually more fundamental and in some senses more important. After all, if the methodological, rational, and systematically logical thinking of the philosophers is somehow or other in conflict with physics, then it would seem, at first glance anyway, that physics is in a lot of trouble. Surely physicists must think rationally and systematically while employing logical relations. If this is so, why has there been such conflict, or, as may more often be the case, why have so many representatives of both disciplines chosen to ignore each other? What may be even worse is when philosophers promote the false idea that philosophy should do nothing more than play a subservient role to the sciences. While it is true that philosophy is important for science, it also has other goals that lie outside the bounds of science.

While many pioneering physicists in quantum theory were openly philosophical, they often rejected, or saw significant limitations with, *contemporary* philosophy, which is an important challenge that we must investigate. Possessing only a superficial or naïve understanding of the issue exacerbates this confusion, as we find in the works of physicist Stephen Hawking. He and co-author Leonard Mlodinow proclaim at the beginning of their recent book that 'philosophy is dead'.[163] Such a remark is the academic equivalent to the sort of pop-culture sensationalism we would expect to find on a supermarket magazine stand.

The authors also claim that 'scientists have become the bearers of the torch of discovery in our quest for knowledge'.[164] The irony here is that any claim about knowledge eventually falls under the purview of philosophy, specifically (though not exclusively) in the field known as epistemology, which deals with such questions as: What is knowledge? How do you know that you know? What counts as genuine knowledge? and so forth. Therefore, if philosophy is dead, Hawking will be unable even to tell us what it means to discover knowledge. In fact, to *discover* knowledge, as he states that scientists (and only scientists) are able to do, necessarily implies that the knowledge that they are capable of discovering already exists; otherwise they would not be able to discover it.

But to say that knowledge already exists is to hold a realist position of one sort or another and, in fact, Hawking and Mlodinow do claim that their whole approach hinges on 'model-dependent realism'.[165] Their version of realism, however, seems to slip immediately into antirealism, which is thoroughly analyzed in Chapter Seven.[166] All that matters here is that we recognize that any claim about being any kind of realist (or antirealist) is a purely *philosophical* claim. In other words, Hawking declares philosophy to be dead and then miraculously (yet furtively) brings it back to life. Let us set aside these sorts of attention-seeking remarks about the demise of philosophy and investigate far more interesting challenges.

A brief (but revealing) history

In order to understand the tensions between philosophy and physics, especially with respect to deeper metaphysical questions, we must first know something about analytic philosophy. On the one hand, analytic philosophers are exceptional at reasoned analysis (within narrowly set parameters), and studying with them can sometimes feel like a sanctuary in a world full of antirealist agendas. On the other hand, they have often been their own worst enemy, contributing to their own irrelevancy, which is arguably linked to the recent closure of several university philosophy departments.

Many philosophy departments now wield such little relevant power in the universities that they end up retreating even further, leaving the philosophers never knowing when their coveted tenure could suddenly be cut short. Only when finally faced with the immediate threat of extinction do they urge the philosophical community to sign petitions and write letters to save them. Sometimes they achieve temporary success, and sometimes they do not. Two relatively recent cases concern philosophy departments that have employed those responsible for the two largest philosophy email lists in the world. The department of philosophy at the University of Liverpool narrowly escaped closure,[167] but 'the Louisiana Board of Regents cut the philosophy major at the University of Louisiana at Lafayette'.[168]

Our brief excursion into the history of analytic philosophy is just as likely to confuse as clarify; however, the confusion will actually result from further clarity. For example, many analytic philosophers tend to dismiss the Platonic tradition and ancient philosophy in general, despite the fact that such philosophers tend not to be well versed in ancient philosophy and so do not really understand what it is they are dismissing. More interestingly, two of the most important founders of analytic philosophy, Bertrand Russell and Gottlob Frege, were both originally Platonic realists with respect to mathematics and logic, which is a fact that seldom receives appropriate attention in today's philosophy classrooms. For philosophers whose claim to fame is seeking intellectual clarity, they are quite poor when it comes to clarifying their own internal inconsistencies.

Russell and Frege were searching for unconditioned truth in mathematics and logic, truth that resides beyond the bounds of space and time, while the logical positivists, who had a very important impact on the development of analytic philosophy, wanted to eliminate all metaphysics and essentially reduce all knowledge to mere subjective experiences. One would assume, therefore, that a Platonist would find good company with Russell; however, he held a multiplicity of positions, such that 'few claims about his views can be made without qualification as to time'.[169] In fact, Russell ended up rejecting his earlier Platonism,[170] which, as already mentioned, was the opposite case for Einstein, who had moved to a Platonist position just in time to help him develop his pioneering work in physics. Thus, one of the most influential philosophers in the last century took a wrong turn away from Platonism, while one of the most important physicists in all of history took the necessary right turn toward it. Consequently, my criticisms of analytic philosophy are aimed more at what it has become rather than its original intentions. From this perspective, one may say that in some ways I am actually more of a traditional analytic philosopher than many of those who currently self-identify with this tradition.

Preeminent analytical philosopher Willard Quine and his followers really do not have anything significant to add to our explorations here, since the analytic movement has gone in the opposite direction of the most important developments in modern physics. It is not that philosophy must follow science, but if your philosophical position is completely at odds with current developments in science, then a serious examination is required to uncover the problem. The problem may be in the sciences, or in your interpretation of the sciences, or it may be in your philosophical assumptions. In any case, there is no point in indulging the analytic tradition when it rejected the founders of the most powerful scientific theory in history, and especially when Quine assumed that 'science can replace philosophy'.[171] With some of the most influential philosophers arguing that philosophy should be replaced by science, is there little wonder that various philosophy departments have come under threat of closure? Did the leaders of analytic philosophy really intend to eliminate philosophy?

Frege is generally considered to be the 'grandfather of analytical philosophy'[172] who, despite being a Platonist (at least a logical/mathematical Platonist), helped pave the way for logical positivism and analytic philosophy.[173] Sociologist Randall Collins reveals the irony that 'Frege's Platonism, when broadened into an epistemology, gave rise to the imperious claims of logical positivism',[174] while at the same time opening 'the path for [philosopher Ludwig] Wittgenstein and the recognition that a language or symbol system contains multiple levels'.[175] By neglecting their own history (until relatively recently and, even then, quite selectively) and by ignoring the complex historical interrelations between science, philosophy, religion, and politics, many analytic philosophers have been able to utilize key ideas from Platonism while at the same time pretending to despise Platonism and the history of philosophy. Of course, not all analytic philosophers reject all ancient philosophy; some are still staunch supporters of a generally Platonic view and believe that truth is objective. Nevertheless, even those who are somewhat sympathetic to Platonism are still highly restricted in their style of philosophical presentation, which, by necessity, precludes deeper metaphysical and mystical implications.

Dummett argues that, contrary to the standard view concerning the development of so-called 'Anglo-American' philosophy, Russell and G. E. Moore were at best only 'great uncles' of the analytic philosophical movement, because the real impetus came principally from the German philosophers.[176] Dummett seems to be correct here as concerns the beginnings of the movement, but Collins offers a more balanced perspective: 'modern logical philosophy comes from the convergence of two lines, German and British. Russell, who brought them together, was the product of a British network going back several generations'.[177]

Analytic philosophy has been around for just over a hundred years, during which time it has taken deep root in North America, the United Kingdom, and elsewhere. It has had tremendous influence on the development of aspiring young philosophers, especially by strongly discouraging them from thinking metaphysically and by shunning anything akin to mysticism. The fact that some analytics are becoming interested in their own history is at least a step in right direction, but there is still a general underlying assumption that nothing worthwhile actually occurred before Frege. There is also now a certain subclass of philosophers specializing in 'analytic metaphysics', but they are still quite restricted in style and content. It is unfortunate that they are pressured to say that they are doing *analytic* metaphysics instead of simply saying 'metaphysics'. The 'analytic' label is somehow or other meant to give them more respectability.

Despite the influential role of logical positivism (sometimes referred to as the 'Viennese Circle') on analytic philosophy, the positivists were actually *anti*-philosophy. In *The Unity of Science*, Rudolph Carnap, a strong proponent of logical positivism, proclaimed that 'the Viennese Circle Does Not Practice Philosophy' and that they '*pursue Logical Analysis, but no Philosophy*'.[178] Carnap was explicit in his dismissal of philosophy, which should lead any reasonable person (other than a professional philosopher, apparently) to ask an obvious question: if the positivists admittedly did not do philosophy, then why didn't they remove themselves from philosophy departments?[179]

Consider the following hypothetical scenario, which will help to illustrate this point. Medical doctors practice an art and rely upon the sciences. They use various scientific technologies, from x-rays to the latest pharmaceuticals, yet they also require intuition, sensitivity, and empathy—at least ideally. Some would argue that the latter characteristics fall outside the realm of science and into psychobabble. Imagine a small group of x-ray technologists who claim that they will save the medical profession from such non-scientific nonsense, because it is x-ray technology that tells us whether or not a patient has a broken collarbone, tuberculosis, or a serious dental problem. Empathy, sensitivity, and intuition are not required, nor is it necessary to ask patients what they feel about their own body or state of mind, as all such things are nonsense when compared to the precision of scientific measurements.

Now imagine that these x-ray technologists claim that they do not practice medicine and refuse to engage with typical medical questions, such as asking patients how they feel or where it hurts, and instead reduce all medical practice to x-raying. These technologists then take over the medical schools and ban all those who practice the 'old-fashioned' nonsensical art of medicine, and the coup is complete. This is essentially what has happened in academic philosophy, whereby a minority of logical positivists, who excelled in devising and implementing important logical tools, overran traditional philosophy. They expunged the most meaningful and important philosophical questions and modes of thinking in much the same way as our hypothetical x-ray technologists overthrew the medical doctors.[180]

Heisenberg admits that there were (or appear to have been) elements of positivism in relativity and quantum theory, for the physicist Ernst Mach had some influence on both Einstein and Heisenberg, yet he adds that Mach's influence should not be overestimated.[181] This caveat is important. After all, Schrödinger claims that Mach's views were very close to the orthodox dogma of the Upanishads,[182] and Philipp Frank, another physicist in the Vienna Circle, notes that despite his positivism, Mach has also been 'proclaimed as a champion of the idealistic philosophy in modern science and as a leader in the struggle against materialism'.[183] Stanley Jaki, who held a doctorate in both theology and physics, added that 'Mach could even speak of Buddhism as a religion most germane to science'.[184]

Heisenberg is often referred to as a positivist,[185] and at some points he does sound like one, such as when he says that we derive empirical laws from observations. But his apparent positivism was related to his rejection of materialism, such as his denial of atoms being physical things or objects.[186] Observations obviously play an important role in helping us discover the laws of physics (sometimes called empirical laws), but the problem is that positivists believe that these laws are mere fictions that somehow or other prove convenient in predicting observations. Heisenberg, however, believed in the nonphysical reality of the mathematical laws of physics.

I am not saying that the positivists had no influence on Heisenberg whatsoever. My point is that he was not a positivist because he refused to accept their self-defeating, anti-philosophical limitations. Both Bohr and Heisenberg agreed with the positivists that we should attempt to give the most accurate and precise clarification of whatever it is we are talking about, insofar as we are able, but such demands are not unique to the positivists; the desire for analytical precision goes back at least to the Greek founders of philosophy.[187] If there is still doubt about Heisenberg's position, allow him to speak for himself: 'the Copenhagen interpretation of quantum theory is in no way positivistic'.[188] He also calls positivism a 'pointless philosophy'[189] and rightly claims that the 'insistence on the postulate of complete logical clarification would make science impossible',[190] which effectively puts an end to the positivistic dream.

Positivism is inherently self-contradictory at its most basic level. For example, positivists reject metaphysics, yet they seek the *unity* of the sciences, even though the very concept of 'unity' has no physical reality and is purely metaphysical. Comte even went so far as to write that 'the first characteristic of the positive philosophy is that it regards all phenomena as subjected to invariable natural *laws*'.[191] However, if a law is invariable, then it never changes, which makes it eternal. Thus, all phenomena would be subject to nonphysical eternal laws.

Finally, Mach's empiricism (or whatever metaphysical view he held), which was an important ingredient of positivism, sought 'the comprehension of as many facts as possible by the simplest possible system of propositions'.[192] The irony, of course, is that seeking unity, simplicity, and invariable natural laws makes the positivists appear to be straightforward mathematical and scientific Platonists.[193] Yet the positivists ignored the deep metaphysical thinking of physicists such as Bohr, Heisenberg, Pauli, Einstein, and Schrödinger, and abhorred the Platonists. They pretended to be able to reject metaphysics and believed that 'concepts such as "mass," "force," and "atom" are merely convenient fictions for simplifying observations',[194] all the while unwittingly adhering to several key aspects of Platonism.

This irony is exacerbated by the fact that, as already mentioned, Frege and Russell (at least at some point in the latter's career) were Platonists, a vital point that is too often neglected by their academic devotees. Dummett notes that 'on Frege's view, thoughts and their constituent sense form a 'third realm' of timeless immutable entities which do not depend for their existence on being grasped or expressed',[195] which sounds like straightforward Platonic realism. Similarly, philosopher Peter Hylton points out that during his Platonist period, Russell insisted that 'mathematics is wholly independent of space and time'.[196] Russell even believed that the mind could potentially have 'direct and unmediated contact with abstract objects', and that 'the absoluteness and objectivity of truth requires the objectivity and independence of propositions'.[197]

This notion of the mind having 'direct and unmediated contact with abstract objects' clearly denotes a mystical experience whereby the mind accesses the nonphysical realm. Thus, Russell, a key founder of the analytic tradition, adopted a Platonist position that included the necessity of some sort of mystical experience at its epistemological foundation. Russell was essentially correct on these particular points, despite focusing on propositions rather than higher aspects of reality, and it is unfortunate that his earlier views are dismissed in favor of what philosopher James Baillie refers to as his later 'mature' philosophy.[198] Russell did not mature philosophically in his later years; he turned away from the road to truth. It is interesting to note, however, that Russell was a friend of Aldous Huxley and would sometimes join him on his visits with Jiddu Krishnamurti, the man once deified by the Theosophists.[199] Perhaps some sort of a spiritual longing remained within Russell after all.

It is also interesting to note that several members of the Vienna Circle, such as Mach and Frank, among others, were actually physicists and mathematicians. It is common practice among professional philosophers to refer to all the members of the Vienna Circle as philosophers, yet those physicists, such as Planck and Heisenberg, who held Platonic assumptions about objective truth (and who just so happened to bring us quantum theory) are explicitly *not* referred to as philosophers. In fact, there has been a consistent and seemingly concerted effort to deny these physicists their rightful place in philosophical discussion. This prejudice can no longer continue, especially when so many anti-philosophy academics have taken over the philosophy departments. As Collins shows, Russell and Wittgenstein were actually 'hostile to philosophy', but they ended up in the attention space of philosophers because that is where the most interest was to be found concerning meta-mathematics.[200] Unfortunately, Russell's prejudices impaired the analytic philosophers' ability to comprehend the depth of the implications of quantum theory.

Staying out of touch (five false assumptions)

In his 1998 APA (American Philosophical Association) sponsored address at the XXth World Congress of Philosophy, Robert Audi outlined several major problems facing academic philosophy today, which I have summarized below. (These five problems are *not* the same five false assumptions that we will examine in this section, but they provide an important starting point.)

1. philosophy is weak in undergraduate education in the US and elsewhere;
2. philosophers are not engaging with the public sufficiently;
3. diminishing research support (often because of (2));
4. concern about an insufficient balance between philosophical grounding and highly specialized or interdisciplinary work, and too little engagement with or even study of other fields; and,
5. a provincialism manifested in, for example, 'stereotyping the views or methods of other philosophers or in positive indifference to alternative perspectives'.[201]

Audi's criticisms indicate some of the underlying and often neglected problems in academic philosophy. Norris would add that 'analytic philosophy of science has taken a number of wrong turns as a consequence of its becoming so far out of touch with developments elsewhere'.[202] Such issues have an important impact upon philosophy, including what sorts of questions we think are worth asking and which topics are worth pursuing (and publishing), what style and tone of writing we should adopt, and which philosophers we should be quoting from. My critique here, however, is more tightly focused on the five false assumptions in analytic philosophy, mentioned in the introduction to this chapter.

It should be noted that although the assumptions I critique here are usually associated with analytic philosophy, there are some philosophers who consider themselves to be 'analytic' but who do not hold all of these assumptions. Likewise, some philosophers may hold one or all of these assumptions yet maintain that they are not analytic philosophers. Hence, my critique is more specifically aimed at a particular set of assumptions that have hampered fruitful dialogue between physicists and philosophers concerning fundamental conceptual difficulties in quantum theory (and in other capacities). Nevertheless, these assumptions are more likely to be held by analytic philosophers, and so aiming my criticisms at analytic philosophy in general (while recognizing the likelihood of various exceptions) is clearly warranted. It is also the case that the assumptions critiqued here may be useful in other philosophical contexts (perhaps in the philosophy of language), but the fact remains that they are still antithetical to a deeper reflection on fundamental philosophical issues in modern physics, and therefore they must be criticized and set aside here.

1. *The history of philosophy is irrelevant*

Analytic philosophers generally admit that a limited examination of the history of philosophy can be a useful pedagogical tool, especially when introducing philosophy to undergraduates. But Sorrell also makes explicit the fact that 'there are philosophers of the past who are used for target practice— that is, whose ideas are currently widely rejected, and who are referred to mainly as sources of deep delusion or fallacy'.[203] Sorrell continues: 'often analytic philosophers are casual in their use of historical figures. For example, there may be a good basis in Plato's texts for associating him with Platonism in mathematics, but no one interested in Platonism in mathematics cares whether what is called 'Platonism' fits those texts'.[204] In other words, they have no concern for history.

This rejection of the history of philosophy as a worthy pursuit was vociferously championed by Carnap and culminated in Quine, who, as Rorty had good occasion to experience, 'made a point of reading as few of the canonical texts as possible, and he recommended this practice to his students at Harvard'.[205] Ironically, *all* the original founders of analytic philosophy have now passed into the history of philosophy, and so by Quine's own decree we should not study them. This is not a trivial point. If we take their anti-history approach seriously, then we should no longer read Russell or Frege, or even try to figure out what they meant. Baillie notes that Frege's *Über Sinn und Bedeutung* 'had long been translated as "On Sense and Reference." However, in recent years, opinion has been that "Meaning," rather than "Reference," is closer to what Frege intended by "Bedeutung."'[206] Since 'meaning' and 'reference' have different definitions in English, it is very significant that opinion has changed concerning which word best approximates the original German, and such further understanding is only possible through historical, sociological, and philological study, all of which are usually neglected by analytic philosophers.

Similarly, if we want to understand Russell's philosophical views, we need to understand the historical context in which they are placed. We must also understand Russell's understanding of the prior philosophers to whom he was reacting, such as Kant, which means that we need to understand Kant. Understanding Kant requires going still further back in history, and so on with Kant's predecessors. If the history of philosophy is irrelevant, then we should no longer cite any works written prior to our own. Wittgenstein, in fact, did just that. In the Preface to his *Tractatus*, he writes that 'what I have here written makes no claim to novelty in points of detail; and therefore I give no sources, because it is indifferent to me whether what I have thought has already been thought before me by another'.[207] If those many academics who esteem Wittgenstein are to take him seriously, then, they should never bother quoting him or even studying him (or anybody else for that matter).

We must not slavishly submit to prior authoritative voices just because they have attained canonical status. What matters, always, is what is actually true. Plato, Nāgārjuna (the famous third-century Buddhist philosopher), and Russell can all be helpful guides on our quest for truth, since they themselves blazed many philosophical trails, and we should learn all we can from them, including from their misadventures. But we must also do the philosophical work ourselves as we forge new trails, which hopefully will inspire and guide others as well. It is ironic that analytic philosophers who usually eschew the past, and are apparently more concerned with arguments than philosophical 'authorities', so often practice the opposite of what they preach. Philosopher John Cottingham makes this point well: 'as an editor of a fairly 'mainstream' journal of modern analytic philosophy, I am constantly struck by the number of submitted articles that to all intents and purposes begin and end with 'Thus spake the master'.[208] The master, in such cases, is not Aristotle or Plato, but Wittgenstein or some other currently worshipped philosopher. Cottingham continues: 'so we have irony here: though many analytic philosophers take a derogatory attitude to history of philosophy, it turns out that the faults they attribute to it are often glaringly manifest in much of their own so-called cutting-edge work'.[209]

The very meaning of the word 'history' is itself ambiguous. Strictly speaking, the very act of reading the sentence prior to the sentence you are currently

THREE: The Curtain Falls for Analytic Philosophy

reading is now a part of history. Each moment that passes becomes part of history, and thus, according to the viewpoint that rejects the study of history, we should not even study what we ourselves have just written or said. Despite this blatant rejection of the history of philosophy, Dummett writes that 'it is important to analytical philosophy that it understands its own history, seeing itself in the context of the general history of philosophy during the nineteenth and twentieth centuries: especially is this true at a time when it is undergoing profound changes'.[210] Surely it is one of the bigger hypocrisies in the history of philosophy for analytic philosophers, who have for so many years openly despised the history of philosophy, to start studying a limited version of their own history.

Apparently the reason we should not study the ancient philosophers, however, is that they were wrong about everything—each and every one of them. In actual practice it seems that a philosopher's ideas are only relegated to the dustbin of the history of philosophy if they are no longer considered *relevant*. But if this is true, then questions raised in quantum theory entail that the thoughts of the ancient Platonists are relevant to modern science (and in many other areas), and so by this account, Platonism should *not* be considered part of the history of philosophy. On the contrary, Platonism would then be considered more relevant than any current philosophical stream. In any case, only by actually reading the ancients and understanding their arguments could Quine possibly have had rational grounds to disagree with them. But, as Rorty states, he emphatically made a point of *not* reading the ancients.

Philosopher Ryan Nichols writes that 'many of the best contemporary analytic philosophers have not made detailed historical studies of the views they oppose (or for that matter, any historical studies at all) and some, as we have witnessed, express a thinly veiled contempt for such projects'.[211] Hilary Putnam provides a prime example. Putnam, a very influential analytic philosopher, was the 2007 recipient of the University College Dublin Ulysses Medal, 'the highest honour that the university can bestow', and was praised as being 'one of the most influential and prolific American philosophers of the past fifty years'.[212] Putnam proclaimed that it would be absurd to try to 'believe what philosophers who lived two hundred or two thousand years ago believed'.[213] What is most interesting here is that Putnam and his wife, who had both been raised as atheists, later became practicing Jews. In fact, in 1994, the same year he published his views about dismissing ancient (and even two hundred year old) philosophers, he 'celebrated his belated bar mitzvah'.[214]

To be clear, I am obviously not implying anything derogatory about Judaism. Throughout this book I am, in fact, emphasizing the importance of ancient philosophical/spiritual traditions. But it is important for us to recognize the significance of Putnam's inconsistency (or, some may say, hypocrisy). He has been turning many students and colleagues against ancient philosophy, while at the same time immersing himself in an even more ancient philosophical and religious tradition, a tradition that just so happens to have developed intimate connections with Platonism.[215] One of the reasons he rejects the ancients in philosophy is that he believes that 'they lived under wholly different conditions and faced wholly different problems', and that such 'a return is impossible in any case'.[216] If philosophers two thousand or even two hundred years ago lived under such wholly different conditions than we do, and if this is to be taken as a reason for rejecting them, then surely the same reasoning would apply to Judaism (and to all religions, and all of science, too). The fact is, it is important to study ancient philosophy and Judaism, and history in general. Unfortunately, what many professors preach in class is very often quite at odds with how they actually live their lives. This is why it is so important to return to the ancient philosophical goal of striving to live one's philosophy.[217]

If Putnam were correct, then Schrödinger, Heisenberg, Pauli, Planck, and Einstein were all absurd for holding beliefs and employing methodological thinking akin to ancient philosophers. Fortunately, the founders of quantum physics were not absurd and, as should be obvious, the many problems the ancients faced are similar enough to our own today. We are also no better at living meaningful lives (and perhaps are somewhat worse) than philosophers who lived two thousand years ago.

Quine makes a similar error as Putnam in saying that the history of physics is irrelevant to current scientific research. First, physics graduates today must be knowledgeable about relativity, a theory that is now more than a hundred years old. If Quine were correct that the history of physics is irrelevant to modern physics, then we should not study relativity but should concern ourselves only with the current number of dimensions postulated by string theory. Quine would probably have replied that the 'history of physics' refers only to those theories that have been rejected and no longer constitute part of the body of accepted scientific theories. But such a view would also be disastrous to physics. Copernicus revived a 'dead' heliocentric theory; at various points it was believed that particle theories should be replaced by wave theories and vice versa; and atomism has been rejected, revived, and simultaneously rejected and revived. Furthermore, we can only know that a given theory was in fact proven to be wrong in the past if we actually study the history of physics in the first place, which will help us not to repeat the same mistakes.

Rorty admits that in seeking tenure he realized that 'there was little percentage in being historically minded',[218] which further helps to explain why there has been such resistance to taking seriously the metaphysically inclined physicists who were influenced by, or certainly resonated with, ancient philosophers. If you want a job as an academic philosopher, you had better follow the unwritten rules. The ways in which we should engage with the history of philosophy will continue to be an important topic to debate, but the fact that we need to take the history of philosophy (and science) seriously should no longer be an issue. To deny the importance of history is to deny what one has just said.

2. *Metaphysics (the deep kind) should be eradicated*

That metaphysics should be eradicated was an ideology propagated by Carnap, Russell, Wittgenstein, and Quine, among others.[219] There are many possible areas to explore in criticizing these anti-philosophy 'philosophers', so I will narrow the focus to only a few essential points. First, Russell conceded that he 'cannot admit any method of arriving at truth except that of science',[220] which, despite aiming to eradicate the need for philosophy, is itself a metaphysical statement. There is no scientific experiment that could ever prove the statement that the only method of arriving at truth is that of science, and so his claim would not have any meaning, given that it is not a scientifically proven statement; it is purely metaphysical. It is also hard to imagine why he would risk his career and go to prison for his pacifism when, according to his own criteria, moral reasoning could never provide us with truth. Once again, we see another eminent philosopher whose professional views and personal life are not in harmony.

Einstein, too, had his anti-metaphysical moments, although he gained clarity and aptly rebuked Russell in a review of his book, *An Inquiry into Meaning and Truth*. Einstein notes that Russell at least admits in the last chapter of his book that we cannot get along without metaphysics. However, Einstein adds that 'the only thing to which I take exception there is the bad intellectual conscience which shines through between the lines'.[221] Clearly, Einstein had some issues with Russell. In fact, Einstein, Pauli, and Kurt Gödel, a logician, philosopher, and intellectual giant,[222] used to have discussions with Russell while at Princeton. Russell believed that all three had a 'German bias towards metaphysics', and despite their extensive efforts, Russell claims that they 'never arrived at common premises from which to argue'.[223] Russell does not bother trying to hide his sarcasm and disdain: 'Gödel turned out to be an unadulterated Platonist, and apparently believed that an eternal "not" was laid up in heaven, where virtuous logicians might hope to meet it hereafter'.[224]

Philosopher Palle Yourgrau writes that 'the failings of an entire century are crystallized in the fact'[225] that Russell could not find any common ground upon which to engage in philosophical discussion with two of the most important physicists and arguably the most important logician of the last century. Russell's inability to relate to them was essentially because of *his* own bias against metaphysics, and especially against Platonism. Here we can see a good example of why philosophy would develop in directions that placed it squarely against the developments in modern physics; the physicists were too philosophical for the philosophers.

It is also important to note that even though Pauli and Einstein held different views about the interpretation of quantum theory, they seemed to be in agreement about *how to think* about such issues. Russell's inability or unwillingness to go deep into metaphysical territory made his philosophical approach irrelevant to Einstein, Gödel, and Pauli. Because of their inability to engage with the most perplexing questions at the heart of quantum physics in a way that is actually helpful for physicists, analytical philosophers have drawn the reasonable charge that, as physicist Dominic Dickson puts it, they are 'arriving after the show trying to tell us what we did'.[226]

3. *The linguistic turn provides the only proper methodology for philosophy*

Dummett points out, in essential agreement with Rorty, that the analytics believe a comprehensive philosophical account of thought can only be obtained through a philosophical account of language.[227] However, Frege gives no justification for the linguistic turn: 'it is simply taken as being the most natural way of going about the philosophical enquiry'.[228] It is quite significant that arguably the most important assumption of analytic philosophy had no supporting arguments. Moreover, as Dummett further notes, 'Frege indeed so far reacted against 'theories of truth' as to declare truth to be indefinable'.[229] Russell had similar difficulties, as he, too, was unable to define his most essential concept. He had claimed 'that logic consists of tautologies…in spite of the fact that, as he says, he does not know how to define 'tautology''.[230] (Wittgenstein, too, did not even bother attempting to define one of his most important concepts, which is known as 'forms of life', but this does not really concern us here.[231])

It is not that I disagree with Frege, Russell, and Wittgenstein on this point, for it does seem impossible to define unambiguously our most fundamental concepts; I am making a similar case about the difficulty of defining 'unity'. The problem is that analytic philosophers are so quick to attack anyone whose ideas they oppose by saying that their opponents have not clearly defined their terms. Wittgensteinians tend to be the first to say, 'Well, what do you *mean* by X?' whenever they are confronted with a viewpoint that they do not like. Striving for such clarification is essential, but too often this tactic is employed merely to try to put an end to further inquiry. The easiest response in this case is to say, 'Well, when *you* explain what you mean by *mean*, then I will tell you what I mean by X.' It is encouraging that some analytic philosophers are now studying their own history, which will allow them to see how their founder declared such an essential notion as truth to be indefinable, gave no logical justification for the linguistic turn, and was even a Platonist.

It is also false to assume that all of thought can be expressed in language. Einstein would agree: 'the words or the language, as they are written or spoken, do not seem to play any role in my mechanism of thought'.[232] It is wrong to assume that what cannot be expressed in language cannot really be thought or experienced. The assumption that the limit of language is the limit of thought disallows any sort of philosophical probing into deeper questions beyond what can be expressed directly through language. Imagine trying to explain the taste of biting into a lemon to someone who has never tasted one. It does not matter how many descriptive metaphors or similes or detailed physiological explanations one uses; language will never fully capture this experience (or any other experience), and so we must admit that there exist aspects of reality beyond what can be analyzed linguistically.

Similarly, if language fails us when trying to think about and precisely describe subatomic events, then analytic philosophy will be of little value to the physicists reflecting on the paradoxes and conundrums they face. The linguistic turn, therefore, has rendered analytic philosophers incapable of reflecting deeply on quantum theory. As philosopher Ian Hacking notes, 'attempts at scientific reduction – reducing one empirical theory to a deeper one – have scored innumerable partial successes, but attempts at linguistic reduction have got nowhere'.[233]

We should always aim for clarity in natural language, to whatever degree is appropriate, but we cannot be fooled into thinking that all that language expresses is all that there is to reality. The tools of language, whether natural or logical, may limit philosophical expression in a similar way as the relevant tools limit a carpenter's ability to actualize the architect's plans, but carpenters think beyond their tools, and so do philosophers. Reality is deeper than language.

The quantum level of reality is beyond our potential to experience directly with our normal senses, and so we must struggle constantly even to begin to understand what we are talking about. We must use metaphors and analogies and arguments. We must philosophize. But we must philosophize about aspects of reality that transcend our normal experiential potential, which is what the logical positivists had forbidden. If philosophers held prisoner by positivism cannot even talk meaningfully about tasting a lemon, what hope do they have of speaking meaningfully about quantum physics?

4. *The relation between nonphysical consciousness and matter is not a serious philosophical question*

This assumption, which is also criticized by Norris, holds that any appeal to constitutive acts of consciousness should be ruled out because they involve a retreat to naïve psychologism, and that the logico-semantic approach is sufficient.[234] What this technical jargon essentially means is that analytic philosophers want to keep logic separate from psychology, and so the rules of logic must transcend psychology. In that case, however, the rules of logic would have an objective existence (or reality) independent of the human mind, which is an aspect of Platonism. However, we must also admit that even though genuine logical laws have an objective existence, our own states of mind directly impact our ability to understand, pursue, and apply these laws.

The real concern here is that philosophers of science tend to be materialists, and so they also are likely to reduce conscious awareness to brain function. But a functionalist view of consciousness has been strongly rejected by various physicists who believe that immaterial consciousness demands to be recognized in quantum physics, where the observer has often been seen to play a central role, and current research is not taking us further away from such an understanding. Since in Chapter Nine I argue in detail for the importance of giving an account of the role of the observer, as necessitated by the postulates of quantum theory and by independent argument, all that needs to be noted here is that, regardless of one's conclusion, such questions are embedded in the assumptions of quantum theory and so we need to be open to analyzing them philosophically. Unfortunately, many analytic philosophers do not want to consider such metaphysical questions or, if they do, their style of writing and method of analysis still overly limit the ways in which such questions can be explored. If you want to understand consciousness, then, in addition to all of the scientific data you can acquire, you also need to explore your own inner experience of consciousness to ever deeper and more challenging levels. Such an approach, however, where we need to take seriously our direct experience of reality, is precisely what is forbidden in philosophy departments.[235]

Striving for as precise clarity as possible is laudable and necessary. But philosophers should not limit themselves to mere logical analysis or ignore our relevant states of mind, our emotions, or any reference to a conscious entity. Materialists may deny their own consciousness, or that they even exist;[236] I, however, know that I am conscious. Nevertheless, the word 'I' is unavoidably fraught with perpetual ambiguity, and any attempt to articulate the precise nature of 'I' will always fall short to some degree, regardless of how clever we may be at arranging our words. Therefore, the mystic's injunction to attain self-realization, a genuine experience that transcends discursive analysis, is the only rational option if we want direct understanding of the essence of 'I'. At any rate, quantum theory has made explicit what was already implicit in classical physics: the observer has a central role, which I will discuss further in Chapter Nine.

5. *There are logically and physically separate atomic facts*

The fifth false assumption of analytic philosophy is that the world can be broken down into independently existing states of affairs, where a statement about X has no relation to a statement about Y.[237] The reasoning behind this false assumption is partially related to the fourth false assumption, because both the fourth and the fifth assumptions neglect the fact that quantum theory tells us that the entire universe is holistically interconnected. But as Norris has argued, those who follow the Copenhagen interpretation of quantum theory (which includes most physicists) tend dogmatically to exclude hidden-variables alternatives, such as those offered by physicist David Bohm. But even Bohm stressed the interconnected holistic nature of reality, which entails the falsity of the atomist view that regards all things as purely separate entities. Bohm writes that 'both observer and observed are merging and interpenetrating aspects of one whole reality, which is indivisible and unanalysable'.[238] We can talk about the absolute unity of all reality, but all logical analysis necessarily divides things into component parts. Therefore, if the underlying or ultimate nature of reality is one, and if in order to analyze it logically we partake in logical distinctions between component parts, then we are never actually talking about the one absolute unity itself, but only a reflection of it. Therefore, strictly speaking, Bohm was correct that the one whole reality 'is indivisible and unanalysable'.

In complete opposition to this interconnected understanding of reality stemming from modern physics, the analytic tradition, as clarified by philosopher Mark Sacks, replaced the more traditional concern about the dichotomy between the knower and the world (observer and observed) with 'the dichotomy between language and the world'.[239] These sorts of 'traditional' philosophical questions, such as the relation between the observer and the observed, or the knower and the known, are at the heart of some of the most perplexing questions in quantum physics, and yet they have generally been dismissed by analytical philosophers.

Physicists as diverse as Planck,[240] Schrödinger,[241] Bohr, Bohm, and Ian Barbour have all recognized that studying individual parts will never give us a complete understanding of the whole, and it is the whole that we are ultimately seeking. Bohm says that 'the present approach of analysis of the world into independently existent parts does not work very well in modern physics'. He argues that 'both in relativity theory and quantum theory, notions implying the undivided wholeness of the universe would provide a much more orderly way of considering the general nature of reality'.[242] Harris explains further that in quantum theory, 'physical entities thus have come to be viewed as wholes of integrated and interdependent parts, rather than as separate and isolated particles'.[243]

The idealists, against whom Russell was reacting, held a view that was very much consistent with Bohm's. Hylton explains that, for idealists (in general), 'the only truly consistent way of thinking – that which yields 'absolute knowledge' – is to be found in the metaphysical conception of the world as a single organic whole, every part of which is internally related to every other'.[244] This notion that the entire universe is a single organic whole, where every part is interrelated to every other part such that there is no completely independently existing entity, is a view consistent with quantum theory, Hegelian philosophy, and even Buddhist metaphysics. It is also consistent with Platonic realism.[245] Platonism, however, does not deny the reality of the physical world. Pirsig recalls his frustration when his professor of Oriental philosophy at Benares

Hindu University answered affirmatively when asked whether the atomic bombs dropped on Hiroshima and Nagasaki were illusions.[246] According to the Platonic philosophers, however, brick walls and atomic bombs really do exist. The physical cosmos is not the limit or source of all reality, but physical objects are still real.

One may wish to object that even if quantum theory implies physical inseparability among entangled quantum states and with the observer (indeed, with the entire universe), it does not follow that we cannot logically have distinct, atomistic *linguistic* parts. In other words, there would still be a place for *logical* atomism in analytic philosophy, even if *physical* atomism is untenable. But this objection ultimately fails, for the logical and linguistic symbols that are placed in such a way as to make some sort of meaningful utterance are themselves logically inseparable, and only together do they form the whole system. Let us not forget that speaking and writing are also physical activities, and as such are governed by the laws of physics.

Bohm notes that in the East there has generally been more attention paid to philosophy and religion, and how the greatest attempts have been made there to seek the immeasurable (such as undivided wholeness), whereas in the West we have been more concerned with measuring.[247] However, the Platonists not only had the same desire for the immeasurable undivided wholeness (the One), but they also aimed as far as possible to uncover and explain the rationality of the physical universe. Both approaches have been a vital part of the birth and continual development of modern science. If anything, the Platonists should have done more measuring.

Analyzing analysis

The etymological root of 'analysis', which is found in the Greek word analusis (αναλυσις) [248] originally meant 'a loosening, releasing; dissolution, death'.[249] Such historical meanings may surprise the analytic philosopher, but they would not be out of place for Bohr. Heisenberg recollects Bohr's idea of complementarity applied to biology, where complete knowledge of a cell's molecular structure can only be achieved through methods that kill the cell; therefore, 'it is logically possible that life precludes the complete determination of its underlying physico-chemical structure'.[250] Dissection kills the cell in a similar way that over-analysis can result in the 'death' of deep understanding of underlying unity and of interdisciplinary research. Such a conclusion seems inevitable, for if your goal is to understand life but your methods of study kill the organism and thereby put an end to life, then you can never study life itself.

Similarly, if in philosophy your goal is to study X but your methods force you into ever-increasingly minutely detailed arguments, then they are taking you further and further away from understanding X. This is the same basic point that Bohm was making about the underlying whole of reality being unanalyzable. Defining concepts as clearly as possible is essential both in philosophy and in physics, but completely clarifying concepts is a logically infinite task, because there is no end to analysis. At some point we must agree with Whitehead that 'clarity always means "clear enough."'[251]

What we are really after here is an understanding of the limits of analysis, which can be demonstrated by a straightforward argument. Let X represent any theory, argument, axiom, or assumption. X ultimately requires justification or proof, but you cannot use X to prove X, so you must rely upon a different theory, argument, axiom, or assumption, which we can represent as Y. However, Y will also require further justification, which cannot come from either X or Y, so it must come from Z. Z will then require further justification, and so on. All we are really doing here is being a bit more formal than a child who asks 'Why?' no matter what answer you give to their questions. At some point you will have to admit that you do not know the answer. While we may think that such an observation is just common sense, it actually has much deeper implications.

At some point, we have to hit a foundation that is itself not logically analyzable, or else we must arbitrarily choose a starting point. If we arbitrarily chose a starting point, then we are not being logical. If we cannot logically analyze our starting point, then we have gone beyond the limits of logical analysis. Everything that has been said here about the limits of logic applies equally to the limits of scientific methodology. Consequently, all logical and scientific reasoning ultimately depends upon something that transcends logic and science, something that is more fundamental than either of them, and at this point we are entering directly into mysticism. What is most compelling is that this conclusion follows from basic logical reasoning.

The problem here is not logical or scientific. It is emotional. It is human. We may not like that there is no logical foundation to logic, and so we ignore the whole issue. It is just too unsettling. If you really think about it and understand the implications, it can be quite frightening. We console ourselves by believing that we are such a rational and sophisticated society, but the fact is that you cannot even logically decide which way to drive to work. If you take one route, you will get hit by a bus. If you take another route, you will arrive safely. Even if you are the greatest logician in history, you cannot logically know what will happen in the future, and so even armed with the most relevant up-to-date information about the road conditions, you still cannot logically know for sure which route you should take in order to arrive safely. Therefore, we must also use our intuition, but then we find ourselves immediately transcending logical and scientific analysis.

The real problem with most professional philosophers (and basically all academics) is not that they think too much, but that they do not think enough.

When we are ruthless with our logical reasoning, when we go all the way to the extremes of logical thought, we inevitably reach the limits of reasoned analysis and must then make the jump beyond. It is fine if we are unable or unwilling to make that journey ourselves, but we are not thereby permitted to condemn those who elect to do so. By taking this journey, we can destroy what is weak or false or unnecessarily limiting, but we must do so with the aim of rebuilding in a better, more holistic way. As philosopher Stephen R. L. Clark remarks, 'analytical criticism, even destructive criticism, is often a good thing, but the real aim of philosophy is not to destroy, but to understand and explore'.[252]

If aspiring philosophers want to be considered a part of the analytic camp, then, as Dummett notes, they will need to quote from the most important writers in the field and write in a style that conforms to their unwritten rules in order to be considered part of this tradition.[253] Gödel did not follow such advice and was thereby systematically ignored by the analytic and positivistic philosophers, despite once being an associate of the Vienna Circle. Yourgrau shows that 'more than most academic philosophers, [Gödel] engaged in philosophy in a manner of which Parmenides and Plato would have been proud: asking fundamental questions about the nature of time, being, death, God and the world of transcendent forms, or "ideas."'[254] The philosophers ignored Gödel because he refused to pay homage to their masters, and because he was a Platonist who thought in a similar way as the ancients.[255] Gödel, like Einstein and Pauli, was too philosophical for the philosophers.

Prying philosophers

Although I have emphasized the importance of philosophy for physics, it is nevertheless true that some of the most important pioneering physicists in the last century strongly rejected philosophy. However, they were rejecting contemporary analytic and positivist philosophy, but certainly not the sort of deep philosophical questioning or methodological thinking rooted in ancient philosophy. Let us begin with Eddington's blunt warning: 'it would probably be wiser to nail up over the door of the new quantum theory a notice, 'Structural alterations in progress—No admittance except on business,' and particularly warn the doorkeeper to keep out prying philosophers'.[256] Einstein concurs: 'when the very foundations of physics itself have become problematic as they are now... the physicist cannot simply surrender to the philosopher the critical contemplation of the theoretical foundations; for, he himself knows best and feels more surely where the shoe pinches'.[257]

What could have prompted Einstein and Eddington to urge us to keep out the 'prying' philosophers, especially at the point when they would seem to be most needed, when the 'very foundations of physics itself have become problematic'? Einstein certainly has a point that the physicist may feel 'more surely where the shoe pinches', but the problematic issues, including his disagreements with Bohr, were fundamentally philosophical. They argued about the meaning, interpretation, and understanding of the results of experiments, and they disagreed over future possibilities based on thought experiments and applied metaphysical reasoning. If the positivistic philosophers could have somehow enforced a restriction on the physicists that eliminated any possibility whatsoever of engaging in metaphysical reflection in relation to physics, it really would have spelled the end of the development of quantum theory.

Eddington and Einstein's warning to philosophers to stay out of quantum theory until it had achieved relative conceptual stability was a warning to the materialists, positivists, analytics, and overly skeptical philosophers. Materialism is still the mainstream position in 'American' philosophy,[258] despite the fact that so many physicists, such as Eddington, have claimed that physics has rendered materialism 'dead'.[259] Eddington has also argued against other prevalent philosophical views endorsed by philosophers such as Russell who believed that science is the only method for arriving at truth and that 'science has nothing to say about values'.[260] Eddington, however, recognized that science has its limits, but that what is found beyond its limits is in no way less real or important just because science has little or nothing to contribute to our understanding. He cannot totally agree, however, that science has nothing to do with values: 'if it were literally true, it would mean that, when the significance of our lives and of the universe around us is under discussion, science is altogether dumb'.[261]

While first building or restructuring the theoretical edifice of the new physics, the physicists would also want to keep out any overly skeptical philosophers. It is relatively easy to wave the banner of skepticism, tearing down every utterance made by an opponent, but such skeptics still believe (or act as if they believe) that their molecular structure will not instantaneously metamorphose into that of a pink elephant.[262] It may be helpful to remember that having a healthy critical attitude is not the same thing as lapsing into *skepticism*, which is a metaphysical doctrine that denies that knowledge of anything is possible.

Although physics would not be possible without mathematics, physicist Mendel Sachs says it is unfortunate that physicists turned their allegiance away from the philosopher to the mathematician. Because of the split between philosophy and physics, Sachs states that 'the dogmatic *approach* remained in physics in our own time primarily because physics stayed apart from philosophy and its critical attitude'.[263] As Gibbins confirms, the authors of standard quantum mechanical textbooks are 'quite justifiably anxious to get off the philosophical material and on with the physics'.[264] Although it is a serious mistake to neglect the philosophical problems in physics, we must also respect Eddington and Einstein's concerns. The analytics would only add to the confusion, or even prevent further developments, before the physicists had some idea what they themselves meant by phrases such as 'wave/particle duality'. It is not feasible to demand precise analytical clarification of problematic areas in physics before ever being allowed to use physics in practical applications. But we also cannot forever ignore such problematic issues and hope to make future groundbreaking advances in fundamental physics.

The mutual distrust between analytic philosophers and the deep-thinking physicists is evident in philosopher Susan Stebbing's obvious anger at Eddington and physicist James Jeans. She accuses Jeans of 'cheap emotionalism and specious appeals' and writes that Eddington's desire to be entertaining 'befools the reader into a state of serious mental confusion'.[265] Stebbing writes that 'Jeans, who believes it to be an advantage not to be trained in philosophy or to have an inclination for it if one wishes to draw philosophical conclusions from physics, seems nevertheless to have read both Plato and Berkeley'.[266] Stebbing, who was otherwise an advocate of critical thinking, joined the analytic philosophy camp and rejected the relevant philosophical views of both Plato and the eighteenth century philosopher George Berkeley. She attacked Jeans because, apparently, he was not aware that such views have been seriously criticized and, 'in the opinion of most philosophers, have been decisively refuted'.[267]

It is poor philosophical reasoning to argue that just because a particular view is apparently at odds with the opinion of 'most' philosophers it must therefore be false. Given the prevailing philosophical environment of the time, Jeans was right that it could be a disadvantage to study the kind of philosophy that was hostile to the required modes of metaphysical reflection that unfolded with the advent of quantum theory. But he was fond of Plato and Berkeley, as their metaphysical views and ways of thinking were much more akin to the new physics than anything that Stebbing and her circle had to offer. Hence, Jeans clearly was not against philosophy. He simply rejected the analytics and positivists.[268]

There are many more such examples that we could examine, but let us just look briefly at one more. Schrödinger had gone so far as to argue that consciousness is universal and singular: 'there *is* only one thing and what seems to be a plurality is merely a series of different aspects of this one thing, produced by a deception (the Indian MAJA)'.[269] He even writes that 'the mystics of many centuries, independently, yet in perfect harmony with each other (somewhat like the particles in an ideal gas) have described, each of them, the unique experience of his or her life in terms that can be condensed in the phrase: DEUS FACTUS SUM (I have become God)'.[270] He notes that the point of view he has been arguing for is in harmony with that of the great mystics throughout the ages, what author and philosopher Aldous Huxley referred to as the 'Perennial Philosophy'.[271] It is difficult to imagine an analytic philosophy department studying Huxley's *Perennial Philosophy*, or taking seriously

Schrödinger's philosophical views. It is not difficult, however, to see that the pioneering physicists were not rejecting philosophy wholesale. They were not rejecting Platonic realism, but they *were* rejecting prying positivists and analytics. Such rejection was wise.

Physical chemist and philosopher Michael Polanyi stresses that freedom of inquiry is essential to the sciences.[272] This obvious point, which is far too often forgotten in our education system and by research funding programs, applies equally to philosophy (and any creative endeavor, for that matter). But analytic philosophy has held too many prejudices, especially during the time of the development of quantum theory, regarding the role of language, consciousness, atomic facts, metaphysics, and ancient philosophy. Such parochialism has impeded many philosophers from engaging in the sort of profound metaphysical thinking required to achieve a similar understanding of reality as that attained by many of the founders of quantum physics. From this perspective, and given the most charitable interpretation possible, Hawking's rejection of philosophy makes more sense, so long as we take his rejection to be aimed at the parochialism of analytic philosophy. However, genuine philosophy is not limited to mere logical or linguistic analysis, nor is it limited to any sort of academic fashion. Heisenberg and many of the founders of quantum physics were deeply philosophical, as Hawking must know, but in a way that had far more in common with ancient rather than contemporary philosophy. Consequently, we must agree with Harris that 'for such useful work as [analytic philosophy] may have done, let us applaud it as it bows out'.[273]

FOUR: Metaphysics and Physics

If there is to be any hope of clarifying the conceptual confusion found in modern physics, then an interdisciplinary approach is necessary.[274] Some of this confusion arises out of the act of interpreting experimental data, but much of it also begins at the very foundations of physics. In this chapter, you will first be introduced to a basic understanding of the importance of beauty and symmetry with respect to ancient philosophy and modern physics. Then, after further establishing that we are in good company in defending an interdisciplinary approach, specifically between metaphysics and physics, I clarify three broad types of metaphysics, which I call *pure*, *applied*, and *presupposed*. This clarification is followed by a brief example of applied metaphysics in the context of examining the nature of the Heisenberg uncertainty principle, one of the most important aspects of quantum theory.

Beauty and symmetry

We tend to take mathematics for granted, as though it should just be there to enable us to understand and manipulate the physical world, but seldom do we take the time or effort to reflect on why we are so fortunate. Mathematics is like an alchemical key enabling us to transform the earth's natural riches into astounding technology. Fortunately, we can all develop an appreciation of the beauty of mathematics even if we are not mathematicians. Let us begin our examination with Ian Mueller, a specialist in the ancient Greek philosophy of science.

Mueller notes that the intellectual climate in which Proclus taught in the fifth century CE was 'generally unscientific—or even antiscientific'. Proclus had to defend the value of mathematics from people who argued that it does not teach us about morality or beauty and that it is of no 'practical value in the "real" world'.[275] Although Proclus does say that 'mathematics also makes contributions of the very greatest value to physical science',[276] his main defense of mathematics comes from the fact that it 'familiarizes us with order, symmetry, and definiteness', which are the essential characteristics of moral significance or beauty.[277] For Proclus, 'mathematics ought to be studied for its own sake, or, if an external motivation is needed, in order to purify the soul for higher apprehension'.[278]

One of the reasons mathematics has survived attacks from those who would have preferred to see us drop it altogether is that its study helps to purify us to receive higher insights from, and develop a fuller understanding of, ultimate reality. Mathematics is important not only for building things, but because it can act as a bridge for us to delve further into the metaphysical/spiritual realm. We may wish to reject the 'impractical' reasons Proclus offers in defense of mathematics, but was he really so impractical? Mathematicians and physicists seek and discover symmetry, order, and harmony, which are all characteristics of beauty. As Rowlands writes, 'symmetry (or analogy) has been the driving force of much of theoretical particle physics, as it was, previously, of classical physics, and physicists seem to *expect* to find symmetries in nature'.[279] Another contemporary physicist, Stephen Barr, notes that symmetry not only contributes to the 'artistic unity of a work', but that the 'connection between symmetry and unity is exceedingly important and applies also to symmetry in physics....Symmetry requires all the parts of a pattern to be present, and is therefore a unifying principle'.[280]

After citing one of Plato's Dialogues, *Timaeus* (31c), Katherine Brading and Elena Castellani, specialists in the philosophy of physics, write that 'from the outset, then, symmetry was closely related to harmony, beauty, and unity, and this was to prove decisive for its role in theories of Nature'.[281] Despite the unsurprising fact that the notion of symmetry has taken on an explicitness and greater technical precision in modern physics than in antiquity, Brading and Castellani also realize that 'symmetry remains linked to beauty (regularity) and unity'.[282] They also recognize that it has been 'in the quantum context that symmetry principles are at their most effective'.[283] Consider also Heisenberg's significant revelatory moment regarding symmetry: 'what was there in the beginning? A physical law, mathematics, symmetry? In the beginning was symmetry! This sounded like Plato's *Timaeus*'.[284] A group of physicists have also recently written that 'symmetry is present in many physical systems and helps uncover some of their fundamental properties [and that] continuous symmetries lead to conservation laws'.[285]

Beauty, too, is inextricable from such considerations. Philosopher Aphrodite Alexandrakis writes that 'for the Pythagoreans harmony *is* beauty, and beauty arises from proportion (*symmetria*)'. She goes on to argue for the Pythagorean influence on Plotinus' conception 'of the cosmos as a harmonious whole and his concern with World Harmony'.[286] Many modern theoretical physicists also recognize the role of beauty in their work. Physicist Paul Dirac believed that 'it is more important to have beauty in one's equations than to have them fit experiment'.[287] Another physicist, James Trefil, writes that Einstein's paper on relativity 'was mainly about philosophy', and the reason scientists were not so concerned about the proof of Einstein's ideas from experimental data 'comes from the fact that relativity is beautiful'. Trefil admits that 'beauty' is 'a strange word to apply to what is, after all, a mass of equations, but that's the way physicists perceive it'.[288]

Jaki writes that 'the fruitfulness of special relativity was an invitation to Einstein to unfold even more of the constant beauty of nature and of the exact science of that beauty'.[289] Jaki also clarifies that Einstein's two theories were 'in a sense mislabeled with the word *relative*, because both the special and general theories of relativity were more absolutist in character and content than any other scientific theory'.[290] This is an important point, clarified further by Trefil, who says that the beauty of Einstein's central principle is that it says 'that any two observers, no matter where they are or how they are moving, will see the laws of nature operating the same way. The universe, in other words, doesn't care how it's being observed—it's the same for everybody'.[291]

What is really being said here is that an important aspect of beauty is realized when we discover the unity underlying variety. The laws of physics, for example, remain constant and provide the underlying unity of the plethora of differences in the physical universe, and this same idea also holds true in the arts. The poet and philosopher Samuel Taylor Coleridge, for example, believed that beauty was 'unity in variety'.[292]

Pauli notes the historical debate 'going back to Plotinus' concerning beauty, which he compares to the quantitative thinker who considers the *parts* to be essential versus the qualitative thinker who considers the indivisibility of the *whole* to be fundamental.[293] In relation to beauty, he further explains the 'feeling or intuitive type' of personality, such as Goethe's, versus the 'thinking type', such as Newton's.[294] The *intuitive type* refers to one who is more acutely attuned to or emphasizes 'non-discursive' thought, a direct perception, understanding, or knowing of the whole. This sort of experience is akin to Fröhlich's mind becoming the physical particle, or the way in which Russell had once assumed that the mind could attain direct contact with abstract objects, for both are aspects of direct experience beyond conceptual, logical, or linguistic limitations. The *thinking type* refers to one who is more rooted in or who emphasizes discursive thought, a logical, step-by-step analysis.

Newton and Goethe had aspects of both types, even as they each tended to express more strongly one over the other. Reason dissolves into higher, direct intuitive knowing, but reason is what we must rely on most often during the course of our daily lives as well as in the sciences, even though this intuitive recognition of total wholeness and transcendence underpins all discursive analysis and material reality.

Although both the quantitative and qualitative aspects of beauty are interdependent, beauty cannot be *reduced* to order, symmetry, and definiteness. We could, after all, organize a very orderly bank robbery, or commit genocide with definiteness and precision, but most of us surely would not say that such activities are beautiful. The Greek word *kallos* originally had connotations of physical *and* moral beauty simultaneously, which may seem a bit odd to us today given that we usually make distinctions between such notions, and it is not often that we speak of moral beauty at all. Consider, however, that even though I would be more attracted to a symmetrical face than an asymmetrical one, all other things being equal, I would think that a morally committed person with a less symmetrical face was more beautiful than a person with a perfectly symmetrical face who willingly engaged in atrocious activities. In such a metaphysical hierarchy, beauty comes after the good, and symmetry after beauty. Symmetry is only possible because of beauty and so it seems reasonable to agree with Proclus that symmetry presupposes beauty.

Beauty is an essential guidepost for fundamental physics, and is in fact presupposed by the very possibility of physics. But it does not seem to matter how many supporting quotations I may offer from great thinkers throughout the ages, because those who have not yet experienced, intuited, or reasoned their way to understanding the importance of beauty are not likely to be convinced. Nevertheless, those who fall into this camp cannot ignore the fact that many great scientists throughout history disagree with them. Ideally, philosophy should play a key role in opening our minds to the reality of beauty, and many philosophy departments do offer courses in aesthetics. But there still tends to be a lingering analytic shadow that subtly (or not so subtly) obscures the object of study, discouraging students from actively seeking the direct experience of beauty for themselves. Indeed, as Clark writes, 'analytical philosophy, perhaps, is not very likely to awaken us to beauty'.[295] Fortunately, we are not constrained by the self-imposed strictures of the analytic tradition.

Distinct but inseparable

Throughout the history of the development of European universities, the process of generating new specializations has been the driving point of creativity in academia, since such differentiation allows for the combining of new ideas.[296] Without specialized disciplines, we would not even have interdisciplinary research in the first place. Interestingly, mathematical science was also the originator of interdisciplinary pressures that influenced the development of the major schools of modern philosophy,[297] a trend that culminated in the development of positivism and analytic philosophy, as already discussed in the previous chapter.

Today, it seems almost clichéd to argue for the importance of interdisciplinary research, especially amongst the sciences where we see how chemistry and biology are dependent upon physics.[298] However, even though chemistry depends upon physics, in order to solve chemistry problems we cannot stay strictly within the domain of physics.[299] Despite the growing recognition of the importance of interdisciplinary research, the fact is that in the day-to-day operations of various academic departments, there is still not nearly enough cross-disciplinary integration. It is very rare, for example, for a physicist to work closely with a philosopher, or for them even to visit each other's departments.

Schrödinger and Heisenberg had some fundamental disagreements about their interpretations of quantum theory, but they certainly agreed on the necessity of interdisciplinary research. Heisenberg said that the sciences are compelled to engage with philosophy 'if we wish to make fundamentally important advances and to understand them'.[300] Schrödinger even went so far as to say that specialization was an 'unavoidable evil' but that 'all specialized research has real value only in the context of the integrated totality of knowledge'.[301] In other words, we must strive for a holistic, interdisciplinary understanding of our specialized research, and we must acknowledge that philosophy is fundamentally important for making advances in the sciences. After all, it is impossible to really understand our scientific research without reflective, systematic, rational thinking.

A philosophical interpretation, however, is required not only after the data have been collected but also at the beginning of the experimental and theoretical process. Physics cannot begin to develop in any fundamental way without taking on a philosophical interpretation.[302] Stathis Psillos, a specialist in the history and philosophy of science, also notes that 'observation in science is not just the act of reporting a phenomenon (whatever that means!). It is the interpretation of a phenomenon in the light of some theory and other background knowledge'.[303] But philosophy is not just about interpretation. Philosophical thinking is at the root of the entire scientific enterprise. For example, Heisenberg believed that in the work of Planck 'we can clearly see that his thought was influenced and made fruitful by his classical schooling'.[304] Heisenberg also correctly observed that by studying ancient Greek philosophy we are able to learn how 'to pose questions of principle', which is 'the strongest mental tool produced by Western thought'.[305] He goes even further by saying that the connections between the various sciences are more obvious now than ever before, and that 'there are many signs of their common origin, which, in the final analysis, must be sought somewhere in the thought of antiquity'.[306]

It is essential to seek the common origin or underlying unity of the sciences, especially if we hope to understand how, for example, biology, chemistry, and physics are interrelated. Therefore, if Heisenberg is correct, we should look for this common origin somewhere in the thought of the ancient Greek philosophers. With this view in mind, we can read what philosopher James Lowry writes about Proclus' *The Elements of Theology*: 'there is no major idea in the whole of Greek philosophy which cannot be found in this small treatise'.[307] Even if this claim is slightly exaggerated, it is still true that a major portion of ancient thought is contained within this text, and so it is essential reading for anyone who wants to understand the unity and common origin of the sciences.[308]

The distinctions between metaphysics and physics seem superficially clear until we begin to ask ourselves what we really mean by these terms and try to specify their boundaries with unambiguous precision. No matter what viewpoint one professes allegiance to, there is no way to avoid having presupposed metaphysical assumptions. If anyone disagrees, it is an easy matter to point out the metaphysical assumptions upon which their disagreement rests. Psychologist Erik Erikson makes a relevant comment about the importance of psychology for history: 'biographers categorically opposed to systemic psychological interpretation permit themselves the most extensive psychologizing—which they can afford to believe is common sense only because they disclaim a defined psychological viewpoint. Yet *there is always an implicit psychology behind the explicit anti-psychology*'.[309] The same can be said about metaphysics; there is always an implicit metaphysics behind any anti-metaphysical stance, or behind any theory or belief whatsoever, which can only be (thinly) concealed by pretending not to hold a defined metaphysical stance.

It is quite perplexing how logical positivists and others who despise metaphysics have also claimed to be serving science for, as shall soon become even clearer, the strict elimination of metaphysics would necessarily entail the end of science. While relating a discussion between Bohr, Pauli, and himself, Heisenberg shows clearly how they all reacted badly to the positivists' dismissal of metaphysics.[310] Bohr had told members of the Vienna circle that although he endorsed the positivist desire for conceptual clarity, he also rightly claimed that banning metaphysics would 'prevent our understanding of quantum theory'.[311] Heisenberg also recalls Pauli saying that physicists must philosophize in everyday language, especially at the point where mathematical formula meet experiments. In fact, he believed that '*all the difficulties of quantum theory will be found to reside in this meeting*, a fact most positivists choose to ignore, precisely because their concepts break down at this point'.[312] It is quite extraordinary that the physicists who brought us quantum theory believed that all the difficulties in the theory needed to be explored philosophically in everyday language, precisely at the point where the mathematical laws of physics intersect with experiments. Consequently, it is essential to explore as far as possible the actual nature of the laws of physics themselves, which is one of the main goals of this book.

Bohr responds by adding that we need to employ a variety of concepts when discussing 'the strange relationship between the formal laws of quantum theory and the observed phenomena'.[313] Bohr, then, makes another critical point. He claims that only by thoroughly examining this relationship and drawing out its apparent contradictions 'can we hope to effect that *change in our thought processes*', which is essential for 'any true understanding of quantum theory'.[314] In other words, as already noted, nature does what it does without contradiction, so the conceptual confusion is in our own beliefs and thought processes, and without a change in our conceptual systems we will not be able to understand quantum theory. Of course, using rational arguments in everyday language to examine fundamental aspects of the nature of reality is to do metaphysics. Consequently, combining as much textual evidence as possible with original arguments can help us make the necessary change in our beliefs and thought processes in order to understand the metaphysical foundations of modern physics. One of the most important conceptual shifts is to recognize that the laws of physics necessarily imply the eternal mathematical law of physics, which is real and yet nonphysical.

It was not just the Copenhagenists, however, who endorsed metaphysics. Einstein, Planck, and Schrödinger, who apparently held opposite views about quantum theory to those held by Bohr, Heisenberg, and Pauli, also understood the importance of metaphysics. Schrödinger writes that 'a real elimination of metaphysics means taking the soul out of *both* art *and* science, turning them into skeletons incapable of any further development'.[315] Strict adherence to positivism will prohibit speculative science, which will eliminate creative science that reaches out into hitherto unexplored or currently unimaginable territory, which would in turn be the end of the progress of the sciences, leaving them as soulless skeletons. However, there is no way of knowing for certain how or in what way science will develop, even though we may be able to understand its metaphysical foundation. The indispensability of exactingly detailed experimental research does not eliminate the need for metaphysical reasoning. It sometimes seems difficult to logically distinguish between them, but in practice we can easily point to many obvious differences.[316]

Probably due to a misunderstanding of the nature of metaphysics and of his own metaphysical thinking, in 1912 Einstein signed an anti-metaphysical manifesto,[317] but he later recognized his error: 'I believe that every true theorist is a kind of tamed metaphysicist, no matter how pure a "positivist" he may fancy himself'.[318] He made such statements despite the fact that some of the leading positivists 'begged him—almost put words in his mouth—to state that experimental data were the trigger of his speculations and achievements'.[319] As Jaki argues, once the 'Machist crust' was removed, Einstein came to understand his 'own mental physiognomy' and his 'reflections upon himself bubbled forth in the unabashed freedom of one who at long last had discovered himself'.[320]

Einstein had, at times, been under the spell of Mach and the positivists, but such moments were always underpinned by what he later came to realize were his realist convictions, in essential metaphysical agreement with Planck. Jaki continues: 'it was only while watching himself that Einstein was able to see the true nature of his own creative deeds in science, and this he did with increasing clarity'.[321] Frank and Moritz Schlick, two pioneering positivists, had to doublethink themselves into ignoring Einstein's rejection of positivism. In a letter to his close friend Michel Besso, Einstein had some interesting remarks to make about Mach: 'I do not inveigh against Mach's little horse; but if you wish to know, this is what I think of it: it cannot give birth to anything living, it can only exterminate harmful vermin'.[322]

Frank had a similar experience after delivering a paper to the 1929 congress of German physicists in Prague, where he 'attacked the metaphysical position of the German physicists and defended the positivistic ideas of Mach'. He had taken Einstein to be a positivist until a 'well-known German physicist', who Frank does not identify, made the following rebuttal: 'I hold to the views of the man who for me is not only the greatest physicist, but also the greatest philosopher: namely, Albert Einstein'. Frank recalls how this speaker told of Einstein's rejection of Mach's positivism and his belief that the laws of physics are more than merely a collection of observations. In fact, this speaker, who Frank left unnamed, believed that 'Einstein was entirely in accord with Planck's view that physical laws describe a reality in space and time that is independent of ourselves'.[323]

In 1930, Einstein informed Schlick directly of his rejection of positivism, telling him that he believed that 'physics is the attempt at the conceptual construction of a model of the *real world* and of its lawful structure'. 'You will be astonished', he continues, 'about the "metaphysicist" Einstein. But every four- and two-legged animal is *de facto* in this sense metaphysicist'.[324] Einstein was aiming for the distinction between the laws of physics as depicted in our textbooks and the actual laws found in the structure of reality. However, he was not able to adequately clarify this very deep problem, which we will discuss further in Chapter Eight.

It is astonishing how so many philosophers of science, who often venerate, serve, and attempt to emulate the sciences, have been able to decry metaphysics, while so many of the greatest pioneering physicists (and others in different scientific disciplines) have held metaphysical views akin to those of Einstein and Planck. Planck agrees with Einstein that you cannot be a scientist unless you believe that the world exists in reality, which is to hold the realist position. However, Planck offers the important caveat that no reasoning process could lead us to this knowledge of the existence of the external world. He also writes that this knowledge 'is a direct perception and therefore in its nature akin to what we call Faith. It is a metaphysical belief. Now that is something which the sceptic questions in regard to religion; but it is the same in regard to science'.[325] Atheists who believe themselves to be scientifically sophisticated need to stop cowering behind their idols, such as Dawkins, and engage with Planck, a truly groundbreaking scientist. To ignore the views of such preeminent scientists as Planck in order to obstinately maintain one's faith that faith is irrelevant to science is to be thoroughly unscientific. The next chapter will deal specifically with the relationship between physics and faith, but here we shall consider what Planck meant by 'direct perception'.

What Planck is referring to by 'direct perception' is the fact that we need to have a direct, intuitive recognition or knowing that the external world is real. This recognition or knowledge cannot be logically proven beyond doubt, so it must be a metaphysical belief that requires some act of faith, since our certain knowledge of the external world is actually beyond logical and scientific certainty. In fact, we cannot even do science, as Planck noted, unless we first have the metaphysical faith that there really is an external world. Even those scientists who profess not to believe in the reality of the external world still must act as if they do; otherwise they could not do science. If there were no external world, then the 'world' would be reduced to whatever was contained within any particular individual mind, eliminating the objective character of the sciences and reducing them to nothing more than one's personal fantasy, thereby landing us squarely in antirealist territory. Fundamentalist atheists like Dawkins can only afford to sound convincing when belittling faith at the cost of being unwilling or unable to fully recognize and examine rationally all the relevant facts.

Planck was being thoroughly rational when he said that no reasoning process could lead us to certain knowledge that there really is an external world. If we cannot rely on logic to prove beyond doubt the most seemingly obvious of facts, such as the fact that there really is an external world beyond ourselves, then we must transcend logic. The idea of knowledge being gained through direct perception has been developed to a profound level in the Platonic tradition. As Corrigan explains, for the ancient Greeks in general, and especially for thinkers such as Plotinus, intellect includes ordered desire and direct, immediate understanding, where this 'understanding is not the sort that has to work things out discursively bit-by-bit'. On the contrary, 'intellect's understanding is more like a complete grasp of the whole at one glance',[326] which is what Bohm was speaking of when he talked about the underlying wholeness of all reality.

This view of the intellect is very different from how we tend to consider it today, since we usually assume that using intellect means employing discursive, or step-by-step, logical thinking. The discursive aspect of cognition is essential, but without the desire to understand what is true, and without the intuitive ability that transcends discursive thought enabling us to grasp the whole of the object of our contemplation, our understanding must remain fragmented and partial. Planck's notion of 'direct perception' in the context given above refers to what the Platonic philosophers called direct, immediate, or non-discursive understanding.

Three categories of metaphysics

I will here introduce three broad categorizations of metaphysics. There are no absolute boundaries between these divisions, so we can expect there to be many instances where they overlap, but this categorization is valuable for understanding the nature of metaphysics in general and the relationship between physics and metaphysics in particular. Metaphysics, which includes both ontology and epistemology, is the rational investigation of the most fundamental questions about the ultimate nature of reality, achieved through the reasoning mind alone, so far as we are able. My classification system includes three kinds of metaphysics: pure, applied, and presupposed.[327]

Pure metaphysics involves reasoning our way to ultimate conclusions without relying upon the current prevailing scientific worldview, or with as minimal reference to it as conceptually possible. (Good examples are Proclus' *The Elements of Theology* or Descartes' *Meditations*.) Indeed, just asking the question of why there is something rather than nothing can lead to an entire metaphysical worldview that only tangentially refers to modern scientific knowledge as examples, but which could actually be applicable to any period of time in the past or future. Pure metaphysics can provide a consciously devised worldview that is believed to correspond to reality so far as possible, and so is intellectually satisfying while serving as the background or framework within which our scientific research can most aptly be carried out. However, pure metaphysics may also provide moral or spiritual guidance for individuals and society, as well as broader intellectual understanding.

Applied metaphysics asks fundamental questions within a specific domain of knowledge. For example, physicists talk about electrons as if they exist, but a metaphysician will ask in exactly what way they exist and what 'existence' means as defined by a physicist (whereas in pure metaphysics we ask what 'existence' is itself). Or a physicist may say that in some sense light is both a wave and a particle (or both, or neither), and the metaphysical questions that then follow are: how can it be both simultaneously, or if it is not really either, then what exactly *is* light?

The physicists need to work with metaphysicians to seek clearer answers to such questions if we are to increase our scientific understanding and improve research. Science cannot escape its reliance upon metaphysics, for we must make rational judgments about what we assume to be empirical facts, which is to do applied metaphysics.[328] Applied metaphysics incorporates speculation with the attempt to develop a coherent and true system by bringing together into a rational unity what we have assumed to be scientific facts. It must also question these same supposed facts, and this sort of questioning leads to speculative answers, which may in turn bring us back to empirical testing or may push us further into pure metaphysics.

Finally, *presupposed metaphysics* is not actually about doing metaphysics or necessarily even thinking metaphysically; rather, it refers to our implied fundamental metaphysical belief system within which we try to make sense of the world and ourselves. When my cat cries outside the door for me to let him into the house, he is certainly acting as if he believes that I exist even though he cannot perceive me, and thus even four-legged creatures are living as realists. We all must have a presupposed metaphysical worldview or at least certain metaphysical assumptions (no matter how inconsistent, false, or incomplete they may be) in order to even begin to talk about facts. The danger, however, is that often, if not usually, these views are held uncritically and unconsciously, which means that we can pretend that we do not have any metaphysical beliefs while simultaneously passing our beliefs onto others indirectly.[329]

Frank's position is a striking example of someone claiming to eliminate metaphysical interpretations while adhering implicitly to a metaphysics that guides his own conclusions. He believed that it was possible to 'distinguish an experimentally testable assertion about observable facts from a proposal to represent the facts in a certain way by word or diagram'.[330] We should realize by now, however, that such a project is impossible. There is no such thing as a strictly observable fact, because all observation requires some sort of interpreting apparatus, whether it is a Geiger counter or a human brain and eyes. The sofa I am sitting on really exists in whatever way that it does, and it exists in the way that it does whether or not I or any other creature happens to be perceiving it. However, if I do observe it, then I must observe it through my own observing apparatus (my eyes and brain), in which case my perceptual and cognitive abilities are limited to my own perceptual and cognitive capacities.

I can also use scientific instruments to tell me things about the sofa's various properties, in which case I am obviously relying on instruments which, by their very nature, can only tell me about limited aspects of the sofa, since each instrument will itself have its own inherent function and limitations. Moreover, I, the observer, must not only implement the instruments but also read and interpret the data, and my interpretation of this data will depend on the theoretical background upon which I am relying.

Every theory, we should by now realize, must have explicit fundamental assumptions that are themselves left unproven and, more importantly, there will be implicit metaphysical assumptions that are left unstated because we are usually not even aware of them. Even if we are aware of them, and since we cannot prove every assumption, then we must leave at least some basic metaphysical assumptions unquestioned. Consequently, even if it were possible to follow Frank and believe that we can actually implement his recommendations, his whole program would still be resting upon the presupposed metaphysics of empiricism. One could object that Frank's position has more to do with methodology than with metaphysics, but that objection could only be sustained by neglecting the metaphysical presuppositions of the empiricist methodology itself.

When positivists object to metaphysics, what they really seem to be reacting against is the category of pure metaphysics, especially its mystical branch.[331] It is impossible to object to presupposed metaphysics, because a metaphysical worldview is presupposed by all of us in order to even voice such an objection, regardless of whether or not we are aware of our metaphysical assumptions and their implications. We need to ask questions and describe our views, theories, speculations, and conclusions in natural language, which is also the basic language of metaphysics. Dummett, however, recalls that in Frege we find 'vehement denunciations of natural language',[332] a view that was not shared by Heisenberg, who writes that 'we know that any understanding must finally be based upon the natural language because it is only there that we can be certain to touch reality, and hence we must be sceptical about any scepticism with regard to this natural language and its essential concepts'.[333]

Heisenberg's point may be clearer when stated in this way: 'even for the physicist the description in plain language will be a criterion of the degree of understanding that has been reached'.[334] It is not, by any means, that natural language should replace mathematics or logic, but natural language is where we really test and develop our understanding. It is certainly odd to see the founder of the dominant tradition of contemporary philosophy denounce natural language, the best tool for philosophers, while a pioneering physicist tells us that it is required for understanding and touching reality.

Despite all the problems with positivism and its rejection by such eminent pioneering physicists, this tradition is still alive and, if not flourishing, doing well. We can see its influence in various scholars from physicists such as Hawking [335] and Stenger [336] to philosophers such as Cartwright,[337] Mark Balaguer,[338] and Arthur Fine.[339] While many philosophers are suspicious of metaphysical thinking, many physicists do not consciously engage with metaphysics at all.

In a similar way, engineers generally do not care whether or not Planck's constant is an objectively true feature of reality or merely a fiction, so long as they get the desired results when applying it. But such practical scientists cannot claim to have any real understanding of the foundations of their discipline, which is one of the reasons Einstein was angry when his son chose such a career, saying that 'what he is interested in isn't really important, even if it is, alas, engineering. One cannot expect one's children to inherit a mind'.[340] Einstein was being a bit harsh on his son, as engineering is as difficult as it is vital to our society. Nevertheless, the very nature of the discipline of engineering is such that it does not need to engage directly with foundational issues, especially in relation to the deeper philosophical questions lurking just below the surface of all scientific endeavors.

Smolin has written recently on the poor state of research and depth of understanding in fundamental physics stemming from this neglect of philosophical reflection. He argues that one of the main reasons researchers have become stuck on fundamental problems, whether in string theory or quantum gravity, is that they have neglected the necessity of philosophical thinking that was essential to the development of quantum theory. He concludes that 'perhaps the problems of unification and quantum gravity are entangled with the foundational problems of quantum theory, as Roger Penrose and Gerard 't Hooft think. If they are right, thousands of theorists who ignore the foundational problems have been wasting their time'.[341]

Metaphysics may not give us directly perceivable, tangible results, but neither does pure mathematics. However, the importance of mathematics when it is applied in physics can hardly be doubted and, likewise, without the possibility of metaphysics, we could not even rationally understand our sense perceptions in a scientific way, nor could we formulate pioneering questions that lead us to new areas of scientific exploration. Fortunately there is a slowly growing recognition of the importance of metaphysics in physics, although Gibbins inadvertently offers a good example of the lingering confusion that my tripartite distinction of metaphysics should help to clarify. He says that 'armchair metaphysics is out of date', although 'scientific metaphysics' has now 'come into fashion'. 'The new metaphysician asks: what is there in the physical world, and what is true of what there is in the physical world? The answers are provided by the philosophy of physics, a subject whose metaphysical part sets out to tell us the way the world is, if physics is true'.[342]

Gibbins' use of the pejorative phrase 'armchair metaphysics' would correspond to what I have called pure metaphysics, while his notion of 'scientific metaphysics' is close to my definition of applied metaphysics, except that scientific metaphysics would be but one aspect of applied metaphysics. However, the sample questions he gives are not limited to applied metaphysics, but could be interpreted to be, or easily lead to, pure metaphysical questions, so it seems that he is doing armchair metaphysics after all. In any case, many of our greatest physicists have been 'armchair physicists' in that they themselves never did empirical experiments but relied mostly on thought alone. Einstein was not only interested in 'scientific metaphysics', for he also admitted the importance of the guiding light of Platonic ideals, such as 'Kindness, Beauty, and Truth'.[343] Einstein did not use the phrase 'Platonic Ideals', which actually makes my case stronger, since he was embracing Platonism, possibly without even realizing it or ascribing a label to his beliefs.

Einstein went so far as to say that 'without the sense of kinship with men of like mind, without the occupation with the objective world, the eternally unattainable in the world of art and scientific endeavours, life would have seemed to me empty'.[344] He also disdained what most of us esteem to be important: 'the trite objects of human efforts—possessions, outward success, luxury—have always seemed to me contemptible'.[345] Einstein's real guiding principles, his highest ideals, were truth, beauty, and kindness (or the good). Surely he would have wanted to follow what seemed most important to him, what was more real and fundamental than luxury cars and fancy shoes and experimental observations. Penrose shares Einstein's fuller Platonic vision: 'One can well take the view', Penrose writes, 'that the 'Platonic world' contains other absolutes, such as the Good and the Beautiful'.[346] However, he does not venture into that territory, because he is doing physics and some applied metaphysics. Penrose does not spend much time in pure metaphysics, which is what we have to do if we are to venture into a dialectical analysis of the Platonic realm of the good and the beautiful.

Uncertain uncertainty

There are virtually unlimited avenues of exploration for applied metaphysical reasoning, from questions in ethics all the way to relativity theory. Let us consider a brief example of such reasoning with respect to the uncertainty principle. The basic idea of the *uncertainty principle* is that the position and momentum of a particle, such as an electron, cannot both be precisely known simultaneously. To the degree that we can precisely measure the momentum of the particle, we cannot precisely measure its position, and vice versa.[347] Among other problems, physicists and philosophers conflate the words 'know', 'observe', and 'measure', which causes significant unnecessary confusion.

Physicist and professor of religion Ian Barbour is better than most at explicating and demarcating the various distinctions involving the uncertainty principle (sometimes called the uncertainty relations). However, he has also made significant errors that are representative of the general conceptual confusion ignored by most physicists and philosophers.[348] The first point to note is that the uncertainty principle, which was even reluctantly accepted by Einstein,[349] resulted from Heisenberg's *thought* experiment.[350] Barbour outlines three possibilities for interpreting the uncertainty principle, stating that the uncertainty is either:

1. due to temporary human ignorance;
2. inherent to experimental or conceptual limitations; or,
3. in nature itself.[351]

If the uncertainty is not in nature, then it is due to some limitation on our part. This limitation could be temporary (such as in number 1), or it could be insurmountable (such in as number 2). By a *temporary limitation*, I mean that I may not currently know something that I could possibly know in the future if I had more knowledge or better technology. By an *insurmountable limitation*, I am referring to our inability to know something because it is actually impossible for us to know it, regardless of how much knowledge or technology we may have. I will here set aside any further debate about whether the uncertainty is a temporary or insurmountable limitation, as such a question will require going too far afield. I will, however, argue briefly that it is far more likely that the uncertainty is not in nature itself, and so must be due to some sort of limitation on our part.

It is reasonable to claim that the uncertainty of a particle's position and momentum is a result of some limitation on our part rather than in nature itself. If such uncertainty were actually intrinsic to nature and had nothing to do with our own limitations, then a particle's position and momentum really would be uncertain. But what does it mean to say that the uncertainty is in nature? The uncertainty could not be an inherent aspect of the equation since the equation itself is not uncertain and remains what it is without change. Even if the uncertainty is 'built into' the equation, the equation itself would still remain certain and unchanging, and so we would have to explain how the certainty and unchanging nature of the equation could contain inherent instability and uncertainty.

The uncertainty could also refer to some unavoidable randomness in nature, but if the randomness is a genuine part of nature, then we have to ask why it is that everything is not random. For those who may respond that this randomness disappears at the macro level, they are not really providing any serious explanation. They cannot satisfactorily explain how this disappearing act works, and in any case it is totally unclear what the boundaries are between the subatomic (micro) and the everyday (macro) realms or levels of physical reality, as well as how the two levels are related.

If there really were randomness at the micro level, then we would have to explain how quantum theory has such extraordinary predictive power. If we then respond that this order is due to the fact that randomness occurs within specific limits, then this does not entail genuine randomness. If there are any limits at all, such limits are objective, which then presupposes order. In that case, we would have to explain how a type of ordering principle could contain total randomness, a similar problem as trying to explain the difference and relation between the micro and macro levels. Regardless of which interpretation of the uncertainty principle we adopt, whether we accept that the uncertainty is in nature or due to some type of limitation on our part, it would still be the case that realism is true. If the uncertainty is due to temporary or insurmountable limitations, then this would be an objective fact of the matter, irrespective of whatever opinion we may hold about it. If the uncertainty is an inherent fact of nature, or inescapably built into the equation, then this, too, would be an objective fact of the matter, irrespective of our opinions. In any case, the importance of such metaphysical reasoning is impossible to ignore if we actually want to understand what we are talking about. We are still uncertain about the uncertainty. (See Appendix A for a more detailed analysis of this issue.)

Coming full circle

It is quite fascinating, as Baggott notes, that although the positivists attempted to eradicate metaphysics, deep metaphysical questions have only become more obviously relevant to understanding quantum physics. 'Three centuries of gloriously successful physics', Baggott writes, 'have brought us right back to the kind of speculation that it took three centuries of philosophy to reject as meaningless'.[352] There are exceptions, of course. Some philosophers have continued to discuss these questions with great enthusiasm, while some physicists detest the idea of talking about God or consciousness or anything appearing to be metaphysical, but as a general comment it does not seem too far off the mark.

Without deep-thinking metaphysical scientists leading the way, pure research would eventually come to an end. Without pure research, applied research would soon lose its fruitfulness, and technological applications would be limited to a reshuffling of current knowledge. Within very limited parameters, we could have continued to innovate new products based on our knowledge of classical physics, but without the advent of quantum physics, virtually none of our high-tech industry would have been possible. If we want to discover and ride the next potential wave of genuinely novel scientific development, which must eventually supersede quamtum physics, then we are going to have start reflecting again in ways similar to Einstein and Heisenberg.

Just imagine a deeper scientific understanding that unleashes a whole new array of technological innovation, so powerful that it makes our current technology look like a horse and carriage next to a space shuttle. Unless we provide a place to encourage and incorporate such metaphysical modes of thought and insight, such new developments will remain only in the realm of science fiction. It seems easy to neglect the metaphysical aspects of physics when one has merely instrumental aims of obtaining research grants, but without those metaphysically and mystically inclined pioneering physicists, quantum theory would never have been born and our high-tech industry of today would never have come into existence.

As chemist Peter Atkins notes, 'around 30 percent of the manufacturing economy stems from the application of quantum mechanics'. If we could understand quantum theory better, he continues, 'there would be an extraordinary surge in the economy...for understanding always enhances application'.[353] Consequently, if we want to profit more from our technological industry, we need to understand physics better. Yet if we truly want to make fundamental progress in physics, we need to foster profound metaphysical thinking, which is quite unconcerned with such worldly aims.

FIVE: Faith in Physics

The purpose of this chapter is to help clarify the relationship between faith and physics, which will include three main parts:
1. the rational inconsistency of the nature of physics;
2. the essential presupposed metaphysical assumptions that must be true in order for physics to be possible; and
3. the relevant beliefs of several key pioneering physicists.

It will not be possible in this brief chapter to bring indisputable clarity to the notion of faith, and anyone who expects such an accomplishment is being unrealistic. Let us not forget that the founders of analytic philosophy were unable to nail down with analytic precision many of their fundamental concepts, and often they did not even bother to try. Nevertheless, so far as possible for my specific purposes here, I aim to bring clarity to this notion of faith, which, despite its elusiveness and ability to ignite disagreement, is an indispensable aspect of the sciences. While clarifying the essence of 'scientific faith', there will also be some overlap with theistic faith, both of which are facets of faith that imply and require Platonic metaphysics if we actually want to understand what we mean.

What is faith?

Whatever disagreements a realist may have with psychologist William James, surely he is correct in saying that 'we cannot live or think at all without some degree of faith'.[354] Like most, if not all, of our fundamental notions, we more or less know what we mean by faith, until we really start to talk about it. Perhaps many physicists will be relatively unconcerned about this discussion on the role of faith, but such disregard would be a grave mistake. Many physicists often claim that faith is fundamental to their discipline, and it is with this more restricted sense of scientific faith that I am most concerned, although we must also consider wider relevant implications.

The role of faith is also apparent in other sciences. For example, zoologist Stanley Beck writes that 'without an underlying faith in natural consistent behaviour in which causes as well as effects are detectable, scientific progress would be impossible'.[355] Arthur Ward Lindsey shared a similar view when, in 1948, while chair of the Department of Biological Sciences at Denison University, he wrote that 'science must have faith'. However, he qualifies this statement by saying that the sort of faith required in science 'cannot share the comfort of faith in a loving God who will ease the burdens of life on earth for the faithful and receive them into a gilded afterlife, for it knows too well the inexorable laws of nature'. Nevertheless, he recognizes that science 'must have faith in the Infinite Power that lies behind the magnificent attainments of mankind and in the capacity of men to push on to greater heights'.[356] We can call this sort of faith 'scientific faith', which I will further clarify below.

All realists need to rely on some version of this sort of scientific faith, which Baggott recognizes when he asks why Einstein was a realist despite his only justification for his realism being 'an appeal to faith'. The answer, Baggott notes, was simple: the realist belief in an observer-independent reality has generally been the unquestioned backbone and driving force behind the most important scientific discoveries for hundreds of years.[357] Although Baggott is essentially correct about the requirement of faith as necessitated by realism, the supposed antirealist founders of the Copenhagen interpretation also shared the same sort of foundational faith as their opponents. *All* scientists, whether capable of admitting or articulating it or not, must share at least some similar sort of faith, for it is embedded in the very fabric of the entire structure of their scientific beliefs. Scientific faith does not necessarily endorse any particular religion, although religious faith, when broadly construed, does share some foundational assumptions with scientific faith.[358]

Despite the importance of this discussion of the role of faith, there is little to be gained here from an overly detailed analysis. It is impossible in a single chapter to capture the vast potential for philosophical exploration regarding the nature of faith, and I am also not going to delve into a comparative discussion concerning the related notion of belief.[359] It is important to maintain a tight focus, for otherwise it is too easy to become sidetracked by trying to account for every possible philosophical twist and turn. We can, however, still gain reasonable clarification of the role of faith in physics.

Physicists cannot absolutely empirically prove their fundamental assumptions, such as cause and effect relations, or the underlying uniformity of the universe. They cannot even prove beyond doubt the veracity of the conservation laws (such as energy, mass, momentum, and angular momentum, which will not concern us here). If we cannot logically or empirically prove something to be true but we have to believe it anyway, then we are to some degree relying upon faith. It is true, however, that faith in the law of conservation of energy is quite different than faith in, say, the Catholic Trinity, but a deeper examination reveals that the differences are not as significant as they may at first seem. The differences are really a matter of degree.

Even if we believe that we have a reasonable amount of certainty about a scientific law, the ambiguity concerning what counts as 'certain' or 'reasonable' immediately enters into our investigations, and we are left reaching for something further, which is also to rely upon faith. But we do not blindly believe in the conservation laws. Many experiments do support or imply these laws, and it was by maintaining belief in the law of conservation of energy against physical evidence that physicists were led to search for the neutrino, which was a type of hidden-variable. Physicists need a rational faith (as opposed to blind faith) in the laws of physics, while remaining open to the possibility that these laws could be superseded, but only by a law that is more abstract, simple, and predictively powerful.

We must have faith that the laws of physics will continue to allow us to depend upon them in order to be able to carry on with our physics (and even to be able to walk around without expecting that our entire molecular structure may instantly and randomly be transformed into that of a frog). We cannot even prove what we mean by 'prove', or that whatever we mean by a 'proof' is actually a 'good' proof, nor can we prove that we even need to have a proof for any claim in the first place. But the assumption that the fundamental laws unveiled by physicists will remain valid within their applicable domains throughout the universe must not only be believed (explicitly or implicitly) but must also be true.

I will later disagree with astrophysicist David Lindley over his apparently unintentional slip into antirealism, but he makes other helpful points regarding faith. For example, he writes that 'underpinning all scientific research, but especially research in fundamental physics, is this article of faith: that nature is rational'.[360] Such observations seem to be all that is required in establishing the role of faith in physics, or in any system of knowledge. What else, really, needs to be said? We cannot define with unambiguous certainty our fundamental concepts or assumptions, whether in physics or philosophy, or in any other field whatsoever, nor can we prove them beyond doubt empirically or logically, and therefore we must have some sort of belief in them over and above logic and empirical evidence, which, in its most basic sense, is to require having faith.

I need only respond to my opponents by asking for proof of the argument supporting their disagreement, and then request proof for the assumptions under which they based their original argument, which eventually they will not be able to provide. Once again, we are left admitting that clarity entails being 'clear enough', but the question remains: clear enough for *whom*? I think it is quite clear what Planck means by faith, but for many people (including many physicists), the explicit and implicit faith of the physicists is not at all clear, and is sometimes even psychologically upsetting (especially for atheists).

Whether or not they are conscious of it, physicists need to have faith in certain fundamental assumptions in order for them even to be able to do physics. Such assumptions can never be proven through experimental procedures, because they are the *starting points* from which mathematics and experimental physics are possible. It also appears to be the case that the greatest scientific discoverers who made the most fundamental breakthroughs at the conceptual level had to be motivated by an intense faith that reality is in itself a unity and is intelligible to the human mind, from which it follows that we can discover truths no matter how partial our insights may be. This basic faith opens the doorway to our potential to understand the universe, and ourselves, more deeply.

Based on my extensive reading of the pioneering physicists who are the focus of this book, *scientific faith* includes the following assumptions:

1. the universe is unified and rational;
2. there is a real physical world that exists externally to any sentient creature or any particular physical object;
3. the laws of physics are susceptible to mathematical formalization, and thus, in the Pythagorean/Platonic sense, the physical world seems to be composed of numbers, or follow strict mathematical laws;
4. the relations between the numbers, expressed as mathematical laws of physics, are not physical and must be unchanging in order to account for their usefulness in predicting novel and disparate phenomena;
5. the more powerful a law of physics, the more simple, abstract, and unifying it must be, and so the more closely it resembles the truth of the matter;
6. some kind of trans-rational insight is required to see what others have missed despite possessing the same available data; and,
7. in some way our minds must be in harmony, or have the potential to be in harmony to different degrees, with both the laws of physics and the physical universe, and must be able to conceive of the nature of their unity.

This list may not be exhaustive, but it includes some of the most essential assumptions of scientific faith.[361] These assumptions are prior to physical experiment and must be believed, at least implicitly, prior to theorizing, for there would be no point in theoretical speculation or empirical testing if we did not really believe that relatively stable answers could be discovered. If we assume, as most physicists do, that we can *discover* answers, then we are assuming the philosophical position of realism. Such assumptions are fine, so long as we do not pretend that we do not have them. We also need to assume that there is an underlying rational order in the cosmos and that the current discrepancies in our theories are not actually discrepancies in the universe itself, which is to say that physicists must seek out the underlying or hidden order of the physical universe. More importantly, these assumptions must actually be true, regardless of whether or not we consciously believe them.

Polanyi agrees with Jeans that the 'outstanding landmarks in the progress of science' have always been due to seeing the inherent order in already known facts.[362] Examples abound of scientists who have had profound insights into previously existing data, including Copernicus, Newton, Charles Darwin, Einstein, Heisenberg, Schrödinger, and Dirac.[363] Planck's insight concerning discrete energy was based upon widely available data and, as Polanyi notes, 'he alone saw inscribed in it a new order transforming the outlook of man. No other scientist had any inkling of this vision; it was more solitary even than

Einstein's discoveries'.[364] A lack of understanding of imaginative vision and intuition necessarily entails a lack of understanding of one of the most vital aspects of scientific progress.[365]

Now that we have a better idea of what scientific faith looks like, we may ask ourselves what sort of description may encapsulate the *general* notion of faith. Philosopher Mary Midgley offers what seems to be an overall accurate summarization, saying that faith is the basic trust that we need before we can even ask whether or not anything is true. Faith 'is the acceptance of a map, a perspective, a set of standards and assumptions, an enclosing vision within which facts are placed. It is a way of organizing the vast jumble of data'.[366] Heisenberg, too, thought that faith entailed the deepest trust: 'if I have faith, it means that I have decided to do something and am willing to stake my life on it'.[367] In addition to scientific faith at the metaphysical level, scientists also have to trust the work of other scientists, even though fraudulent research can slip through the cracks of peer review. But this only serves to strengthen the claim of the importance of faith, because scientists must also trust other scientists to uncover fraudulent science.[368]

Contrary to the ill-conceived assumption that faith is a relic of pre-scientific nonsense, Midgley notes that 'in our age, when that jumble is getting more and more confusing, the need for such principles of organization is not going away. It is increasing'.[369] Our need for a rational faith underlying our ability to attain a vision of organizing principles is, indeed, increasing. How can we even ask whether or not a statement is true unless we first believe and trust (have faith) that truth is a viable concept, that the statement being true or not is a matter of whether or not it accurately describes a real state of affairs? Realism clearly presupposes faith in our ability to discover or approach truth, which in turn presupposes that truth is real, even if it is beyond our grasp. Antirealists also have faith that their claims are true or, at any rate, that they are genuinely better or more useful than those offered by realists.

Theologian Paul Helm argues for a web-like notion of interlocking beliefs, claiming that because the lines of justification for beliefs are not linear, 'there is no danger of regress of justification proceeding indefinitely, and therefore no need to invoke a non-inferential stopping place for such regress'.[370] In essence, he is arguing that there are no foundational beliefs required in a web of beliefs, because they are mutually interdependent to one degree or another, which is an odd claim coming from a theist. In any case, he is mistaken for two reasons. First, although beliefs are interlocking, clearly some are more fundamental than others, which implies some sort of hierarchical linearity infused within dynamic web-like interrelationships. For example, believing that God is real (or not real) is more fundamental than believing that your favorite television program will be aired as scheduled next week. Second, Helm argues that what justifies the reasonableness of our beliefs (in his case, our religious beliefs) is that they cohere as a whole.

The *coherence theory of truth* basically says that truth depends upon the degree to which the statements of a theory cohere as a whole, which is to say that they are consistent or reasonably imply each other. However, striving for coherence is a necessary but not sufficient condition to guarantee the truth of a system or theory. A purely coherentist view does not guarantee any genuine relation to reality. For example, any set of completely false statements may be logically consistent, such as the statement that: all bananas are blue; my dog is a banana; therefore, my dog is blue.

We should strive for coherence in any system, theory, or concept, but our set of coherent beliefs must also correspond to the way things actually are in reality. The *correspondence theory of truth* basically says that our statements are true if they correspond to reality. In other words, the claim that 'there is a cat on my desk' is true if and only if there is, in fact, a cat on my desk; otherwise my claim is false. But we cannot be content with a simple correspondence theory either, for a bunch of random statements that happen to correspond to certain aspects of reality may be totally useless. I can say: Canada is located north of the United States; there are many different types of insects; and some people have more money than others. Although all of these statements are true, which is to say that they correspond to reality, there is no relevant connection between them and so they do not cohere. In order to provide a viable scientific theory, our beliefs must correspond to reality *and* be internally coherent so far as possible, and we must also be pragmatic in selecting which truths are most worth seeking. However, achieving the appropriate balance between correspondence to reality and internal coherence is a never-ending challenge, especially since sometimes it can be rational to be logically inconsistent, as we shall see in the next section.

Rationally inconsistent physicists

Eddington clarifies how physicists used classical laws to calculate the energies of the orbits of a hydrogen atom, which involves Planck's constant, even though Planck's constant is contrary to the classical laws of radiation. 'The whole procedure', Eddington notes, 'is glaringly contradictory but conspicuously successful'.[371] How can it be that such logical contradictions in physics do not necessarily impede its incredible success? The main reason in this case is that the contradiction is not in nature, but in our theories. However, what may be even more interesting, and what we will focus on in this section, is the rational inconsistency of the pioneering physicists themselves.

We must admit that while it is essential to take seriously the philosophical views of physicists, it is also very difficult to place them in one particular and completely exclusive philosophical camp. Einstein, for example, has said that science may be defined as 'methodological thinking directed toward finding regulative connections between our sensual experiences',[372] which sounds like he is endorsing empiricism. But he also opposed empiricism and positivism, employing phrases such as 'senseless empiricism' and 'sterile positivism'.[373] Heisenberg recalls Einstein telling him that he did not want to endorse naïve realism, which usually just amounts to materialism, while at the same time expressing the view that Mach's empiricism was also too naïve. Einstein seems to have held a mix of philosophical views in relation to the sciences (as all scientists tend to do), although his realism remains constant.

I am not expecting philosophers to follow the physicists, nor vice versa, but philosophers do need to be able to engage seriously with the physicists' philosophical views, despite their rational inconsistency. Gibbins writes that 'from a great physicist one should not expect too much consistency of the type that philosophers value'.[374] But why should this be the case? Isn't physics a rational or logical activity, and therefore shouldn't its theories be rationally or logically consistent? The same physicists whom I am asking the philosophers to take seriously as philosophical thinkers were anything but straightforwardly consistent. But it was the physicists, not the philosophers, who were initially confronted with the apparent paradoxes of quantum theory, such as wave/particle duality. These physicists had to try to make sense of the data with which they were confronted, which led them into uncharted subatomic terrain and forced them to confront ancient metaphysical questions.

Lowry notes that many Platonists wanted to show that 'different philosophies really only appear to be different', and that 'behind such a desire for agreement is the perhaps somewhat euphoric conviction that truth cannot be contradictory'.[375] Similarly, relativity and quantum theory cannot really be in contradiction, because they are both aspects of the whole. Physics makes fundamental progress when we discover underlying unifying laws, such as those that will eventually show the inherent unity underlying or transcending the apparent contradictions between classical and quantum physics. In the meantime, however, physicists must make do with what they have. Many physicists (perhaps the majority) need to focus on piecemeal work, asking questions and experimenting in highly constricted fields of research. However, some physicists (perhaps a significant minority) need to be seeking the underlying, unifying laws, for it is only when we make such fundamental discoveries that we make genuine scientific progress into new areas. There is no need to suppress one approach in favor of the other.[376]

The point is that in some cases we need to emphasize empirical observation and at other times we need to emphasize deeper theoretical speculation. Being a mathematical and scientific Platonist includes accepting the reality of material entities, the importance of empirical observation, and the necessity of being pragmatic. It also means rejecting the notions that all of reality can be reduced to the physical, that all of physics can be reduced to observation, and that truth can be reduced to usefulness, as the *pragmatist* would generally claim. Denying the metaphysical doctrines of materialism, empiricism, and pragmatism does not entail a rejection of material objects, empirical research, or the importance of being pragmatic.

Physicists could not be completely logically consistent, as Heisenberg well understood, because such a demand would make science impossible, but that does not mean that they were irrational. If physicists were completely logically consistent, then they could never be successful. For example, if Einstein had adopted what appeared to be a positivist outlook in one case and was therefore always required to maintain this outlook in order to be logically consistent, he could not have gone on to further developments in his field. Similarly, if we were not permitted to use quantum theory or classical physics until successfully proving their underlying consistency, we could not have developed the physics that we have today, along with all the subsequent technological applications. We must be able to use what we have to the best of our ability, but we must also continue to seek underlying unity, because it is the seeking of such unity that drives the whole discipline of physics forward.

Consider the following example. The statement 'A = A' is a *tautology*, which means that the statement must always be true by logical necessity. If all of science were to be reduced to such logical consistency, we could never discover anything new, because we could only ever say what we already know to be true. Moreover, the statement 'A = A' tells us nothing about the nature of A, and even the 'equals' sign is highly problematic. If 'equals' means absolute identity, then A can only ever be absolutely identical to A abstractly, but never physically. In other words, the tautological statement 'A = A' can only be true in the abstract, nonphysical realm, because the very instant we apply this statement in any way whatsoever, which means bringing it into time and space, it necessarily loses its guarantee of being true by logical necessity.

If A refers to different objects, they obviously cannot be absolutely identical, because they must at least be located in different places. But even if A refers to the same object, it still takes some amount of time to say (or even think, or write) 'A = A', in which case it is logically and physically possible that, during the time elapsed to say (or think, or write) the statement, the object could have changed in some way. Actually, it *must* have undergone some sort of change, because all physical things are constantly changing in one way or another at every moment to some degree, even if only in their relative space-time coordinates. Strictly speaking, then, 'A = A' can never imply absolute identity in the physical world, which means that as soon as we apply the tautological statement to the physical realm, we immediately lose the logical guarantee of truth. In other words, abstract logical relations can never guarantee deductive validity in the physical world, which is to say that we can never guarantee that, physically, 'A = A'. Necessarily, therefore, physics can never represent a completely logically consistent theory unless it has no concern whatsoever for the physical world.

Even though physicists can achieve great practical success without attaining complete logical consistency, this does not give us license to haphazardly assert nonsensical and blatantly contradictory propositions. We do need to *aim* for complete logical consistency, while acknowledging that in the end it is never possible to actualize in any physical sense. What I want to say turns on the following point: being logically consistent is not the same thing as being rational. It may be rational to believe a conclusion that is not logically certain, and it may be irrational to believe a conclusion that is logically certain. Before providing an example, let me first offer a brief clarification between deductive and inductive reasoning. In a *deductive* argument, the conclusion necessarily follows from the premises, whereas in an *inductive* argument, it is likely or probable that the conclusion follows from the premises. For example, it is not rational to believe the following logically consistent (deductively valid) argument.

1. Tom is a banana.
2. All bananas are pink elephants.
3. Therefore, Tom is a pink elephant.

But it is rational to believe the following deductively *invalid* (though inductively plausible) argument.

1. The sun's rays have reached the earth every day in my past.
2. The known laws of physics predict that the sun will continue to exist tomorrow.
3. Therefore, I believe that the sun's rays should reach the earth tomorrow.

It is not possible to have a deductively valid argument when the conclusion is based upon the past and is projected into the future. Just because something happened in the past does not necessarily mean that it will continue to happen in the future, and even if something did not happen in the past, it may very well happen in the future. This logical fact causes great headaches for those who think deeply about the laws of physics, because there is no logical guarantee that what happened in an experiment in the past will recur in future experiments. In other words, just because my book has fallen to the ground every time I let it go in the past, it does not logically follow that it will happen again the next time I let it go (but you would be wise to bet a fortune that it will). Our faith in the constancy of the laws of physics, a faith that transcends logical certainty, is rationally justified.

There are other serious limitations to logic to consider. As stated in an introductory logic textbook, 'the techniques of formal logic cannot normally tell us which claims and beliefs are true and which are false. Truth is usually a matter of the way the world is, and logic does not tell us that'.[377] Logic cannot even tell us which premises to choose in the first place, and so there is no way to logically begin reasoning logically, because the original premises cannot come from logic itself. Logic cannot even tell us if our premises are actually true. What follows is that there is no logical method to begin using logic. Eddington makes a similar point: 'Reasoning leads us from premises to conclusion; it cannot start without the premises. The premises for our reasoning about the visible universe start in the self-knowledge of mind'.[378] This is a very important point that every philosopher should know but which is mostly ignored, perhaps due to the psychological fear of admitting the role of mind underpinning our ability to use logic in the first place.

Related to the discussion of the limits of logic is the recognition that there is also no scientific method to begin using the scientific method. Apparently forgetting about his belief that only science can give us truth, Russell admits that 'it is quite difficult to think of the right hypothesis, and no technique exists to facilitate this most essential step in scientific progress'.[379] If there is no technique with which to begin science, then how do we begin it, and what are the implications of recognizing the limitations of the scientific method or of admitting that there is no scientific way to begin science? An intuition or insight, coupled with some kind of basic motivation, is the starting point for both logical reasoning and our scientific investigations. Intuition, however, involves some trans-rational awareness, state of mind, or direct perception or understanding. This fundamental role of intuition inevitably opens the way to mysticism as well, and, consequently, many academics who spend much of their lives working discursively can become quite psychologically unsettled by the prospect of looking honestly and directly at this process. Turning away from reality, however, is not scientific.

We aim to present our research as if everything were neat and orderly and logically clear, which is really nothing more than a facade. Joseph Agassi, a philosopher specializing in logic and scientific methodology, reminds us that the apparent orderliness of scientific research veils the hidden reality of the messy problem-oriented workshop. The student who falsely believes that 'science is orderly—inductive, or deductive or both—but not the mess of the problem-oriented workshop' is not yet qualified 'to work in the scientific research workshop'.[380] Those who admit and recognize the messiness of the workshop are much more likely to help clean up some portion of it.

While writing this book I have aimed to present my ideas as logically, rationally, and clearly as possible, with each point hopefully following the last in a way that seems to make sense. But the reality remains that the many struggles involved in formulating these ideas have not always followed a neat logical progression. Many insights have occurred suddenly while doing the most mundane of activities, or they have been the result of some significant life experience. This point is extraordinarily important because, by removing the 'human factor' and ignoring the reality of the messiness of the workshop, and especially by downplaying the fact that there is no scientific method to begin science and no logical method to begin logic, we can pretend that everything is fine so long as we have a consistent argument. The analytic philosopher can rest peacefully at night.

What we have to admit is that physicists, especially when they are breaking new ground, cannot be expected to have complete consistency. However, I am now going to make an apparently opposite claim, though the subtlety of it will steer me away from the charge of contradiction. Despite my above claims, I will add that we still need to correct the false, confused, contradictory, and misleading reasoning, pronouncements, and assumptions of these same pioneering physicists whom I have been discussing. The fact is that these physicists were sometimes blatantly inconsistent in the sense of not maintaining rationally plausible views, and at various times they fell prey to poor reasoning and false or highly problematic assumptions. They also sometimes confused the meanings of important concepts and ended up talking past one another in their arguments, which resulted in unnecessary confusion permeating quantum theory. (Obviously these observations apply equally to philosophers, and to everyone else, too.)

When breaking new ground and confronting baffling data, physicists have to be able to make bold speculations that reach beyond the bounds of 'normal' science. They cannot be forced into the straitjacket of logically tight reasoning from the outset because, first of all, the premises to be used in such reasoning may be false, since they would be rooted in the assumptions that are the reason for the inexplicability of the data in the first place. Second, as already shown, such complete logical consistency would not allow for inductive reasoning or the seeking of novel intuitive or imaginative solutions in the problem-oriented workshop of actual scientific practice. However, the reasoning of the physicists still has to make sense—it has to be rational—which, again, can easily divert us into a different topic concerning a dialectical exploration of the similarities and differences between rationality and logic and what counts as 'reasonable'. The key point here is that being rational does not necessarily entail being logically consistent, as rationality and logic are intimately connected yet also distinct. To use a simple example, it is reasonable for me to believe that my wife will love me tomorrow, but no logical argument can prove that this belief is necessarily true.

Faith-filled physicists

Exploring some of the unavoidable overlap between religious and scientific faith will require a bit of philosophical meandering until the end of this section. Some academics seem incredulous (and can even become quite upset) when it is pointed out that Planck, the founder of quantum theory, believed that faith was essential to doing physics. Planck's scientific faith is not to be equated with faith in papal infallibility or the holy trinity, but what is of relevance is that he had faith in the rationality and unity of the cosmos, which are views that are also foundational to theism. Gödel was also fond of reminding philosophers, much to their general annoyance, that the founders of modern science were not atheists.[381] One may respond that in the context of their time it was not unusual for Copernicus, Kepler, Galileo, and Newton to believe in God, but such a response indicates a lack of understanding of how the development of scientific thought depended precisely upon some sort of theologically inspired thinking.[382] Kepler, for example, said that he 'constantly prayed to God that [he] might succeed if what Copernicus had said was true'.[383]

Not all contemporary philosophers with an understanding of the sciences have rejected theological underpinnings in science. Clark, a notable exception, has argued extensively that 'in ethics and in science alike we must rely on faith, on indemonstrable axioms',[384] which is an unavoidable fact—we cannot demonstrate with absolute logical certainty the veracity of all (or perhaps even any) of our fundamental propositions and assumptions. He argues further that reason itself, or rather our ability to reason and the concomitant assumption that reality should be amenable to our reasoning mind, cannot be proven by the application of reason alone. We must admit with Clark that 'reason rests on faith'.[385]

Louis de Broglie, another pioneering physicist who contributed to the development of quantum theory, makes a related point about pure science untiringly pursuing 'the search of this hidden order'. He adds that 'we are not sufficiently astonished by the fact that any science may be possible, that is, that our reason should provide us with the means of understanding at least certain aspects of what happens around us in nature'.[386]

James makes a related point about the search for hidden order in nature: 'a man's religious faith (whatever more special items of doctrine it may involve) means for me essentially his faith in the existence of an unseen order of some kind in which the riddles of the natural order may be found explained'.[387] James is here intimating the distinction between primary and secondary religious beliefs. In this case, belief in the existence of an unseen order in nature is an aspect of primary religious faith, which is also an essential aspect of scientific faith.

That most of us are not 'sufficiently astonished' that science is even possible at all is generally due to a lack of appreciation and understanding of science. Those who are most deeply immersed in the profoundest aspects of pioneering thought are generally the first to admit their astonishment and ignorance about the subject matter they are exploring. Those of us who pretend to know everything tend to act as if such things are of no real consequence, which reveals just how little we really know what we are talking about.

Those who have reflected deeply on how or why it should be possible that our thoughts are capable of grasping 'the profound relations existing between the phenomena' find it difficult to avoid concluding that it is astonishing, a 'great wonder', as de Broglie put it.[388] Why should there be any relations at all, let alone 'profound' ones, between things, and how is it possible that we can grasp these stable relationships which are responsible for the changing phenomena around us? Logic alone will not provide the answer to this question of the rationality of the cosmos, and all of our experiments *presuppose* that our theories and data will actually be able to tell us something about reality. Why should the experimental method and theorization prove so fruitful? It is not a sufficient explanation to say that our thoughts can find agreement with reality because they just happen to do so.

The predominant philosophy of the Middle Ages took for granted that the universe was intelligible to us,[389] which was an essential assumption for the rise of modern science. Those who have not understood this point have assumed that, for example, Newton's theological conjectures were an 'aberrant and antiquated appendage to the corpuscular philosophy'.[390] But in actuality, the exact opposite is true; his theological propensities were the foundation of his scientific thinking. As Rowlands argues, Newton was not a mechanist but had the 'theological cast of mind to recognize the true abstract nature of scientific thought',[391] and the abstractness of fundamental physics has been growing ever since. The common assumption of the conflict between science and religion is not completely false, but it obscures the profound metaphysical interconnections between them that have persisted (at least implicitly, and sometimes explicitly) as part of the foundational guiding beliefs of many pioneers in theoretical physics.

John Hedley Brooke, a specialist in the history of science, shows how 'in the past, religious beliefs have served as a presupposition of the scientific enterprise insofar as they have underwritten that uniformity. Natural philosophers of the seventeenth century would present their work as the search for order in a universe regulated by an intelligent Creator'.[392] If there is order in the universe, then it is a rational assumption to presuppose that there is some divine intelligence behind that order. Brooke argues that science may have developed without a prior theology but, nevertheless, 'particular conceptions of science held by its pioneers were often informed by theological and metaphysical beliefs'.[393] A key point that Brooke is overlooking here, however, is that although science may have been able to develop without its pioneers adhering to any particular theology, it is still the case that scientific faith was essential for these pioneers. It is even more essential that scientific faith, which is essentially an expression of scientific Platonism, must actually be true.

Planck and Einstein remind us of further necessary requirements for innovative science, such as intuition, creative imagination, and faith in ultimate success and in the underlying order behind appearances.[394] For example, Planck states that Kepler's 'faith in the existence of the eternal laws of creation' was an essential factor enabling him to see the inherent order in disparate astronomical observations. Kepler 'was the creator of the new astronomy', whereas the great sixteenth century observational astronomer Tycho Brahe, who Kepler assisted, 'remained only a researcher'.[395] Planck's comment about Brahe being only a researcher is a bit unfair, in the sense that Brahe's observations laid the foundation for Kepler's theoretical leaps. But in our modern culture we almost always worship the 'practical' person, such as Brahe, and ignore the kind of mystical mindset required for great theoreticians such as Kepler. In reality, we need both types.

Planck argues that there is no logical way to find the most suitable scientific hypothesis, and that only an independent imagination, strong creative power, and accurate knowledge of the relevant facts can allow the mind to immediately grasp the problem or seize 'upon some happy idea'.[396] He also agrees with Einstein that you cannot seriously do science unless you believe that an external world really exists, even though our knowledge that there really is an external world cannot come to us by any process of reasoning. As mentioned in Chapter Four, Planck says that such knowledge is a result of 'a direct perception and therefore in its nature akin to what we call Faith. It is a metaphysical belief'.[397] In fact, the realist belief that there are verification-transcendent truths about the world and the universe, truths that we may never be able to prove or even know at all, necessarily requires a deep faith. It is with good reason that Clark firmly believes 'that there is a reality essentially independent of all human cognition and experience which may nonetheless be known by us'.[398] This faith is necessarily mystical, because the laws of physics and other higher metaphysical principles do not have any physicality, and yet they are real.

Einstein was mystical and, similarly to Clark, he believed that we could have 'knowledge of the existence of something that we cannot penetrate'. But he was strongly opposed to organized religion and what he saw as the naïve belief in personal existence after death. Yet his description of this essential mystery as being the 'profoundest reason and the most radiant beauty' could have come straight from almost any Platonist, and he devoted his life to trying to understand a portion, 'be it ever so tiny, of the Reason that manifests itself in nature'.[399] Without this relentless drive for truth and understanding, fundamental science would come to a halt. 'To this', writes Einstein, 'there also belongs faith in the possibility that the regulations valid for the world of existence are rational, that is, comprehensible to reason. I cannot conceive of a genuine scientist without that profound faith'.[400]

The profound faith that Einstein is talking about, the faith that the universe is regulated by rational laws that we are able to discover through the use of our own reasoning faculties, also requires a further vital component, which is the intuitive flash of insight or direct understanding. As Bohm puts it, 'one may be puzzled by a wide range of factors, things that do not fit together, until suddenly there is a flash of understanding, and therefore one sees how all these factors are related as aspects of one totality (e.g. consider Newton's insight into universal gravitation)'.[401] Ignoring this vital component of scientific progress, which also applies to artistic and business endeavors, leaves us in ignorance about some of the fundamental aspects of scientific discovery, and creativity in general. Bohm also reminds us that these intuitive flashes of understanding 'cannot properly be given a detailed analysis or description'.[402]

Bohm used the same kind of language as Planck to describe this insight beyond logical thought, both employing the word 'perception' in its widest sense. If we consider the views of these three physicists together (Planck, Einstein, and Bohm), we can see that there is general agreement amongst them about the limited applicability of logical analysis. It does not follow that no analysis is possible of the act of insight, the flash of intuitive understanding, but only the foolish would believe that we could actually capture such creativity in an airtight logical package that could be replicated by anyone following the same steps. Surely I do not need to stress the obvious point that genuinely novel creativity does not happen in this manner.[403] But because this point is so obvious and yet so difficult to pin down, it is easy to gloss over it, to turn away to more manageable endeavors that are more amenable to logical analysis and have much less ambiguity or fewer mystical implications.

The truly revolutionary insights in science (and the arts and business), the sort of acts of direct perception or direct understanding such as Planck's that gave birth to quantum theory, have been due to insights beyond logical thought. Even when the insight can be shown to be a logical conclusion from the original data and variables, the conclusion was not obvious to all those who had originally tackled the problem, and so logic alone cannot provide the answer. Only after the conclusion has been discovered can logic be used to lead others to it in a discursive fashion. Knowledge that does not depend on any process of discursive reasoning, which is a 'direct perception', a 'flash of understanding', is essentially what the Platonists referred to as 'non-discursive' thought or knowledge.[404]

Zoologist and Nobel Prize winner Peter Medawar points out that scientific discovery cannot be premeditated, for not even the greatest scientists can say with any certainty that they will make an important discovery.[405] Medawar's obvious point sometimes seems to be totally forgotten by research funding bodies, especially when they neglect the importance of non-discursive insight or intuition for scientific discovery. This is the problem with critical thinking or problem solving in general. We are seeking a solution that is not obvious, and only a genuine moment of creative insight will reveal the answer, which later is checked for its logical or empirical support. If you leave out this creative step, this intuitive leap into the unknown, you are ignoring one of the most vital, indispensable aspects of the scientific process. No presentation of scientific methodology can be considered adequate if it cannot provide some sort of an account of this trans-logical creative intuition.

Heisenberg, for example, was deeply influenced by Plato's Timaeus in his youth and had a mystical experience of the 'central order' of things, which was to affect his later thoughts 'profoundly'.[406] Heisenberg had originally thought that the Timaeus was 'completely nonsensical' and kept 'wondering why a great philosopher like Plato should have thought he could recognize order in natural phenomena when we ourselves could not'. While attending a Youth Assembly offering various political speeches, he grew frustrated that the viewpoints expressed there 'were no longer directed toward a unifying center'. This disharmony became increasingly painful for him, to the point where he was 'suffering almost physically'. After several hours of such speeches and agony, suddenly a hush descended over the crowd as a violinist on a balcony began playing Bach's Chaconne. At that moment, something extraordinary happened to Heisenberg: 'all at once, and with utter certainty, I had found my link with the center....There had always been a path to the central order in the language of music, in philosophy and in religion, today no less than in Plato's day and in Bach's. That I now knew from my own experience'.

Heisenberg now understood Plato's idea that the underlying structure of all matter is of geometric forms rather than things, and that such forms appear to refer to the 'atom's structure in time and space, to the symmetrical properties of its forces, to its ability to form compounds with other atoms'. These structures are not material objects, and so will not be amenable to normal descriptions of things, but they must be susceptible to mathematical treatment. It required a direct personal experience (the flash of understanding, a direct perception) transcending simple discursive reasoning in order for Heisenberg to understand Plato's notion of unifying order and nonphysical geometric forms as the basis of physical reality. This he knew with 'utter certainty', which is quite significant coming from the man who gave us the uncertainty principle.

Moreover, it was this experience that profoundly affected his later thoughts, deeply influencing his way of understanding quantum theory. Here we can see the roots of the Copenhagen non-materialism that put more emphasis on the underlying geometric symmetries represented in the mathematical laws of physics, which have more reality than the physical world because they are representative of the eternal central order of the cosmos, what Einstein referred to as 'Reason' and what Pauli had called the 'cosmic order'. Analytic philosophers can protest as much as they want about this way of speaking and thinking, but it can hardly be denied that such beliefs and experiences have given us quantum theory and all of physics.

Eddington even writes that 'the physicist now regards his own external world in a way which I can only describe as more mystical, though not less exact and practical, than that which prevailed some years ago'.[407] Schrödinger writes in a similar vein: 'I have therefore no hesitation in declaring quite bluntly that the acceptance of a really existing material world, as the explanation of the fact that we all find in the end that we are empirically in the same environment, is mystical and metaphysical'.[408] They are both correct. The deeper we logically analyze a fundamental concept, such as truth, without the concomitant intuitive understanding needed in order to guide our reasoning process, the slipperier this concept becomes. The more we fear not being able to grasp analytically its full meaning, the tighter our cognitive grip squeezes, which inevitably has the unfortunate effect of extinguishing both our discursive apprehension and our potential for non-discursive insight, like a blanket smothering a flame.

We may not be able to unambiguously define the nature of truth, as Frege also realized, but we still can, to varying degrees, intuitively grasp it. Although this intuition, this direct perception, can never be fully actualized in discursive form, we still need reason both as a springboard and as an anchor. Recognizing the inherent limits of reasoning sets us free (as terrifying as that may be), while the power of reason provides us with the potential for actualizing as closely as possible what we have come to know non-discursively, which is what grounds us in the everyday aspect of reality. All of science and art engages in this dialectical play between creative insight and discursive unfolding, whether in mathematics or painting, or in whichever way we choose to express ourselves.

We must be courageous and continue exploring unknown metaphysical terrain that overlaps and is contiguous with our empirical evidence. Failure to do so will impair our own intellectual development and hamper further significant scientific progress. Deeper metaphysical territory may be less suited to many explorers, but that does not at all detract from its reality and importance. It is in this area that we are best able to understand rationally the intuitive flash of insight or understanding that, as Bohm noted, is not amenable to detailed analysis. We can study such insight rationally, and we also want to understand the role played by the brain and the rest of our physiology,[409] but genuine insight is attained beyond the limits of reason and empirical data. This mystical moment of insight is not just for artists and spiritual aspirants, but is also fundamental to creative progress in the sciences. This does not make it irrational; rather, it makes it trans-rational. Irrationality is an enemy of science. Rational mysticism reveals the foundation of science.

Even with such insights as those experienced by Bohm and Planck, we are still, to a tremendous extent, dependent upon discursive reasoning. However, we cannot be afraid to admit along with Polanyi the fact that in the end we must rely upon ourselves, upon our own convictions, no matter how much reasoning we offer in support of our beliefs.[410] This is not giving in to antirealism, but rather it is admitting the inescapable role that we ourselves must play in all such considerations. Even if we completely disagree with Schrödinger, Einstein, Planck, and Heisenberg and think that their views are absurd, it is still the case that these mystically inclined physicists have provided us with the most powerful scientific developments in history. Is it a mere coincidence that the greatest pioneering physicists tend to be considered 'absurd' by the standards of those who are incapable of comprehending them?

The eighteenth century philosopher David Hume provided strong arguments against our knowledge of causes and effects, or at least of there being any way to logically prove that there are such things as causes and effects beyond mere constant conjunctions. What he thought, essentially, is that the best we can say is that X tends to be associated with Y, which is to say that Y appears to be the cause of X, but all that really happens is that X and Y tend to be constantly conjoined. However, there is no way to prove logically that Y causes X.

For example, when I flick the light switch in my bedroom, the light turns on. It appears that the flicking of the switch causes the light bulb to shine. However, as we already know, there is no way to prove logically that the next time I flick the switch the light will come on again. The best we can say is that the flicking of the switch and the shining of the light bulb have constantly been associated together in the past. For Hume, this recognition led to skepticism, which is not, in fact, a logically necessary conclusion. I hold a similar view as Hume about the inability of logic to prove cause and effect relations, but I then go in the opposite direction, where I recognize that the laws of physics are real and that the light bulb will always shine when I flick the switch, despite the failure of logic to prove this with certainty. The only time it will not shine is when there is another scientific explanation for this outcome. For example, the light bulb has burned out, or there is a power outage, and so forth.

Contrary to Hume's attack on reason, Whitehead writes that 'scientific faith has risen to the occasion, and has tacitly removed the philosophic mountain'.[411] Heisenberg adds that taking Hume's denial of induction and causation seriously would 'destroy the basis of all empirical science'.[412] Schäfer also notes that Hume's attack on causality is logically justified when limited merely to observing external events, but that when the self-conscious mind is involved causality is clearly established, even though such principles of inference 'are non-rational and non-empirical in the sense that they cannot be verified by an observation of physical reality nor derived by a process of reasoning'.[413]

Hume was right to raise doubts about the nature of logic and reason, but he was wrong to reject the faith required in order to make his anti-reason arguments reasonable in the first place. He necessarily had faith in reason and in his abilities to argue reasonably, whether or not he openly admitted this fact. After all, he believed his own arguments to be reasonable and true. Jaki is correct, therefore, that 'for Hume, reason was a welcome ally only when its sharp thrust served his purposes'.[414] In other words, Hume's arguments against reason required him to have formidable reasoning abilities, and so his arguments against reason could only make sense to the degree that they were reasonable, undermining his skepticism towards reason.[415] Hume needed faith in reason in order to use reason against both faith and reason.

Barr makes the following relevant point: 'even the atheist, precisely to the extent that he is rational, has a certain kind of faith. He asks questions about reality in the expectation that these questions will have answers and that these answers will make sense...It is a faith that reality can be known through reason'.[416] Atheists who are not fundamentalists and are therefore willing to put their own beliefs to the test of rational examination must admit that they are relying upon reason to tell them something about the ultimate nature of reality; namely, that God does not exist. The only way they can make their case, in other words, is to have faith that reason can unveil aspects of reality. But then we have to ask the atheist why reality is such that reason has this sort of power, a question that immediately takes us deep into the realm of metaphysics, which quickly slides into theologically relevant territory. I am not going to discuss any further here the plethora of misinformation surrounding the science and religion debate, where abundant confusion can be found on both sides. Suffice it to say that many leading pioneering theoretical physicsts have at least been comfortable with a mystical understanding that is harmonious with Platonism.[417]

All physicists, whether they are religious, spiritual, atheist, or agnostic, are implicitly relying upon the hidden or underlying unity that accounts for the integrity of the vastly disparate phenomena we find in the universe. But our current approaches are reaching their experimental limitations. For example, Lindley has shown how cosmology and particle physics have become conjoined where, in their mutual search for fundamental forces and particles, they have been forced to recognize the current limits of empirical testing within their respective fields. Increasing the size of particle accelerators will do comparatively little to take us further into the sub-nuclear realm, and no one was around to witness the beginning of the universe (if there even was a beginning in the way that we think we can imagine it). As Lindley writes, 'full understanding of the fundamental particles and forces of the birth of the universe will be achieved, if it is ever achieved, only through a long and indirect chain of reasoning from what happened a long time ago, under circumstances we can barely imagine, to what is before us now'.[418]

This 'indirect chain of reasoning' is a good example of applied metaphysics, which presupposes predictable cause-effect relations from the beginning of the universe, a universe governed by eternally stable laws that we can discover. We should now have a better appreciation of the fact that such assumptions require scientific faith in order, for example, to be able to conduct an experiment that attempts to replicate the conditions of the Big Bang. Physicists can only believe that an experiment conducted today can help us understand the initial conditions of the universe at the moment of the Big Bang by assuming that the same laws of physics that apply now also applied then. The physicist must believe in the possibility of this indirect chain of reasoning and in the truth of cause-effect relations. But Hume would be right (in a qualified sense) that there is no logical certainty to be found in such beliefs. Nevertheless, it is completely rational to have faith in these assumptions.

While I have just shown how numerous physicists admit the importance of faith in their discipline, it goes without saying that there are many who disagree, or believe that they disagree, because they do not like the word 'faith' and its usual theistic implications. There are important differences between theistic and scientific faith, but what we need to admit is that there is also unavoidable overlap between both types of faith at their foundation. In other words, it is certainly not an irrational position to maintain both primary theistic faith and scientific faith, and Platonic realism makes sense of the underlying unity of both. As we have seen, many eminent physicists, from Kepler to Planck, have admitted the importance of faith in physics, which is a significant challenge to the 'militant' atheism[419] of Dawkins.

Fanatical physicists

Sometimes these pioneering physicists found themselves far from the path of the so-called detached, objective, unemotional, unbiased scientist extracting facts from the physical world. It is not unusual for many people, including scientists, to cultivate such a misconstrued image, but nothing could be further from the truth. I am arguing against the narrow conception of what constitutes accepted scientific inquiry, showing how the founders of quantum theory were thinking and acting far beyond such conventions of laboratory measurements or the testing of hypotheses. Their beliefs, attitudes, and actions, including their 'fanatical' moments, played important roles in the development of quantum theory, and are thus genuine parts of pioneering science. Let us look at some examples.

Bohr had referred to Schrödinger's and Einstein's views as 'appalling' and 'high treason'.[420] Einstein had his own nasty rebuttal: 'this theory (the present quantum theory) reminds me a little of the system of delusions of an exceedingly intelligent paranoic [sic], concocted of incoherent elements of thought'.[421] Heisenberg also recalls how Bohr was acting as 'a remorseless fanatic, one who was not prepared to make the least concession or grant that he could ever be mistaken'[422] when arguing with Schrödinger, who happened to be lying sick in bed at the time. Given the opportunity, Schrödinger readily vented his frustration: 'if all this damned quantum jumping were here to stay, I should be sorry I ever got involved with quantum theory'.[423]

Heisenberg, too, expressed great anger towards his colleagues, which is made abundantly clear in his 1926 letter to Pauli: 'the more I think about the physical part of Schrödinger's theory, the more disgusting I find it'. He goes on to write that he considers Schrödinger's views to be 'Mist' (translated variously as junk, rubbish, crap, and bullshit).[424] Although Bohr and Heisenberg were close companions, they also had disagreements and strong emotional reactions to each other's differing opinions about certain aspects of quantum mechanics. As Baggott tells us, 'Bohr put Heisenberg under intolerable pressure—so much so that harsh words were exchanged on all sides, and at one point Heisenberg was reduced to tears'.[425] Surely these actions and attitudes overthrow the common misconception that scientists are detached, dispassionate, purely objective observers of nature.

However, if these pioneers did not hold such conviction, such powerful faith, it would not have been possible for them to produce their great accomplishments. It does not follow that scientific faith is therefore true just because this personal faith in one's own abilities is necessary to impel the physicist forward. The elements of scientific faith hold true independently of what we personally happen to believe. But this sort of personal faith is intimately wrapped up in scientific and theistic faith, and it is not so easy to demarcate them in one's actual life, however well we can draw logical distinctions on paper. Of course, various people fanatically believe that they are right about this or that cause or ideology, even though they may be totally deluded. Mainstream contemporary psychology, however, is not theoretically well equipped to distinguish between the fanatical actions of someone like Bohr and a genuinely psychologically ill person.[426] But this lack of comprehension of such an important psychological topic does not invalidate the significance of Bohr's convictions or the actions, attitudes, and beliefs of these pioneers in modern physics. It just means that psychologists have a lot of work to do to elevate their understanding in this regard.

These pioneering physicists found themselves in the middle of a very messy workshop, a workshop that seemed to have been all but demolished and was being rebuilt by fanatical, self-styled prophets. But let us not forget that the experimentalist, the inventor, the innovator, and the salesperson selling you the latest high-tech gadget all rely entirely upon the efforts of these fanatics. In other words, your computer, your mobile phone, and, for good or bad, our nuclear power are all made possible because of these fanatical pioneers. It is true that a different kind of creative intelligence is required to turn theory into experimental application, and yet another kind to create technological products that (hopefully) benefit us. But none of these things would be possible without the mystically minded pioneers, which was just as true for the foundations of classical physics four hundred years ago. It would be irrelevant if 99% of the human population were to believe that Einstein and Bohr were totally insane. It is irrelevant if the majority of scientists are atheists, or agnostics, or theists, or if they believe in directed panspermia, invisible pink elephants, or an unreal earth.[427] What is relevant is that a significant percentage of the key pioneering theoretical founders of modern physics have held foundational scientific beliefs that explicitly or implicitly imply and presuppose Platonic realism. This fact is enough to warrant serious attention to the Platonic tradition and to the arguments and information I have presented here.

Most physicists have unwittingly followed the advice of Simmias, one of the interlocutors in Plato's dialogue the *Phaedo*. Simmias tells us that if we cannot be taught or find out where truth is for ourselves, then we should 'at least take the best possible human doctrine and the hardest to disprove, and to ride on this like a raft over the waters of life and take the risk'.[428] The raft for most physicists is usually the Copenhagen interpretation. Despite its tremendous success in many areas, its unresolved conceptual difficulties have played an important part in the continuing sense of unease in physics and in providing the impetus for the development of alternative approaches. Some physicists even claim that there are no quantum jumps or particles,[429] which would have made Schrödinger happy.

Some experts will still say that I have not adequately discussed all the factors relevant to a discussion on the nature of faith, especially in relation to science and religion, accusations that I readily accept. However, these experts in the humanities are not likely to complain when a simplified explanation of quantum physics is offered for their benefit, at the expense of ignoring further subtle, complex discrepancies that are of great importance to experts in this scientific field. The essential point that we need to take with us from this chapter is that all scientists must explicitly or implicitly have faith that the universe is rational, which entails that it operates according to stable laws, a view that is also a part of theistic faith. Platonic realism makes sense of both primary theistic and scientific faith, while clarifying their underlying unity.

SIX: Expanding Realism

The foundational intuitions of realists and antirealists are so fundamentally different that it almost seems impossible for one side to conceive of how the other can maintain such beliefs.[430] On the one hand, we must remember that our intuitions are the starting point for all our arguments, and that it is very difficult to convince others purely with rational arguments when they hold opposing foundational intuitions. On the other hand, we must always be prepared to give up our false beliefs. The purpose of rational argument is not so much to convince others as it is to test our own beliefs to see if they are true, and then let them go if they turn out to be false. However, one foundational belief that we must refuse to forsake is the belief that we can actually be wrong. This may sound like a strange thing to say since it is so obviously true that we can be wrong, but its obviousness escapes the antirealists.

Outlining the struggle

If there really is no objective truth or reality, then it is impossible for me to be really wrong about anything. In fact, one of the motivating reasons behind the thinking of those who promote antirealism seems to be some inner need to never be wrong. Realists, however, know that they can be wrong. My critics may tear this book to pieces, or they may show that I have made hundreds of errors. None of this, however, is of any great importance. What does matter is that at least I cannot be wrong about one thing—and that is the fact that I can be objectively wrong, and so can you and everyone else, too. Therefore, since antirealism necessarily implies that we cannot be objectively wrong, we must reject at the outset the basic assumption (or desire) behind antirealism.

Scientists usually (if not always) assume realism in their work. Barbour points out the obvious fact that dinosaurs, for example, 'are held to be creatures that actually roamed the earth, not useful fictions with which we organize the fossil data...Even the physicists, who more than the others have been forced to examine their concepts, still speak of the *discovery* (rather than the *invention*) of the electron'.[431] Being a realist was, and is, necessary for our evolution. As an example, suppose that instead of accepting the objective fact that the saber-toothed tiger standing in front of them was truly dangerous, your prehistoric ancestors chose the more comforting belief that it was really as harmless as a kitten and tried to take it back to their cave as a pet. In that case, you probably would never have been born. But most people seldom, if ever, think about how difficult it is to justify their realist beliefs, which we all tend to take for granted, such as the belief that my car still exists when I stop perceiving it. There are several philosophical difficulties that are inherent to our naïve realism, difficulties that can really only be overcome by ascending to Platonic realism.

SIX: Expanding Realism

Philosopher Gerald Vision notes that 'from its inception to the present, philosophy may be viewed as a series of struggles between various realisms and anti-realisms'.[432] There remains pervasive confusion surrounding this struggle, especially concerning quantum theory. My arguments and classifications are not free from problems either, but they do allow for a deeper understanding of how quantum theory is actually realist, which, as we now know, is contrary to the common story told in academia. However, there are more than thirty-five types of realism,[433] and so to help simplify things, I offer three different categories—broad realism, abstract realism, and factual realism, all of which are aspects of Platonic realism.

Norris also provides a helpful overall characterization of the general beliefs of realists and antirealists, which I have paraphrased here. Realists tend to believe that:

1. there is a mind-independent reality whose properties must be discovered (rather than merely invented);
2. there are verification-transcendent truths that exist beyond our epistemological limitations;
3. mature scientific theories are true descriptions of physical reality; and,
4. theories should explain the phenomena via a causal (depth-ontological) account.

Antirealists deny all of the above, due to their belief that (1) all properties of 'reality' are merely internal relations, rather than reflections of a mind-independent reality; (2) it makes no sense to talk of truths beyond our best knowledge; and (3) 'truth' is just our current best explanation. They also deny the fourth item in the above list because they are skeptical of ontology as being a return to naïve, pre-scientific metaphysical habits of thought.[434] I will explain these difficult ideas throughout this chapter, but let us begin with a more fundamental question.

What is reality?

The first major challenge for realists (and for everyone else) is to clarify what we mean by the word 'reality'. This is basically an impossible task, but we have avoided it long enough and now we must at least make the effort. *Reality* is one, and yet it has three general aspects: that which exists, that which has being, and that which is the foundation of both existence and being. Admittedly, this definition may not seem to be of much help at first glance, so let us explore it further.

The realm of existence (or becoming) is the realm of space and time, where things come into and go out of existence, and while they exist are constantly changing in one way or another, to one degree or another.[435] As already mentioned, things that constantly change never remain what they are, since each and every moment they become something different to greater or lesser degrees, which is why everything that exists is only relatively real. Given this first part of the definition of reality, there is a trivial sense in which every living creature seems to contribute to creating reality, because every living creature in some way interacts with the environment and helps to change it, which is to make changes in the realm of becoming.

But we need to speak more accurately here, because even in this trivial sense, no creature ever actually *creates* anything, even in the realm of becoming. We can only work with what is already here, such as taking materials from the earth and applying techniques based upon scientific laws that allow us to alter and shape these materials into something like an automobile or a cathedral. In a similar way, a bee gathers pollen and makes honey, thereby altering the environment and, in some way, altering reality in the realm of becoming; before, there was no honey, and now there is. Even at the lowest level of reality we still never really create reality; we can only shape or transform certain aspects of it in ways that are constrained by the laws of physics.

We can still speak of things being objectively true or false in the realm of becoming, such as saying that my car is blue or that I enjoy honey. These are *objective* facts even though they may not necessarily be *absolutely* true. This point may seem a bit odd at first, so let us examine it further. If I say that I like blue, there is still inherent ambiguity surrounding each word in the statement. The most difficult ambiguity concerns the meaning of the word 'I', which is used to refer to a particular individual. By asking what we mean by the word 'I', and without pusillanimously abandoning the investigation when presented with the first comforting pseudo-answer thrown at us, we will be entering into the philosophical labyrinth of the nature of personal identity.

What, after all, does 'I' refer to? Does it mean my brain, my body, my soul, my personality, or a combination of all of these? Such questions lead to innumerable sub-questions, like asking what it would really mean to say that 'I am my brain'. What part of my brain am I? Or am I all of my brain, and if so, does that mean that there are as many different 'me's' as there are parts of my brain? If not, then how many 'me's' are there? And how many parts of my brain are there? Do we count all the subatomic particles in my brain as parts? How is my conscious awareness to be accounted for purely by seeking correlated brain states? What, even, *is* a brain state? Whatever answers may be provided to these questions, they can certainly be pushed into ever-increasingly difficult territory with very simple but vitally important further questions. The same situation applies to the words 'like' and 'blue'. What does it really mean to say that we 'like' something, and which *shade* of 'blue' do you like, anyway? When you speak of 'blue', do you mean a particular wavelength of light or your personal perception of what you believe blue appears to be?

Such questions are endless, and we can only pretend to have the absolutely true answer by remaining ignorant. We may, however, have actually hit upon an absolutely true answer in the realm of becoming, but we could never absolutely know that we have done so while remaining solely in this realm. The realm of becoming is inherently fraught with change and ambiguity. It is directly tied to the endlessness of discursive reasoning. The ambiguity and constant state of change at this level of reality does not make everything in the physical universe an illusion, as some philosophical/spiritual traditions have claimed, but it does mean that things in the realm of becoming are not stable, and so they are inherently *less* real than what we can discover in the realm of being.

The realm of being is only accessible by the mind. I am not here using the word 'mind' to refer to discursive thought, the endless logical maze of analysis. I am using the word mind in the ancient sense of *nous*: pure intellect or consciousness, that which can apprehend or know eternal truths and know that it knows such truths. This part of us is what can achieve sudden flashes of understanding of the whole, or gain direct perception of an eternal truth, as Bohm and Planck discussed. It may seem tempting to deny the reality of the realm of being, but if we do, then we must admit that there is no logical necessity inherent to our denial. In fact, if there is no realm of being, then not only are the laws of physics fictitious, but so are the laws of logic, and if the laws of logic have no objectively real status, then there is no way for my opponents to offer an objective logical argument in their defense. The only way they could do so would be to rely upon the reality of some stable, unchanging truth; otherwise, I would simply charge them with subjective prejudice, and we would have no reason at all to believe them. The only way that they could provide an objectively true argument would be to rely upon some eternal idea or truth, which would be to admit the reality of the realm of being. Gödel was certainly correct to approve of Russell's early belief in logical realism, where logic is just as concerned with the 'real world' as the sciences, although it is more abstract and general.[436]

The only option that allows us to reject the realm of being is to say that truth amounts to nothing more than one's idiosyncratic opinions or perceptions. If this is the case, then we have no objectively good reason to believe anybody about anything. In fact, we would never need to learn anything, because whatever we happened to perceive or believe would already be our reality. This conclusion also logically follows from the views espoused in Plato's day by the sophists, who would teach people how to argue without being concerned with finding truth. One of the most famous sophists was Protagoras, who taught that 'man is measure of all things'.[437] As Socrates pointed out, if Protagoras is correct on this point, then whatever I believe to be true is just as true and just as false as anything else. In other words, if I believe that elephants can fly, become invisible at will, and do calculus, then my belief is just as true as any other belief.

Alternatively, if Protagoras is correct, and if Socrates believes that Protagoras is *not* correct then, by Protagoras' own doctrine, Socrates must be correct, which necessarily entails that Protagoras is incorrect. This example represents the fatal flaw caused by remaining stuck in the realm of becoming, which is where we find the antirealists. Some antirealists, however, would even deny the realm of becoming, which is to deny the reality of the very sentence that you are currently reading.

While the realm of being is home to the eternal mathematical law, as well as the prior metaphysical principles such as symmetry and beauty, the realm beyond being is the ultimate source of all reality, which is the absolute One.

More fun challenges

In this section we will examine the four basic realist beliefs as characterized by Norris.

1. *There is a mind-independent reality whose properties must be discovered (rather than merely invented)*

In the introduction to this chapter, I listed four basic realist beliefs, each of which faces difficult challenges. The first claim, that there is a mind-independent reality, is not as obviously true as most realists suppose. Claiming that reality is mind-independent may imply that reality is independent of my mind, indicating that my mind is not in or part of reality and is, therefore, unreal, which is not what realists want to assert. My mind must also be part of reality, which would mean that, strictly speaking, my mind is not independent of reality. However, antirealists make the logical mistake of arguing that just because my mind cannot be outside of reality that, therefore, all of reality depends upon my mind, which usually ends up meaning that I have created reality with my mind. Clearly, the reverse is more reasonable; my body depends entirely upon the rest of reality in order to have existence, yet the rest of reality seems not to need my body. In other words, if the rest of reality (excluding my body) did not exist already, or at least certain aspects of the rest of reality (such as oxygen, water, the laws of physics, etc.), then I could never have become physically manifest.

A similar analogy would be the fact that while humans are an intricate part of the global ecosystem, the rest of nature (minus humans) would seem to get along just fine without us, whereas we are completely dependent upon the rest of nature (which is why haphazardly polluting our environment is so insane). By admitting that my mind is also part of reality, I cannot strictly say that reality is mind-independent. However, this does not in any way give support to the antirealist claim that reality is a mere mental construct with no intrinsic or objective existence. All it says is that realists need to keep pushing themselves further and questioning their own assumptions, and that they need to be more careful with their wording.

What the realist really wants to say on this issue is that reality is not limited to any particular mind, or even to the aggregate of the minds of all sentient creatures in the entire universe. For example, I cannot verify whether or not you are eating an apple at the exact moment that I am writing this sentence; nevertheless, it will still be the case that either you are or are not doing so. Consequently, it is a fact that you either are or are not eating an apple even though I cannot know this fact, and so reality is not limited to my personal experiences, which is to say that reality extends beyond the boundaries of the cognitive and perceptual limitations of my mind. My mind is not the limit of reality, which is what realists should say instead of claiming that reality is mind-independent.

Let us consider another example. When I leave my office, I have no direct observational means of verifying that my computer still exists. I can only use applied metaphysical reasoning based on past experience and the known laws of physics to argue that my computer still exists in my office, unless there is some rational explanation for its disappearance, such as if someone were to steal it. Now, suppose that you happen to reject the idea that your computer still exists in your office after you have left and can no longer perceive it. In that case, you should not hesitate to report it missing to the police every time you leave your office. If you do not do so, you are behaving as if you really believe your computer will still be in your office when you return, which is to be a realist in this basic sense. If you claim that it does indeed disappear when you leave the office and that you will bring it back into existence upon your return, then unless there are some magical properties about your office, you should be able to make your computer manifest *anywhere* at will. Since I will assume that you cannot perform such a magical feat, I can safely say that reality is not limited to the experiences of your particular mind.

Of course, if realists simply mean that reality does not depend upon my mind for its existence, and yet my mind is still a part of reality, then that would be acceptable. However, we would still have to explain the nature of mind, the nature of reality, how mind is part of reality, and how reality can be known through my mind and yet we still know that reality is what it is regardless of how I perceive it.

2. *There are verification-transcendent truths that exist beyond our epistemological limitations*

The second challenge for realists concerns the ontological status of verification-transcendent truths that exist beyond our epistemological limitations, but which we are nonetheless able to discover. In other words, what is the actual nature or status of something that is beyond our ability to know or perceive, such as a law of physics? It seems that many realists do not consider this to be a genuine problem, but rather take for granted that such truths can simply be discovered. However, we cannot discover something that is *not*; we can only discover what already exists or already has being. In other words, if something does not already exist in some way, then it is impossible for me to measure or observe it in the first place.[438]

But we need to be able to talk in some way about the ontological status of truths that are yet to be discovered, as well as those truths that have already been discovered. Otherwise, the antirealist need only ask us for an account of the supposed truths that we can discover, and when we are unable to respond they may gain confidence in saying that our 'truths' are nothing but social constructions. So how can the realist respond?

There are at least three different sorts of truths that we can discover: conventional truths, physical truths, and abstract truths. A *conventional* truth includes our arbitrary rules, such as stopping at a red light instead of green light, as well as the rules that follow logically from the original rules of any game. Simply put, the rules for a game entail what we are permitted to do while playing that game, but these rules are arbitrary, and therefore are only conventional. If we discover a new rule implied by our initial rules, we can say that we have discovered a conventional truth. However, the very fact that we can say that a new rule logically follows from the initial rules implies that there is something about the rules of logic that are not bound by the rules of the game, and that the rules of the game cannot contravene the rules of logic. Therefore, the rules of logic are not conventional, and thus they require an ontological account.[439]

One could object here and say, for example, that the rules of logic themselves are arbitrary, but such an objection cannot seriously be maintained, at least concerning the most fundamental axioms of logic. In order to even formulate an argument, antirealist or otherwise, we must first rely upon fundamental axioms of logic, whether explicitly stated, known, or unknown. These fundamental axioms of logic in turn require an ontological account. If the rules of logic were truly arbitrary, we would never be justified in saying that any argument or belief is illogical, because then everyone would be permitted to construct any sort of 'logic' that they wished to suit their own particular purposes. However, to construct any sort of logical system in the first place, even a new logical system that could somehow entirely replace what we currently take for granted as fundamental logical axioms, would require the implicit acknowledgement of some way of coherently binding together these various new axioms and propositions. If nothing about any logical system needs to make sense, then there is no point in bothering to try to be logical; we may as well say that that 'all elephants are pigs' and 'all pigs are green bananas', therefore 'George Washington was a fruit fly'.

SIX: Expanding Realism

But it gets worse. Take any commonplace example you wish and you will find logic lurking everywhere, ready to pounce and force itself upon you. Antirealists who wish to reduce all logic and rationality to mere constructions or arbitrary fictions end up relying upon the very assumptions that they deny to realists in order to make an attempt at any sort of coherent statement.[440] If there is no non-arbitrary stability underlying the propositions and premises used in their arguments, then there is no objective reason to assent to their conclusions. If the antirealists then respond that I do not understand that they are denying objectivity in the first place and that therefore there are no objective standards, then I simply need to ask them how, in that case, they can say that I am actually wrong. In other words, if there is no objective truth of the matter, then I cannot be wrong to claim that there is objective truth, nor can I be wrong about anything at all. Therefore, the antirealists can never say that realists are really wrong. They can only say that they happen not to like what realists are saying, and there is nothing logical about saying that.

The second sort of truth concerns physical objects and is part of the wider category of factual realism, which I discuss in the next section. The notion of *physical truth* basically says that I can find objectively true properties about physical objects and about how the properties of different objects interact. For example, smashing object A against object B brings about result R. To put it more concretely: if someone hits me hard in the face with a large rock, the result will include physical pain (assuming I have 'normal' nerve function). If the skeptical antirealist claims that there are no physical truths, or at least that we could never have knowledge of such physical truths, we can easily test this claim. Such antirealists need only allow me to hit them in the face with a large rock to see whether or not pain results. If, after being struck, they still insist that even though they know that pain resulted in this particular instance, they still do not know (as Hume would have to say) that the next time I hit them with a rock the result will be similar, then I can test that claim too, for as long as they wish to maintain that they do not know any physical truths.

The material reality of the rock has objective properties that necessitate specific results when struck against a human face, which also has certain objective material properties, and we can discover what the physical truths are about these objects and how they interact. We can then accurately predict the results of other similar interactions without having to experience them firsthand each and every time. However, a detailed scientific investigation of any particular physical truth necessarily leads us further away from physical reality and into the realm of abstract truth.

The third kind of truth concerns the abstract realm and is part of the wider category of abstract realism, which I also discuss in the next section. An *abstract truth* is a truth about some aspect of the abstract realm, such as a law of physics, which is real and yet nonphysical. I will discuss the nature of the laws of physics in more detail in Chapter Eight, so here we need only give it brief consideration. Either the laws of physics are merely 'convenient descriptions', or they are real abstract laws that cannot be broken. Sometimes it may appear as if a genuine law of physics can be broken, but this can only occur where a hitherto unknown law eventually accounts for anomalies that the prior law cannot explain because it is beyond its applicability. If the laws of physics are just descriptions, they cannot have any non-trivial predictive ability. 'My sweater is black' and 'the sun is bright' are examples of descriptions, but neither of them is useful for making predictions in the way that the laws of physics are. A black sweater may help conceal me at night, and if the sun is bright then I should wear my sunglasses while driving. In this trivial sense these descriptions allow me to make certain predictions, in much the same way that any statement about anything might enable me to make a prediction (provided I exercise sufficient creativity). But none of these predictions are remotely similar to the way in which quantum theory allows us to make extraordinarily accurate predictions about widely disparate phenomena under varying conditions. Modern physics is so powerful precisely because it touches upon the truth of fundamental abstract laws that physical reality must obey.

3. *Mature scientific theories are true descriptions of physical reality*

The third challenge for realists is their general claim that mature scientific theories are true descriptions of physical reality. As a realist, I cannot bring myself to fully accept this claim, and yet I certainly do not side with quantum antirealists such as Gibbins, who states ominously that 'quantum mechanics is not known to be true. It is a truism of the philosophy of science that no generally applicable physical theory ever could be'.[441] We cannot position ourselves on a middle ground between realism and antirealism, either. Rather, we need to expand realism beyond the limitations and falsity of materialism. My realist position is that given our perceptual and cognitive limitations, it does not seem to be possible that we could ever know with total logical or empirical certainty that any given theory is absolutely true, even if considered only within its applicable limits. Empirical facts never correspond *exactly* and *absolutely* to theory. As Einstein said, 'as far as the propositions of mathematics refer to reality, they are not certain; as far as they are certain, they do not refer to reality'.[442] Of course, he was here referring to *physical* reality.

The mathematical laws of physics, although distinct from any particular physical phenomenon, nonetheless are the foundation of all physicality, and so there can never *in reality* be any discrepancy between these laws and physicality. But we still find some discrepancy, however minimal, between a theory and our measurements, which is due to our own cognitive, epistemological, and physical limitations. Moreover, a law that we discover and put in our textbooks may still only be approximating the actual law that it closely represents, which will result in further discrepancies between our theoretical predictions and the relevant experimental data. When we make a prediction based on a theory that is approximately true, even if it is 99.9% approximately true, the possibility of finding discrepancies remains. Philosopher of science Karl Popper offers a good defense of objective truth, which implies approximate truths: 'the very idea of error, or of doubt (in its normal straightforward sense) implies the idea of an objective truth which we may fail to reach'.[443]

My position on this point is very similar to that offered by James, who believed not only that there is truth, but also that 'it is the destiny of our minds to attain it'.[444] I am also in *qualified* agreement with him that we cannot infallibly know that we have attained truth even if we have in fact done so. Where I differ from him is in my belief that it may be possible to know that we know, but only at the non-discursive level of direct perception, the flash of understanding, or mystical insight. The claim that 'we cannot infallibly know that we know' is itself a claim implying infallibility (i.e. we can infallibly know that 'we cannot infallibly know that we know'). What we have to recognize here is that we can only know something for certain when it is eternal and unchanging, because anything that is finite and changing can never be known with certainty.

Socrates is famous for apparently not knowing anything for certain, but this impression is misleading. He did not know with certainty things that could not be known for certain, such as things in the realm of becoming. But he did know for certain some truths that could be known for certain, because they are part of the realm of being. We can, therefore, agree with Socrates that I know that I do not know for certain the diversity of disparate phenomena, but what I can know for certain is whatever is metaphysically prior to the material realm of becoming.

An example used by Socrates was that he knew something to be beautiful to the degree that it participated in absolute beauty.[445] What he means by this is that he may not be able to say for certain to what degree, if any, something here in the physical realm is beautiful, but what he did know for certain was that if anything is beautiful at all, it is only beautiful to the degree that it participates in absolute beauty. He believed he had infallible knowledge that absolute beauty was real, even if he could not say with total certainty what exactly this absolute beauty was, for any description of absolute beauty involves the partialness of language and discursiveness of thought. In terms of physics, what knowledge I have about particular physical phenomena may be more or less accurate but never absolutely certain, whereas whatever knowledge I have of the unchanging mathematical laws of physics has the possibility of being certain.

The upshot here is that the realist claim that mature scientific theories are true descriptions of physical reality needs to be rephrased more accurately. Quantum theory has forced us to realize what was always implicit in classical physics, which is the fact that the very nature of theorizing and the constantly dynamic changing universe, coupled with our cognitive and perceptual limitations, necessarily implies that we are limited in our ability to represent physical reality with absolute accuracy.[446] Consequently, the realist position with respect to mature (or any) scientific theories should be put in this way: the degree of truth that a scientific theory attains depends upon the degree to which it accurately represents the relevant aspects of the genuine abstract lawful structure of the universe, and the entities presupposed by more accurate theories should be considered to be real or close to real representations of the actual entities themselves.

4. *Theories should explain the phenomena via a causal (depth-ontological) account*

The fourth aspect of the general realist beliefs, which concerns the antirealist's skepticism of ontology as being a return to naïve, pre-scientific metaphysical habits of thought, has already been addressed in our discussion of how ontology is not limited to, or eliminated by, epistemology. In other words, the limit of my knowledge, beliefs, or experiences is not the limit of reality. All scientific theories presuppose causality in one way or another, and we should aim to uncover and explain the relevant aspects of causality as they pertain to a particular theory. To deny causality is to deny the possibility of scientific explanation.

We are all realists

The struggle between realist and antirealist philosophies at the heart of quantum physics was brought to the fore by the well-known Einstein-Bohr debate. As Baggott writes, it was 'one of the most important scientific debates ever witnessed', for Einstein was directly challenging Bohr about the meaning of quantum theory.[447] He continues: 'at stake was the interpretation of quantum theory and its implications for the way we attempt to understand the physical world. The outcome of the debate would determine the directions of the future development of quantum physics'.[448] The main emphasis of the debate, especially as it developed over the years, has been declared to be a clash between realism and antirealism, a saga that physicists such as Penrose and Hawking have continued.[449] However, I am not going to discuss the debate itself, but focus on other related, though often neglected, issues instead.[450]

Realism in physics has often come to be associated with materialism; that is, with strict physical causal relations between separate, independent, and determinate particles of matter. As Gibbins writes, 'realism in the philosophy of quantum mechanics means the idea that quantum systems are really like classical particles. Everything points against it'.[451] Antirealism, conversely, has come to be associated with indeterminism, uncertainty, inseparability, and immaterialism.[452]

SIX: Expanding Realism

Through a series of confused arguments and equivocations on the word 'realism', many philosophers and physicists have falsely assumed that quantum theory is antirealist. What causes the most trouble is when one then extrapolates from quantum antirealism, which is essentially the denial of materialism, to antirealism in general, which denies the very conditions and assumptions presupposed by physicists (including the Copenhagenists) in order to do physics in the first place. In other words, antirealism in general denies that there is any such thing as objective truth and reality, but, if that were true, then physics would not be possible, because everyone could create any physics they wanted without any constraints whatsoever. It is therefore essential to reclaim the essence of Platonic realism, which claimed that what is more real or has more being is that which is unchanging, such as the laws of physics. Platonic realism does not view objects as mere collections of qualities, not if such a view entails that material things do not exist and are instead mere illusions. Platonic realism denies materialism, but it also denies immaterialism.[453]

While many of the pioneering physicists rejected materialism, the twentieth-century analytic tradition of philosophy of mind was, as philosopher John Searle notes, resolutely materialist.[454] The analytic philosophers and the quantum pioneers were like two ships passing in the night, except that one was heading for the rocks, the other for the open sea. Dawkins is a good example of a scientist perpetuating the confusion surrounding materialism, which has been rightly challenged by Ward, who notes that Dawkins tries to deceive 'the reader into thinking that all respectable scientists are really materialists—which is as false a belief as most that one can think of'.[455]

Trying to convince those who adamantly refuse to consider that there is more to reality than what they can kick with their foot may be more difficult than squeezing oneself through the eye of a needle.[456] But perhaps Plato was a bit harsh when he called materialists 'terrible men',[457] and 'very stubborn and perverse mortals'.[458] After all, we are all stubborn and perverse mortals to some degree, and we can all act terribly. Nonetheless, those who cling to mud as if it were the ultimate foundation of the cosmos are unable to grasp the higher aspects of reality.[459]

We are now in a better position to examine more directly how quantum theory is realist. It is true that Bohr and Heisenberg made certain claims that seem to be antirealist, and I will explore, clarify, and, where necessary, argue against such views in the next chapter. Here, however, I will show how they were overwhelmingly realist, but in order to do so we first need to clarify three forms of realism that are relevant to this discussion: broad realism, abstract realism, and factual realism, all of which are aspects of Platonism.

Broad realism simply states that we can be objectively wrong about what we happen to believe or assert,[460] for if there were no truth about anything then I could say any absurd thing whatsoever and not be wrong. 'Plato currently resides in New York City', and 'ducks taught mathematics to Einstein' are statements that are not just false according to some conventional standard; they are objectively false. Antirealists also believe that realists are objectively wrong. They believe that it is an actual fact of the matter that realism is false. But if it is actually the case that realists are wrong then, by necessity, at least the most fundamental aspect of realism—broad realism—is correct, because there would be at least one objective fact of the matter independent of what anyone happens to believe or know, that fact being that realists are wrong. Of course, if this is the case, then realists are actually correct and the foundation of antirealism instantly crumbles.[461]

Bohr was a broad realist, believing that he was *really* right and that Einstein was *really* wrong. But he was much more of a realist than what is implied by this basic category. He believed that the results of quantum mechanical experiments could repeatedly confirm wave-particle duality. He was also committed to the realist belief that reasoning based on thought experiments could lead to true conclusions.[462] In other words, Bohr, the supposed antirealist, believed that the experimental results of quantum theory could be replicated and that applied metaphysical reasoning could tell us something about the nature of reality, both of which presuppose objective reality.

It is impossible to do science or to have a proper dialogue or philosophical argument with anyone who denies broad realism, for it is impossible to hold anyone or any scientific theory accountable if truth is nothing but a fleeting entity capable of change from moment to moment. As Polanyi correctly observes, 'any effort made to understand something must be sustained by the belief that there is something there that can be understood'.[463] Similarly, Clark has argued that 'it is wholly irrational to speak of scientific progress while at the same time we disdain all knowledge of realistic truth'.[464] After all, we cannot discover or understand something that has no reality whatsoever, and it is impossible to say that science has, or has not, made progress unless we are realists about the objective fact of the matter. Despite the ambiguity surrounding the concept of 'progress', it can hardly be doubted that we have greater scientific understanding today than we did two thousand, two-hundred, or even twenty years ago.

The fact is, we are all realists of one sort or another. Philosopher Roy Bhaskar says that 'it's not a question of being a realist, or not a realist. It is a question of what kind of realist you are going to be – explicit or tacit'.[465] The real challenge comes when we attempt to defend a particular kind of explicit or tacit realism. So, while antirealists may have raised some important questions for realists, what antirealists really need to do is to qualify their own implicit realist position. For example, they may believe that subatomic entities do not really exist, but if they are correct, then it would really be the case that such entities do not exist; there would be a real fact of the matter and, therefore, they would be adhering to broad realism.

Abstract realism (a term I adopt from Sachs but develop further) refers to the reality of the laws of physics, as well as the higher metaphysical principles, such as symmetry and absolute beauty. Heisenberg was clearly an abstract realist, as all physicists must be to some extent, explicitly commending Plato and Pythagoras in opposition to the doctrine of materialism numerous times in his book *Across the Frontiers* (1974).[466] In fact, Heisenberg argues for 'the objective character of mathematics' and claims that true or valid mathematical principles will retain their truth and validity for any sentient being in the universe.[467] He also states that we can only *discover* the basic forms, such as these mathematical laws of physics or chemical constitutions, which 'possess a genuine objectivity' and which must also 'depict reality'; we 'cannot simply construct them'.[468]

Such discoveries are of eternally valid laws. For example, regarding the laws of the lever formulated by Archimedes over two thousand years ago, Heisenberg writes that 'we can have no doubt that, at all times and all places, they retain their validity'.[469] The important caveat is that 'we are by no means able to claim that all phenomena can be described in terms of these concepts'.[470] Similarly, Newtonian mechanics is true (or approximately true to a significant degree) insofar as we recognize and do not transgress its inherent limitations, but within its limitations it is universally valid. The same reasoning must also apply to quantum theory.

That the laws of nature hold true for life anywhere in the universe, Heisenberg continues, 'is not just a theoretical opinion, for we can see in our telescope that the same chemical elements exist there as they do with us, that they enter into the same chemical combinations and emit light of the same spectral composition'.[471] He also says that there must be at least three universal constants, such as Planck's constant, which are *'independent* constants of nature'.[472] Even Bohr believed that the abstract laws of physics and logic were true independent of human belief or knowledge. He told Heisenberg that 'as much as all living organisms are constructed in accordance with the same laws of nature, and largely from approximately the same compounds, so the various possibilities of logic are probably based on fundamental forms that are *neither man-made nor even dependent on man'*.[473]

Bohr also told Heisenberg that 'we have good reason to assume that quantum-mechanical laws can be proved valid in a living organism just as they can in dead matter',[474] entailing that abstract quantum mechanical laws apply equally to everything in the universe (within their applicable domain). Recall how Pauli, too, believed in a cosmic order that is independent of the world of phenomena. It is truly astounding, then, that these three physicists have been used as the foundation of essentially all antirealist claims deriving from quantum physics given their *overwhelmingly* realist commitments.

The Copenhagen enemies—Einstein, Planck, and Schrödinger—were even more obviously abstract realists. As already mentioned in Chapter Two, Whitehead notes that Einstein was part of the Platonic/Pythagorean tradition. Schrödinger, too, although he disparaged Plato's apparently failed social and political philosophy, believed that the reason for Plato's fame was 'that he was the first to envisage the idea of timeless existence and to emphasize it'.[475] He accepts Plato's belief that mathematical relations hold 'irrespective of our inquiry into them' and that 'a mathematical truth is timeless, it does not come into being when we discover it'.[476] Norris has said much the same; the truth or falsity of the theories that undergird our present-best science is 'decided by the way things stand in physical reality or in a realm of nonphysical objective truths, for example, those of mathematics that are wholly unaffected by whatever we might think or be able to establish concerning them'.[477] Even Frege, the founder of analytic philosophy, who was also a mathematician, stated that 'the mathematicians cannot create things at will, any more than the geographer can; he too can only discover what is there and give it a name'.[478] In such cases of discovery, we have discovered an abstract truth.

Given the importance of the fundamental nature of mathematical relationships, we should also briefly consider the metaphysical position known as *structural realism*, which, according to my classificatory scheme, is a species of abstract realism according to one interpretation, but a qualified antirealism according to another. According to John Worrall, a specialist in the philosophy of science, 'on the structural realist view what Newton really discovered are the relationships between phenomena expressed in the mathematical equations of his theory'.[479] In other words, Newton discovered certain mathematical laws (relations, or structures) of physics that express how material entities relate to one another. Psillos claims that structural realism ultimately fails because it is not realist enough. It 'remains silent about the entities and processes that are described by mathematical structures' because they are beyond 'what can be quantitatively described'.[480] As we already know, such structures are not physical, even though they capture the relationship between physical entities or processes. However, if our metaphysical system implies that we cannot say anything meaningful about the entities and their behaviors as described by such mathematical laws, then our system is in need of revision.

Eddington seems to have been a qualified structural realist in the sense that he thought the laws of physics are real but that physicists could not directly penetrate beyond the symbolism of the equations to the entities themselves. He believed that we could translate all of physics into 'Jabberwocky', a nonsense verse poem by Lewis Carroll, and that so long as all of the numbers remain unchanged, physics would remain intact. 'Out of the numbers', Eddington continues, 'proceeds the harmony of natural law which it is the aim of science to disclose. We can grasp the tune but not the player'.[481] However, Eddington's view, and structural realism in general, logically entails that material entities or processes exist even if we can never know exactly what those entities (or processes) are. While structural realism is an important example of mathematical and logical aspects of abstract realism, this category of realism does not itself account for physics.

The idea of having constant, unchanging relations with varying terms is a fundamental aspect of Platonism.[482] Siorvanes also notes that for Proclus (and in Platonism generally) 'truth is a relation, not a thing',[483] which at first glance makes Platonism sound like structural realism. But physical phenomena, according to scientific Platonism, are still real; they are just lower on the scale of reality in the metaphysical hierarchy. For example, we do not look to mud for moral guidance or to predict the motions of the planets. We look to abstract ideas, such as justice, or to the laws of physics.

Even though our scientific investigations may begin with a physical examination of the properties of mud, eventually, if our scientific understanding is going to progress, we must achieve a more fundamental abstract understanding. To revise an analogy offered by Heisenberg,[484] by flying high above our local terrain we obtain a greater understanding of the ecological relationships on the ground, and learn how to live better lives within our own particular geographical area. With our higher, abstract, universal understanding, we can better understand the relationships between particular phenomena, be they ecological, material, sociological, or even the contents and patterns of our own minds.

With respect to the reality of scientific theories and scientific entities, we find a confusing mix of views. Some theorists are realists about theories, but not about entities, while others are realists about entities, but not about theories. In the first case, I may believe that our scientific theories and laws are true representations of reality, but that the entities that are part of those theories either have no physical reality or are merely constructs that result from the interplay between our theory and the relevant technology. In the second case, I may believe that entities, such as electrons, the moon, or my car, are all real, but that the theories that make sense of these entities are merely fictions that help us to predict observations.

Bohr was an entity realist since he believed in the reality of atoms, while he is often assumed to have been an antirealist about scientific theories because quantum theory does not give us a literal picture of the subatomic world, but rather a 'symbolic representation'.[485] But Bohr was not an antirealist about theories, either. The theories of quantum mechanics are not pictorial or literal (in the physical sense), but they do represent actual nonphysical mathematical laws and metaphysical principles. Bohr was simply rejecting materialism.

Whether or not our theories completely or only partially represent the actual mathematical laws and physical entities as they are in reality, entities still exist and the universe still operates according to mathematical laws. Realists may be expecting too much when they claim that our scientific theories describe with absolute certainty entities as they actually are in physical reality. As realists, we have to admit our fallibility. The antirealist denial of all entities existing independently of our theories and measurements, however, does not at all follow from this limitation. We may not be able to know for certain that our theories provide absolutely accurate representations of the entities postulated by those theories, but we can be sure that there are entities that we can potentially discover, and that our theories can more or less accurately tell us exactly what those entities are. A structural realist or antirealist may deny the reality of atoms, or at least deny that we can have any knowledge of atoms, but they would then have a difficult time explaining the power unleashed by atomic and nuclear weapons.

We shall now move on to *factual realism*, which first requires a clarification of the meaning of 'facts'. Some facts do appear to be theory-*dependent*, but in actuality all genuine facts are theory-*independent*. The notion of 'facts', in the way that I, and most people, use the word ultimately reduces to 'the way things are'. As discussed further in the next chapter, an example of a theory-dependent fact would be to say that there are X number of planets in our solar system, depending on how we define a planet. But our classifications do not literally create planets; we do not bring something instantly into existence by naming it. In other words, although there are eight planets in our solar system, the new classification scheme that redefines Pluto as a dwarf planet instead of a planet does not eliminate or create any new objects in the sky. Pluto is still there, regardless of what we call it.[486]

Consider the following simple but illustrative example. My cat Tyson sometimes jumps up onto my desk en route to the windowsill, where he enjoys a bit of relaxing time until he is ready to go outside in the morning. This mundane example contains a profound metaphysical lesson. I am (fairly) sure that Tyson has no way of conceptualizing or understanding what a desk really is, such as it being a convenient place to put my computer and hold various documents. I really have no idea how he perceives and cognizes my desk, other than it being an object that helps him attain his ends, and even then, how can I really be sure that he is 'thinking' about 'attaining ends'? After all, I do not know what it is like to be a cat.[487]

Despite the apparently unbridgeable gulf between our two perceptual and cognitive worlds, there is still the fact that *something* exists that withstands the weight of both my computer and Tyson, regardless of whether or not there are an infinite number of possible interpretations and ways of understanding what that something is. My desk is whatever it is, which is to say that it has objective scientific properties, regardless of how I or any other sentient creature perceives it or what we believe about it.

Similarly, you may feel cold while sitting in a room while I feel warm, but the objective fact remains that the actual temperature is 18 degrees Celsius. Thus, while factual realism allows for both objective and subjective physical truths, we cannot reduce the objective properties of things to our subjective experiences of them. However, the subjective physical truth of your experience of feeling cold still depends to some significant degree upon the objective physical facts of your physiological makeup.

I must now briefly anticipate certain antirealist arguments from the next chapter in order to discuss a relevant issue raised by physicist John Wheeler, who writes that 'useful as it is under everyday circumstances to say that the world exists 'out there' independent of us, that view can no longer be upheld'.[488] Wheeler is correct that our knowledge is constrained by our measurements (and intentions, goals, desires, etc.), but it does not therefore follow that there is no objective reality. Just because my body does not exist independently from the earth (and the rest of the universe), it does not follow that the earth cannot exist without my body. I could not have been born here if the earth had not already existed, which means that the earth was here before I was. (It should probably be a bit embarrassing telling non-specialists that part of my job is trying to convince various academics—including some scientists—that the earth existed before they were born.)

Wheeler also states that 'no elementary phenomenon is a phenomenon until it is a registered (observed) phenomenon'.[489] This statement is a goldmine for antirealists, because if it were really true, then nothing could exist until I happened to look at it. At some moments Bohr seemed to agree with this viewpoint, but when we examine his views more closely we see that he did not. He writes that 'the use of phrases like "disturbance of phenomena by observation" or "creation of physical attributes of objects by measurements" is hardly compatible with common language and practical definition'.[490] Such antirealist notions are not compatible with basic rationality, either. The very fact that we perceive anything at all seems to imply that there is *something* 'out there' and that it was there before we noticed it. We need only ask Wheeler what he would say to a person who runs him over with a car and then claims that such an accident could not have happened because the driver did not observe Wheeler in that moment, meaning that Wheeler could not have existed.

It is true, however, that I may believe I perceive a pink elephant 'out there', when in fact I am hallucinating. We then find ourselves right back at the edge of the trap set by the skeptical antirealist, who can always retort that 'you could just be in virtual reality, and so whatever you believe to be real is not actually out there; it is just in your head'.

The first problem with such a response is that my head would have to exist objectively, or else I would not be able to have a hallucination 'in my head'. Second, I can only be wrong that my hallucinations accurately represent reality if, in fact, there is a real objective fact of the matter that my hallucinations are not accurately representing, which is to vindicate realism. In other words, I may very well be in virtual reality or hallucinating, but then there would necessarily still be an objective reality 'out there' that is causing me to live in my own dream world, which, again, vindicates realism.

However, some facts, such as wave/particle duality or the orbit of an electron, seem to depend upon the theory that we choose to adopt. Even Einstein told Heisenberg that 'it is the theory which decides what we can observe'.[491] A theory does not create reality, but it does constrain which facts receive our attention. But facts are still only true if they actually correspond to the way things are, even if we are ignorant regarding this correspondence. If it turns out that a supposed fact is actually false, then we have to say that it never really was a fact in the first place.

There are an infinite number of possible theory-independent facts concerning the physical universe, all of which we cannot possibly observe or even contemplate. Our theories, goals, and cognitive and perceptual abilities allow us the possibility of coherently selecting what is of most relevance to us. But that which is selected for consideration by such theoretical or practical goals and limitations already exists, and that which we have chosen to ignore within such limitations also still exists. For example, my conscious decision of which book to select from the shelf does, in that moment, limit all of my potential choices to one actual book. But my choice neither creates the book nor eliminates the reality of the other books on the shelf. Similarly, the standard analogy of a fishing net that only catches fish of a certain size reminds us that there are still other fish in the sea than the ones we happen to catch; likewise, there are still theory-independent facts beyond the theory-dependent facts that we happen to be able to (or want to) notice. The net is like our theory, selecting which fish we are able to catch, or which theory-independent facts we are able to observe. The inability to understand such a simple distinction has been one of the causes of the great and unnecessary confusion that has arisen in the attempt to understand some of the apparent paradoxes of quantum theory.

As good as it gets?

At certain moments, Bohr and Heisenberg seemed to suggest that although we may be able to progress further *within* quantum theory, we could progress no further *than* quantum theory. At other times they clearly did not think this was the case, but I will here consider those occasions when they did appear to make such a suggestion. For example, Heisenberg claimed that 'nature works only in such a way as not to violate quantum mechanical formalism',[492] which may seem to imply that quantum theory is the final theory of physics. It is true that nature, within specific parameters, cannot violate quantum theory, in the same way that, within still narrower parameters, it cannot violate classical physics. But other aspects of reality remain beyond the powers of quantum theory to explain. Nevertheless, some of their pronouncements had a note of finality, suggesting that there was no going beyond quantum theory, that it was as good as it gets.[493] Such misplaced confidence is reminiscent of Planck's professor, Philipp von Jolly, who told him that physics was essentially a complete science with little prospect of further development.[494] But there is no end to the possibility of deeper understanding in physics, or in any field or aspect of life. The only way physics as a discipline could come to an end would be if we were to stop pursuing such research. Fortunately, Planck ignored von Jolly's misguided advice on this matter.[495]

Baggott writes that Bohr 'denied that quantum theory has anything meaningful to say about an underlying physical reality that exists independently of our measuring devices. It denied the possibility that further development of the theory could take us closer to some as yet unrevealed truth'.[496] At first glance, Baggott's analysis gives the impression that such a claim would entail that we cannot go deeper than quantum theory. Perhaps it is true that quantum theory cannot be developed any further and that, as it stands, it cannot tell us about nature itself, but in that case we would only be recognizing the objective limitations of the theory.

In a similar way that quantum theory surpasses yet incorporates Newtonian physics, we must also discover a deeper and more powerful theory that embraces yet goes beyond quantum theory. Bohr's view also entails that there is an underlying physical reality, but that quantum theory may not be able to tell us about it exactly as it is in any physically literal sense. Nevertheless, it is still telling us something about this physical reality, since it enables us to understand, explain, predict, control, and transform aspects of it to an extraordinary degree. Our current scientific success, however, is nothing more than a few pebbles on the endless shore of potential knowledge.

SEVEN: Imploding Antirealism

Quantum theory has been assumed to be antirealist, often because of the confusion between materialism and realism, but also because of an even more elementary mistake in reasoning. Antirealists assume that just because there must be a limit to what physics can tell us about reality, it must follow that there is no reality beyond such limitations. Other antirealists go even further by saying that there is no objective reality at all. Many scholars have fallen into this subtly enticing antirealist trap, perhaps mesmerized by its underlying promise of bestowing godlike powers upon us. In other words, antirealists would like to be able to create their own version of reality just by wishing it to be so, and to pretend to be able to ignore aspects of reality that are inconvenient or uncomfortable. The result is that they feel entitled to force 'their reality' on others.

Many antirealists create the misleading impression that they can draw significant support for their position from quantum physics. However, even if it is true that physics cannot say anything at all about physical reality apart from our means of measurement, it does not follow that there is no physical reality beyond our measurements. This mistake in reasoning is so obvious that it is difficult to comprehend how so many otherwise intelligent people have missed it, yet all quantum antirealist claims ultimately rest upon such misguided assumptions.

In this chapter, I first discuss the general character of antirealism before analyzing a few comments made by Bohr and Heisenberg that seem to be antirealist. I then argue against extreme and ambiguous antirealism. Throughout this chapter, the key point to remember is that ontology is not eliminated by epistemology, which is to say that the limit of our knowledge is not the limit of reality.

What, really, is antirealism?

We cannot discuss in detail all of the different versions of antirealism,[497] as we need to focus on what is most relevant for our purposes here. It will be helpful, however, to give brief consideration at the outset of our discussion to both relativism and postmodernism. *Relativism* generally refers to the belief that every truth claim is conditioned by our society and particular circumstances, goals, beliefs, and cognitive and perceptual limitations, and so no one can ever make a universally true claim. Relativists therefore say that there is no such thing as a universal truth. The obvious rebuttal is to point out that the claim that 'there is no such thing as a universal truth' would itself be a universal truth, which necessarily makes relativism false.

Relativists are correct to point out how so often our understanding of what is true or false, or right or wrong, is conditioned by our society. But they are wrong to say that, therefore, there is no such thing as objective truth. Relativism ends up saying that no claim is any more true than any other claim; if I say that 2+2=17, my math teacher cannot say that I am wrong. In fact, relativists can never say that my view is false, because relativism can never say that anything is objectively false. But as Polanyi recognizes, 'there is no purpose in arguing with others unless you believe that they also believe in the truth and are seeking it'.[498] How can we have a serious philosophical dialogue with relativists when they deny that there is any such thing as objective truth and yet have no hesitation to claim that I am really wrong not to be a relativist?

Those who follow *postmodernism*, despite often rejecting the label, generally assume that reason cannot tell us anything about reality, and they deny that there is universal or objective truth, just like the relativists. Postmodernists also tend to reject dichotomies and hierarchies, and they scoff at any talk of truth. The postmodernist Jean Baudrillard puts it this way: 'truth is what should be laughed at. One may dream of a culture where everyone bursts into laughter when someone says: this is true, this is real'.[499] If Baudrillard were on trial for committing a serious crime that he did not really commit, I am sure he would not wish for the jury to laugh at the evidence proving his innocence.

Postmodernist professors also attack hierarchies, while at the same time accepting a salary that gives them the hierarchically based authority to decide whether or not their students will pass their course. These postmodernists also believe that their ideas are better than many ideas that predate postmodernism (or have since replaced postmodernism), which means that they are making a dichotomy between their ideas and their opponents' ideas. It also implies a hierarchy whereby they believe that their own ideas are superior to their opponents'.[500]

The most important version of antirealism to consider at this point is idealism, but we need to recall the distinction mentioned in Chapter Two between ancient and modern idealism. Ancient idealism assigned a greater reality to the nonphysical, unchanging, intelligible realm responsible for the order of the physical cosmos. Modern idealism, on the other hand, entails that reality is nothing other than a construct of the human mind (or perhaps another sentient creature's mind). Thus, ancient idealism is actually Platonic realism, while modern idealism is a form of antirealism.

A. E. Taylor defines idealism as the doctrine that all reality is 'mental', whereas realism is the 'doctrine that the fundamental character of that which really is, as distinguished from that which is only imagined to be, is to be found in its independence of all relation to the experience of a subject'. He continues his explanation by saying that 'what exists at all, the realist holds, exists equally whether it is experienced or not'.[501] However, there are two separate claims being made here, although they appear to be identical, and once we see their subtle distinctions we will be better able to appreciate another reason quantum theory has been falsely assumed to be antirealist.

The claim that reality is independent of all relation to the experience of the subject is significantly different than saying that reality exists (or is) whether or not it is experienced. The second distinction, that reality exists (or has being) equally whether it is experienced or not was endorsed by Bohr, Heisenberg, and Pauli, which would make them realists. As I have shown, they believed that nature could not violate the laws of quantum mechanics, laws that they believed themselves to be discovering, and their views, when taken as a whole, overwhelmingly presuppose Platonic realism.

It is worth pursuing in more detail Taylor's first distinction, which, without proper understanding, can be seen to evoke the claim that quantum theory is antirealist. Since quantum theory has shown us rather forcefully that all aspects of reality are interdependent, it may ultimately seem impossible for reality to exist (or be) independent of all relation to the experience of the subject, the perceiver, or the experimenter. Since the observer and the observed have been shown to be intermingled in quantum physics in the sense, for example, that the experimenter's choice can play a direct role in the outcome of an experiment, it is easy to see why antirealists have claimed that realism is dead when realism is defined as implying the absurd position that nothing has any relation to anything else.

Obviously, we must in some way be in relation to what it is we are studying; otherwise, there would be no way of obtaining any information about it. Consider the following simple but important example. I can use a thermometer if I want to determine the temperature of water in a glass, but when I put the thermometer into the glass of water, the heat of the thermometer will interfere with the temperature of the water, and so the measured result is the temperature of the water + thermometer (+ glass + table + floor, etc.). We cannot find the temperature of the water in isolation from our methods of measurement, and we cannot know the temperature unless we ourselves look at the reading on the thermometer, which requires an act of consciousness. We cannot even know the temperature of the water in complete isolation from the totality of all physical reality, of which the water is a miniscule aspect, and we do not need to evoke quantum physics to reveal this interdependency.

The experimenter imposes the distinctions, limitations, and boundary conditions for the practical purposes of the experiment. However, they are not logically defensible demarcations, because there is no logically necessary reason to exclude any potential variable in any experimental situation. Given the holistic nature of reality, every part of the universe must necessarily be considered as part of every experiment. In practice, we obviously have to limit our variables to the few that are most immediately relevant to our purposes, but there is still no logical necessity to such limitations.

Many philosophers and scientists have believed that reality could be known with absolute objectivity by an impartial experimenter, which is a false metaphysical assumption. Quantum theory has emphatically shown that physicists, in their capacity as physicists, cannot know physical reality with absolute objectivity, not if such objectivity implies that the discovered aspects of reality have absolutely no relation whatsoever to the experimenter. However, this claim is completely different from saying that there is no objective reality. Just because we may not be able to know completely and with absolute certainty some aspect of objective reality, it does not at all follow that there is no objective reality.

The claim that reality can be found in independence of all relation to the experience of the subject is false according to quantum theory, as well as according to pure metaphysical reasoning, for all knowledge is only known when appropriated by an intellectual subject capable of such cogitation. However, it does not follow that we are creating that which we come to know. I can only acquire knowledge from a book if I read and understand it, but I did not create that knowledge, nor does my act of acquiring it eliminate the knowledge found in in other books I have yet to read. Similarly, we can discover certain bits of knowledge in the 'book of nature', but the totality of all such knowledge is not limited by the particular paragraph or chapter we happen to read.

Another subtle point can be made here by way of the following example: I have just moved a pencil from one spot on my desk to another and, in so doing, had an immediate effect upon my physical environment, as every living things does in every moment. (We could perhaps include non-living things in this example as well.) Therefore, the claim that the experimenter impacts the experiment is rather trivial, because every creature impacts its environment in some way at every moment. The point, however, is that if every human on earth were to suddenly die, the earth would still be here doing whatever it does (and it would probably soon be in a much healthier state, too). In other words, the earth was here before we were, and it can exist without us, although our existence here depends very much upon our relationship with the earth.

The supposed antirealism of the Copenhagenists basically amounts to nothing more than the recognition of the limits of the physicist's ability to describe physical reality in absolutely precise terms. That being said, some of their comments do seem to indicate that there is no quantum reality beyond our measurements, which would be a denial of certain aspects of realism. However, their rejection of a pictorial representation of quantum aspects of reality was really a rejection of materialism, not a rejection of Platonic realism. Nevertheless, if they really intended to make antirealist insinuations, then they were wrong, and such views are incompatible with their overwhelming realist commitments. Unfortunately, antirealist interpretations of these sparse comments have led to even greater wholesale antirealist proclamations in our day, which the Copenhagenists never would have accepted.

Apparent antirealists

Baggott writes that, according to the Copenhagen interpretation, 'it is not meaningful to regard a quantum particle as having *any* intrinsic properties independent of some measuring instrument'.[502] The related notions that a quantum particle has no intrinsic qualities distinct from our measurements, and that deterministic description is rendered impossible even in principle, upset Einstein. He demanded that 'every element in the physical theory must have a counterpart in the physical reality'.[503] Einstein was correct in saying that the Copenhagenists were wrong to claim that quantum particles have no intrinsic properties apart from being measured. Even given their own assumptions, the Copenhagenists cannot make such an assertion about the nature of quantum aspects of reality, because they have assumed that they either cannot make any claims about quantum reality or that there is no quantum reality. Therefore, they would not be permitted to say anything about whether or not particles have intrinsic properties apart from being measured.

Einstein was also right to reject the Copenhagenists' dismissal of absolute determinism, but he seemed to be asking for the impossible by demanding that *every* element in physical theory have a counterpart in physical reality. Theory and observation never correlate absolutely and, even if they did, it seems impossible to know with absolute certainty when they do. Einstein may have forgotten his own words, as mentioned in Chapter Six: 'as far as the propositions of mathematics refer to reality, they are not certain; as far as they are certain, they do not refer to reality'. As his own theories relied heavily upon mathematics, which could not refer directly to physical reality to the degree that the mathematics was certain, his requirement that every part of physical theory must have a counterpart in physical reality would not be possible. Nevertheless, he had the correct intuition that something was amiss with the Copenhagenists' interpretation.

Bohr infamously said that 'there is no quantum world. There is only an abstract quantum physical description. It is wrong to think that the task of physics is to find out how nature is. Physics concerns what we can say about nature'.[504] We can adopt the charitable interpretation that he was merely saying that physics has epistemological limits and cannot make ontological claims. However, in that case Bohr did not follow his own advice, because he made many ontological claims. He was right to stress that how we conduct our measurements has an influence on the result of an experiment, but he was wrong to assume that there are no intrinsic properties just because when we measure something, only then do we 'see' its properties. Either the properties are intrinsically part of the quantum particle (or wave), or we are creating the properties (and the particles/waves themselves) out of nothing. Alternatively, there may be no determinate properties, because each particle/wave already harbors all possible properties from which our measurements select one or some limited number.

If it is true that quantum particles have no intrinsic properties until we measure them and, further, that there is no way for quantum theory (or perhaps any theory) ever to penetrate any deeper into reality and discover that these particles do, after all, have intrinsic properties independent of our measurements, then this fact would still be a fact independent of what any of us happens to believe about the matter, which therefore supports realism. However, there is no way to prove empirically that quantum particles do not have intrinsic qualities until we measure them. Both the belief that they do and that they do not have intrinsic properties before we measure them cannot be purely experimentally decided, and recognizing this fact reveals the metaphysical nature of this confusion. Given his other realist commitments, it is safe to assume that Bohr did not believe that we actually create quantum particles from absolute nothing with our measurements or observations. He believed that *something* was there prior to our measurements or observations, even if that something is indeterminate and unknowable in itself. But either we measure *something* or *nothing*; we cannot measure *nothing*, so we must measure *something*, even if we cannot ever know what that something is in and of itself apart from our measurements.

When Bohr claimed that we could not know quantum reality because there *is* no quantum reality outside of our measurements, he was both partly right and seriously wrong. It is not wrong to demonstrate our epistemic limitations, but we are neither philosophically nor scientifically justified in making the ontological claim that there is *no* quantum reality. Consider the example of the glass of water discussed previously in this section. The only way we can directly measure the temperature of the water is by placing a thermometer into the glass, but this act will alter the temperature of the water (unless the thermometer is *absolutely* the same temperature as the water, which we can never know).

The metaphysical assumptions of science do not allow us to say that the epistemic limitations inherent in interfering with the water temperature we are measuring entails that the water has no actual or definite temperature until we measure it. The water has some specific temperature at some specific time regardless of the interference of the measuring process. Although antirealists want to prevent realists from making claims about reality beyond our direct experience, they will turn around and immediately make such claims themselves. In this case, claiming that the water has or does not have any particular

145

temperature prior to our measurement is to make a claim about reality beyond our direct experience and knowledge.

Besides making clear the distinctions between ontological and epistemological assertions, my example shows the problematic nature of applying the Copenhagen assumption (that subatomic particles have no intrinsic properties until measured) to macroscopic objects or events.[505] It is bizarre to say that the water in the glass has no particular, intrinsic temperature until it is measured, even though the views of some physicists would imply such a thing.[506] The water must have a particular temperature at a particular moment whether or not I measure it.[507]

The claim that the water never has any definite temperature could mean either that the water has no temperature at all (which is surely a strange assertion), or that its temperature is constantly changing at every moment. Strictly speaking, the temperature of the water probably is changing, however slightly, from moment to moment and, therefore, we cannot know exactly what the temperature of the water was prior to measurement or what it will be after measurement. In fact, the water temperature may vary slightly between the water's surface and the bottom of the glass. We can only know what the temperature *was* at the particular moment when we made the measurement, and even then we only know the temperature of the water + thermometer (+ glass + rest of the universe), which forces us to admit the holistic nature of physical (and nonphysical) reality.

We can sometimes slip into antirealism unwittingly when we defend a holistic account of reality. The erroneous slip occurs when we shift from the metaphysical realization that no physical thing could exist unless it is in relation to other things to the false conclusion that therefore nothing at all has any existence. It is true that, in one sense, nothing could have a temperature unless it was in relation to other things, and therefore nothing ever has an exact temperature of its own independent of everything else. However, at the very least, the temperature of the water (plus the rest of the universe), despite constantly changing from moment to moment, must have some particular temperature at some precise moment in time. The water will not suddenly evaporate, freeze, or disappear into oblivion without some further rational scientific explanation that could account for such an event. There are objective limits, and such limits are defined by the intricate interplay of innumerable factors all abiding by the same laws of physics, and likewise for the subatomic particles that make up the atoms and molecules of the water. When a particular particle (or wave) is measured, it is not suddenly going to transform into a fire truck, or any other random object; it will have a particular manifestation at a particular time, which can be predicted with extraordinary accuracy, and it will surely have been and continue to be something similar to whatever it was prior to and after measurement.

The metaphysical notion of wholeness, of the inseparability of subject and object, or experimenter (or measuring device) and the system being measured, has mistakenly led to the antirealist conclusion that there is no reality independent of the mind of the observer. But admitting that everything in the universe is in some way interdependent does not mean that electrons have no properties or do not even exist until we measure them. Bohm says that wholeness is not simply an ideal toward which we should strive; rather, what is real *is* this wholeness. 'It is just because reality is whole that man, with his fragmentary approach, will inevitably be answered with correspondingly fragmentary responses'.[508] No theory could ever be absolutely true, but only ever partial, since every theory, statement, or description can only partially describe or explain any phenomenon.

Bohm is actually being a realist by claiming that there is ultimate reality underlying our limited theories and experiences, but if all of physics is partial, and if this is all the knowledge that we are capable of, then we could never know that our knowledge is partial, nor could we know that there is an underlying reality of unbroken wholeness. Only by gaining some glimpse of the whole could we recognize that our knowledge is partial. Bohm thinks that our partial, fragmentary, perhaps illusory theories are never actually true or false, but only clear in certain domains,[509] yet they can still 'point to or indicate a reality that is implicit and not describable or specifiable in its totality'.[510]

One may believe that reality is not undivided wholeness, but rather a chaotic motley of chance mutations and colliding objects, but Bohm would say that such a view is wrong. There *is* a reality, and this reality is undivided wholeness. How he *knows* that this is reality, however, is a different question. Probably following Krishnamurti's influence (or at least his encouragement),[511] he wants to say that by understanding the fragmentary nature of our thought itself we can come to understand or perceive the reality of wholeness underlying our fragmented thought. Although we may be able to experience or know directly this undivided wholeness of reality, as soon as we reflect upon this experience, or talk about it, or build a theory upon it, we must necessarily create a fragmentary description. The absolute unity of perceiver and perceived (or observer and observed) may be possible, but as soon as any thought about this experience-knowledge whatsoever occurs, there is necessarily fragmentation. What is most interesting here is that the notion of the unity between the observer and the observed, or the knower and the known, is not only found in Eastern spiritual/philosophical traditions; the Platonic philosophers also recognized it.[512]

Heisenberg, too, made some comments that are hard to defend against the charge of antirealism. For example, although it may be true that the laws of physics formulated mathematically in quantum theory no longer allow us to speak of (physical) nature in and of itself, but only 'nature exposed to our methods of questioning',[513] it does not follow that there is no reality beyond what physics can tell us, as Heisenberg obviously realized. But he also stated that it is no 'longer possible to ask whether or not the particles exist in space and time objectively'.[514] In this case, Heisenberg could mean that:

1. quantum theory will never allow us to know anything about the objective nature of these particles;
2. these particles do not exist in physical space and time until measured;
3. these particles do not exist at all; or,
4. these particles do not exist in physical space and time at all, but are still real.

Concerning (1), if we cannot know anything about the objective nature of such particles, then physicists would have to stop making any statements about their objective nature. If we accept (2), then we have to ask how we could measure something in the first place if it is not in space and time, and how something outside of space and time could suddenly come into space and time to be measured. There are no easy answers to such questions. If we accept the antirealist position of (3), where we assert that these particles do not exist at all, then it would be essentially impossible to explain just about anything we are able to do in science. Heisenberg seems to have meant option (4) where, as already noted in the previous chapter, he believes that the 'smallest units of matter are in fact not physical objects in the ordinary sense of the word', but are actually geometric forms or structures.[515] In other words, the smallest units of matter are not actually physical, but they are real geometric structures that can only be represented mathematically.

Perhaps Heisenberg should have said that the smallest units of matter are still physical, but that they can only be represented accurately in the language of mathematics because they behave according to mathematical laws. It is also true that, to the degree that we can mathematically represent any physical object, no matter how large or small, we are talking about it with scientific objectivity. The issue becomes blurry when we discuss macro objects, because we can perceive them directly with our senses, and so we experience various qualities in relation to them, such as hot or cold, sweet or sour, green or blue, and so forth. But each object that we experience has underlying objective scientific properties that are what they are regardless of how we experience them. In addition, Heisenberg should have said that we cannot know with absolute certainty and absolute objectivity the totality of all possible aspects of any micro or macro object, but that we can still have objectively true knowledge of limited aspects of such objects. However, as discussed in the previous chapter, he did admit that the same chemical elements exist on other planets, and he also believed that certain areas of physics are universally valid for all time. Such beliefs are part of Platonic realism.

Heisenberg and many other physicists have also more or less claimed that we cannot describe what happens in between two observations of an electron, and that it is meaningless to assert that it has any characteristics at all if we cannot measure them.[516] It has never been clear what is meant by such pronouncements, especially since there are important differences between claiming that:

1. physics cannot describe what happens between any two observations;
2. nothing whatsoever happens between any two observations; and,
3. nothing *determinate* happens between any two observations.

Option (1) says nothing more than that there are objective limitations to what physicists can achieve (in this case, they cannot describe what happens to an electron between observations), which is a relatively trivial point. Option (2), however, makes no sense at all. If *absolutely nothing* happens between any two observations, entailing that even the passage of time has ceased, then the physical universe would be in a sort of temporary suspension until we happen to measure the second event. But there could only be a second event if, in fact, something happened (or changed) since the first event, so *something* must be happening between the observations.

Alternatively, we would have to say that only this particular electron enters into a timeless and motionless state after the first observation, which is to say that it must suddenly go from being in time and space, and therefore in some kind of motion or change, to being completely outside of time and space. This is the only possible way that absolutely nothing could happen to an electron in between observations. However, in order for a second measurement of the electron to occur, it must reenter time and space; it must move or change from being inside time and space to being outside of it, and then reenter time and space again when we make the second measurement. In other words, *something* has happened in between measurements. If we really cannot make claims beyond what we can measure, then we cannot claim that something or nothing happens between measurements. Yet simple applied metaphysical reasoning shows us that, obviously, something is indeed happening in between measurements, even if we cannot know or explain what exactly is happening.

Option (3), the claim that nothing *determinate* happens between any two observations, seems to indicate that the electron is behaving randomly in between observations. But if we are not justified in saying that the electron is behaving deterministically between observations then, for the same reasons, we are not justified in saying that it is behaving indeterministically (or randomly) between observations. Even if it is behaving randomly, its randomness is still highly limited within specific parameters, and we would still have to explain how random behavior gives nonrandom results in physical reality, which physicists cannot adequately explain.

The most likely case is that Heisenberg simply wanted to assert that we cannot know absolutely what is happening between observations and that, as physicists, we can only say what the case is when we make a measurement, even though we cannot logically or empirically separate what is being measured from what is doing the measuring. But all these claims are logically trivial and no different in principle than what is necessitated by my previous example of measuring the temperature of water in a glass. In the end, the realist must accept the limitations of our empirical knowledge of physical reality through the focused lens of physics, yet antirealists are wrong to assume that such epistemological limitations then apply to reality itself. Physics is not the only route to knowledge, and physicists and positivist philosophers who assume that it is continually make metaphysical assumptions and dogmatic proclamations beyond the boundaries of physics, making their positions as internally untenable as they are incompatible with everyday confrontations with physical reality.

A final example of how some physicists continue to hold inconsistent assumptions concerns the belief that the following two postulates are both true:

1. the goal of physics is not to describe nature as it is in itself, but as it is exposed by our questions; and,
2. the uncertainty of the uncertainty principle is inherent in nature itself.

These two statements are clearly incompatible. The reason for this incompatibility is that (2), the claim that the uncertainty is inherent in nature, is a claim about nature itself. It should be obvious that if (1) is true, we can never make *any* claim about nature itself, so we cannot consistently say that nature really is inherently uncertain without violating the first claim that physics cannot describe nature itself. In any case, describing nature as it is exposed by our questioning necessarily implies that there is underlying nature that is what it is, even though we are only able to access certain aspects of it depending on the kinds of questions we ask. This view is congruent with Bohm's notion of the reality of unbroken wholeness underlying our fragmented perceptions.

We can agree that physicists are describing nature as it is exposed to their questions, where the trick is to ask the right questions, since our epistemological limitations are unavoidable. But this claim is no more significant than saying that I can describe a painting in many different ways, perhaps as a fellow artist, an art critic, an artistically uneducated person, a capitalist, a physicist, or a chemist. If I ask questions about the chemical structure of the paint, I will get a certain kind of answer, but if I ask for the selling price of the painting I will get a much different answer. Such answers are not incompatible, although they may still be right or wrong (or more or less accurate) independently of each other. Moreover, these different answers, which follow from their respective questions, do not imply that the painting does not exist. The painting exists as it is regardless of what questions we ask or whether or not we happen to be admiring it.

The basic philosophical assumption of realism is not materialism, but that there is objective truth independent of our beliefs, opinions, and perceptual limitations. Consequently, by equivocating on the word 'realism', antirealists can use the first meaning to claim that physics proves realism to be false (since physics proves materialism to be false), and then shift to the second meaning, thereby denying that there is any objective truth about anything. By denying that there is any objective truth about anything, they can then conclude that there is no objective reality, which can lead to the claim that there is no reality at all.

This sort of unsound reasoning used by antirealists is analogous to making the following racist argument: we must discriminate between poisonous and edible berries, and so we must discriminate against Asians. It is easy to see the main flaw in this reasoning, which occurs when we move between two different meanings of the word 'discriminate'. In the first case, it is necessary to discriminate or differentiate between poisonous and edible food. In the second case regarding Asians, we have changed the meaning of 'discriminate' to mean holding unjust prejudice against a group of people.[517]

Physics cannot tell us absolutely everything about physical reality, but it can illuminate a portion of it. Eddington makes a similar point: 'if our so-called [scientific] facts are changing shadows, they are shadows cast by the light of constant truth'.[518] Absolute truth is constant, but our understanding of it is changing and (hopefully) deepening. The deeper we explore the foundations of physical reality, the more we leave the familiar world of the physical and enter into the nonphysical, changeless, and abstract. Such a journey is bound to be unsettling to anyone brave enough to endure it. It was certainly unsettling to the pioneers of quantum physics. Their tremendous insights and discoveries came at the price of some relatively simple philosophical confusion that has had enormous implications on society when unleashed in the form of antirealism. It is unfortunate that these quantum pioneers did not receive the philosophical assistance they needed at the time, but we are now in a better position to clear up some of this confusion. The key point to understand is that although Heisenberg and Bohr made a few comments that opened the way to sliding down the irrational path of antirealism, they were overwhelmingly Platonic realists. Their rejection of materialism was not a rejection of realism, and antirealists who still pretend to believe that they can create any reality they want just by snapping their fingers can no longer rely on support from modern physics. In reality, they never could.

Diehard antirealists

Lindley provides a good example of how extreme quantum antirealism has developed from the Copenhagen interpretation: 'there is no longer any meaning to be attached to the idea of a real objective world; what is measured, and therefore known, depends on the nature of the measurement'.[519] Lindley seems to have forgotten, however, that he also admits the importance of a real objective world: 'some things in the end can be determined only empirically, by looking at the world and figuring out how it works'.[520] Thus, he denies the reality of an objective world that he also appeals to in deciding how things actually work, thereby negating his antirealism. Antirealists tend to expedite the inevitable implosion of antirealism as soon as they begin to speak. With that in mind, let us now examine the views of two extreme antirealists who have travelled so far down the antirealist path that one wonders if they can ever find their way back again.

Steeped in Heidegger, or at least a certain interpretation of Heidegger,[521] philosopher Karl Rogers denies the reality of electrons.[522] He claims that it does not matter whether an electron is 'real and out there', because its reality cannot be 'divorced from the socio-technical processes in which it is stabilised and utilised'.[523] In other words, unless we live in a society that has the technological means to study electrons, electrons do not exist. However, if electrons exist only when we measure them, then either they do not exist anywhere except where they are currently being measured, or they do *now* retroactively exist everywhere as soon as a machine has measured them, which would imply some incredible magic. Either option is going to face far tougher challenges than those that could ever possibly haunt the naïve realist.

Concerning whether or not electrons have any 'reality outside of the technological framework', we need only consider the familiar notion of radicals in biology. 'Radicals are compounds that have a single electron, usually in an outer orbital. Free radicals are radicals that exist independently in solution or in a lipid environment'.[524] There are several 'dietary free radical scavengers', such as vitamin E, ascorbic acid, carotenoids, and flavonoids. For example, Vitamin E is an 'efficient antioxidant and nonenzymatic terminator of free radical chain reactions, and has little pro-oxidant activity.... The chemistry of vitamin E is such that it has a much greater tendency to donate a second electron and go to the fully oxidised form'.[525] There are also several 'disease states' associated with free radical injury, including cervical cancer, alcohol-induced liver disease, diabetes, aging, Alzheimer's disease, multiple sclerosis, and Parkinson's disease.[526]

Biological cells are made up of molecules, which are made up of atoms, which consist of numerous subatomic entities (whether conceived of as waves, particles, both, or neither), and these entities include electrons. Since no biological or chemical process can violate a law of physics, then physics is ultimately at the foundation of biology and chemistry, and physical explanations include the assumption that electrons exist independently of 'socio-technical processes'. The evolution of life has required millions of years,[527] which means that cells have been around for a very long time, and cells, according to our best biological and chemical theories, consist of many subatomic particles such as electrons. If electrons did not really exist outside of our ability to detect or create them with current technology, then no life would exist anywhere in the universe.[528]

Consequently, if Rogers is correct, then either no life exists anywhere, or life could only have come about retroactively once an electron emerged within our newly devised technological framework. But the only way we could have allowed for the emergence of a technologically-dependent electron is if we existed prior to our ability to create the relevant technology, which would require our biological bodies, and therefore electrons, to exist before electrons and our bodies came into existence. In other words, if Rogers is correct, then no humans could have existed before developments in modern physics, which is to say that no humans could have existed in the past until humans existed in the present. So, you had to exist today before your great-grandmother could have existed in the past.

Rogers believes that it is irrelevant whether or not electrons are 'real and out there' fundamental particles,[529] but I hope it is obvious to most of us that it *is* relevant whether or not they are. If they were not real and out there, then no life could exist anywhere and, apparently, no one could ever experience any disease states, such as aging, that are associated with free radical injury until we had first created the technology to measure electrons or allow them to emerge. If Rogers is correct, then no disease associated with free radicals could have existed prior to our ability to measure electrons. To cure these diseases, then, all we would need to do would be to destroy the relevant technology and stop talking about electrons so that these free radicals stop existing. In other words, if we just un-invent the electron, or choose to remain ignorant about it, then we will never age. We would also retroactively stop existing.

Rogers could reply that electrons are simply abstracted from experiences, the same way he assumes that mathematics is abstracted, and so electrons are not *causes* of anything. In this case, talking about electrons would be a useful basic description only possible within a particular socio-technical context. We could then project this description backwards merely for explanatory purposes, without ever believing that electrons really exist or existed. Such a response, however, is untenable. It is possible that our current understanding of what electrons and other such entities are may change, or that electrons may not be *exactly* as we think they are, but all reasonable evidence and rational arguments point to the fact that something very much resembling what we call electrons must exist, and they must exist independently of the measuring apparatus and of the observer, even if our epistemological limitations imply that we cannot have absolute certainty about the ontological status of electrons until they are measured. At the very least, even if our theories of electrons are totally false, there is still something existing 'out there' that enables us to create our modern technology.

Electrons as they are now conceived (or something very much like them) are real, and they were real before any humans were roaming the earth. We may only be cognizant of their reality within the confines of the technological (and, as Rogers should have added, the theoretical) framework, but this framework does not bring electrons into existence. On Rogers' account, it would be as if, to alter the metaphor, the shining of our flashlight in the dark actually brings our keys into existence, which then suddenly brings our car, and roads, and stoplights into existence as well.

If we deny the socio-technological-independent reality of electrons, then there is no good reason to stop us from denying the reality of atoms, molecules, cells, animals, and planets. As Barbour noted, dinosaurs are not commonly considered to be merely convenient fictions to explain the fossil data, but since they were made of cells, molecules, atoms, and subatomic particles, which apparently have no existence outside of our relatively recent technological framework, then dinosaurs must be fictions. The consistent antirealist (if such a person could actually exist) may deny that any life ever existed anywhere until we measured electrons, and I too can play such games and deny any reality to anything other than myself—or even deny reality to myself—but, again, this inane skepticism becomes pointless, more pointless even than positivism. Rogers' antirealism would be the end of not only science, but of the whole universe.

The second extreme antirealist we will consider is B. Alan Wallace, a specialist in religious studies who spent many years studying Tibetan Buddhism and often acting as interpreter for the Dalai Lama. His erudition and spiritual wisdom, however, have not prevented him from offering untenable antirealist conclusions based on only a couple of passages from Bohr and Heisenberg and, ironically, even from Einstein. While completely ignoring the obvious Platonic realist commitments of his Copenhagen mentors, Wallace offers us what he calls the 'centrist view', which relies upon Buddhist metaphysics. It is impossible, obviously, to do justice to his Buddhist view in such a short space, but fortunately we need only consider two of the most relevant points here.[530]

First, Wallace takes the general Buddhist position that nothing at all, not even minds or the laws of physics, have any inherent existence. The argument, in essence, runs as follows: everything has arisen in interdependency from the void (Emptiness, 'Sunyata'), therefore nothing exists independently of or separately from anything else, therefore everything exists only in 'conventional' reality but nothing exists in 'ultimate' reality, and so there is no objective reality. This conclusion may not seem to follow, and indeed it does not, but this is, nevertheless, the essence of the argument.

According to Platonism, however, this less than perfect world of constant change really does exist, and it really does matter what we do here while we are alive, not only within the here and now but also in preparation for what comes after death.[531] The subtle but crucial distinction is that in Platonism this everyday reality is not a mere *illusion* but, rather, we are *deluded* when we believe that this everyday reality is all there is and that there is no greater reality behind the appearances. In Platonism, the ultimate nature of reality is the One itself, which is what it is independent of all things, be they physical or nonphysical.

While Wallace admits that most mathematicians hold a Platonist conception of mathematics,[532] he still criticizes Platonism as *reification*, which means attributing a real existence to fictitious concepts. In other words, to reify is to believe that something abstract or intangible is actually real. But if all of mathematics is a fiction, then so are the laws of physics, and so the foundation of physics would also be a fiction. In fact, on the Buddhist view, it would seem that everything in the universe, including the self and even Emptiness itself, is nothing more than reification.[533] The next chapter deals specifically with the nature of the laws of physics, so here we need only point out that there is no logically necessary reason to say that *only* something physical can be real and, in any case, we cannot even define unambiguously what is meant by the terms 'material', 'physical', or 'tangible'.

Wallace also latches onto the apparent antirealist comments made by Bohr and Heisenberg, neglecting their overwhelming realist commitments, to argue that there is no objective reality. Such misappropriation contributes to his fall into antirealism. For example, he writes that 'those who continue to adopt a realist interpretation of quantum theory continue in the age-old attempt to conceive of physical reality as it exists independently of our systems of measurements'.[534] He argues that 'in the absence of any system of measurement, we have no evidence of waves, particles, potential, or anything else'.[535] He even says that 'an electron existing as an independent entity is in principle unknowable; therefore this independent entity does not exist as potentiality, for it does not exist at all'.[536] Wallace has now gone even further than Rogers by completely denying that electrons exist at all. Such fashionable postmodern pronouncements are not a sign of sophistication, however; on the contrary, they indicate a departure from reality.

As realists, we have to admit the possibility that electrons may not actually exist in the way we think they do. Our theories may only be making relatively accurate predictions about observations, and everything about our theories may have nothing to do with physical reality. Even if that were the case, that fact would itself be an objectively true fact, and it would still be true that physical entities, from the smallest to the largest, exist whether or not we can measure them or have any knowledge of them. To deny this obvious fact is to deny that you have a body or that there is a universe. Any antirealist who denies that they have a body or that they are living in a universe should have no qualms with stepping in front of a speeding bus.

Wallace also claims that 'phenomena are brought into existence through the processes of verbal and conceptual designation'. In fact, 'their very nature is defined by language usage: the relationship between an object and its attributes is determined by the way we speak and think about them'.[537] If Wallace is correct that phenomena are brought into existence through the process of verbal and conceptual designation, then phenomena could exist only for those creatures with adequate intellectual and linguistic capacities. Thus, he should be able to avoid being hit by a bus if he stops thinking and talking about buses and eliminates the word 'bus' from his vocabulary.

These antirealists who try to draw support for their views from quantum theory would do well to listen to the wise words of the theory's chief originator. Planck writes that 'reason tells us that if we turn our back upon a so-called object and cease to attend to it, the object still continues to exist'. He adds that the laws of nature 'are in no way affected by any human brain. On the contrary, they existed long before there was any life on earth, and will continue to exist long after the last physicist has perished'.[538]

Amorphous antirealism (or, an anti-antirealist antirealist closet realist)

Vision argues that some of the most well known antirealists, such as Rorty, have been successful not only in enlisting many converts, but 'more importantly in the fact that they have managed to define the issues in their own terms'.[539] But we need not accept their whimsical rules. As we shall see, Rorty discards any distinction between truth and falsity, reality and appearances, and knowledge and opinion. It is, to put it mildly, disconcerting that so many otherwise intelligent people have fallen under his sway. (In Appendix B, I provide a detailed critical analysis of Bas van Fraassen, another amorphous antirealist who is somehow simultaneously a philosopher of science, an antirealist, and a Catholic.)

The first point to note about Rorty is that if we really believe that he genuinely meant what he committed to writing, then he has no way of objectively differentiating between the most benevolent actions and the most horrific. In fact, his attempt to demolish any distinctions between appearances and reality, or knowledge and opinion, would be the end of science and all rational discourse.[540] If we really take Rorty seriously, then we really *cannot* take him seriously. His philosophical view amounts to saying that we should do whatever makes us happy and not bother thinking deeply about anything that makes us feel uncomfortable. But shouldn't we feel uncomfortable with a philosophical view that makes it impossible to tell people not to rape, murder, or mutilate if such things make them happy?

Rorty shares some similarities with the ancient Skeptic and Cynic philosophers, although I am skeptical that he engaged in sexual intercourse or masturbation in public, or lived in a barrel, as some famous Cynics did.[541] These ancient philosophical schools may have helped some people cultivate a courageous way of life, but they were not able to assist us in scientific understanding. It was not their goal, and neither is it Rorty's. But his epistemological pronouncements about the sciences, which are rooted in his whole approach, bring him directly into the purview of this book. For example, he draws no epistemological line of distinction between theoretical physics, literary criticism, and astrology,[542] which means that we should be able to rewrite the 'story' of physics any way we choose. If he is correct, then I should also be able to consult an astrological chart from any random newspaper in order to write like Shakespeare or make the next great discovery in theoretical physics. Literary criticism may be no more or less important in our lives than theoretical physics, but it does not follow that the two disciplines are epistemologically identical.

Rorty also goes so far as to (pretend to) demolish distinctions between knowledge and opinion, and truth and appearances, stating that all such 'dualisms' should be 'dissolved'.[543] First, it should be noted that the term 'dualism' is widely misused and abused, but there is no need to go into such controversies here, except to note that, by his own criteria, Rorty has created another dualism. He believes that dualisms should be dissolved, which creates a duality between his views and those who disagree with him, whereby he believes that his own views are superior.

If there are to be no distinctions between knowledge and opinion, as Rorty insists, then we may as well consult a construction worker when we are sick and a medical practitioner when we want to buy plane tickets to Spain, or get a literary critic to build a nuclear reactor. I would be interested to see if any followers of Rorty would accept my opinion of how to fix the broken engine of the airplane they are about to board, which happens to be that we should deliver a swift kick to the fuselage and fill the fuel tanks with protein powder. If there is no distinction between knowledge and opinion, then my opinion is just as knowledgeable as that of any aeronautical engineer. Whoever refuses to fly in a plane repaired according to my suggestion clearly values knowledge over opinion, and so they cannot be a follower of Rorty.[544]

Rorty also claims to believe that the most useful philosophers of the twentieth century 'are those who have argued that the Greek distinction between appearance and reality can, and should, be set aside'.[545] He wants to replace this distinction with descriptions that we happen to find useful to suit our own purposes. If Rorty is correct then, since I am lazy and hate to study anything, it would suit my purposes not to bother trying to understand calculus and simply create my own unique description of mathematics. I could, for example, describe derivatives as apple pies, and if Rorty were my math teacher he could not say that my description is wrong. I would be guaranteed an A+.

If Rorty were to reply that, obviously, some ways of describing derivatives in calculus are better or more accurate then others, then he has just admitted the realist position. We cannot, after all, say that anything is a better or more accurate description of any particular thing unless there is some sort of objective truth about it. If I describe my car as if it were a peanut, but you describe it as if it were a truck, then your description would be more accurate than mine. Whoever disagrees is welcome to get into their peanut every day and drive to work. Similarly, Rorty would also have to admit that a description that may best suit the purposes of a serial killer would be just as morally right as any other description.[546]

The consequences of taking Rorty's views seriously would be disastrous. His philosophical view would force him to say that killing babies for fun is just as morally permissible as feeding the homeless. If Rorty is correct, then each and every one of us can also make up our own laws of physics, laws that must be equally as true (and equally as false, and equally as meaningless) as any laws discovered by the greatest physicists. Yet, quite unbelievably, Rorty also claims that he is not a relativist, nor does he think that anybody really is one.[547] Rorty is at least correct that no one is truly a relativist, given that relativists really do believe that they are right, which they cannot do if they are to maintain their relativist stance.

When you are a relativist, you do not have to be consistent, and so Rorty can, in fact, be a relativist about everything while denying that he is one, because relativists do not have to be consistent, rational, or truthful. A relativist can say with a perfectly straight face, 'I hate Asian people, but I am not racist'. One of the reasons our society is such a mess is precisely because many of us actually do take relativism seriously, but only when it seems convenient for us to do so, when it is 'useful for some purpose'. Rorty is such an antirealist that he is in some sense even an anti-antirealist, although that really just amounts to being an extreme antirealist. He is apparently an *anti-representationalist* and *anti-essentialist*, which basically means that he rejects any foundation for anything, even though his own views have a foundation, which I will discuss shortly. But first let us see how his approach to philosophy would have rendered the development of quantum physics impossible.

Rorty claims that the controversy between idealists (antirealists) and realists is 'pointless',[548] which is to say that all the arguments between Bohr and Einstein, which were essential to the development of quantum theory, were a colossal waste of time. As Rorty notes approvingly, Wittgenstein refused 'any longer to be tempted to answer questions like "Is reality intrinsically determinate, or is its determinacy a result of our activity?"'.[549] In other words, Rorty has just agreed with Wittgenstein that one of the most important questions in quantum theory was too philosophical for their anti-philosophical agendas. This parochialism renders them and their followers incapable of discussing the deepest and most significant philosophical issues that have confounded the pioneering physicists in quantum theory.

Ironically, Rorty has also characterized the realist versus antirealist debate as one of the 'pseudo-problems' that have plagued philosophy for long enough and that should be dissolved.[550] However, a 'pseudo-problem' is something that only *appears* to be a real problem while in *reality* it is not. This means that Rorty has just made a distinction between reality and appearances, between what is true and false, between a *real* problem and a *pseudo*-problem, even though his own philosophical stance forbids any such distinctions. And yet, he also states that 'the anti-essentialist has no doubt that there were trees and stars long before there were statements about trees and stars',[551] which is precisely part of the realist position that many antirealists pointedly deny.

Rorty also states supportively that those philosophers, such as James, who 'are dubious about the "correspondence theory of truth," nevertheless have no sympathy with the notion of nature as malleable to thought, or with the inference from "one cannot give a theory-independent description of a thing" to "there are no theory independent things."'[552] It is an amazing twist of reasoning (or intellectual bullying) to say that any questions about realism versus antirealism are meaningless while simultaneously holding an implicit, naïve realist stance and promoting extreme antirealism, even while denying being an antirealist or a realist and the whole time mocking rationality.

Rorty has also claimed that 'we will see Dummett's notion of philosophy of language as "first philosophy" as mistaken not because some other area is "first" but because the notion of philosophy as having foundations is as mistaken as that of knowledge having foundations'.[553] In other words, Rorty believes that Dummett is objectively wrong, which further illustrates my point about Rorty's inconsistency (or hypocrisy). If Rorty is correct that there is no such thing as true or false, right or wrong, or good or bad, then he cannot say that Dummett is wrong. But he says it anyway. Interestingly, about a decade later he admits (or pretends to admit) that he cannot say that anybody is wrong. He states that 'it would be inconsistent with my own anti-essentialism to try to convince you that the Darwinian way of thinking of language—and, by extension, the Deweyan, pragmatist way of thinking of truth—is the objectively true way. All I am entitled to say is that it is a useful way, useful for certain purposes'.[554] The problem with Rorty bothering to aim for consistency, such as not wanting to say 'X and not-X', is that he has now accepted the standards of logical reasoning dating back to the ancient Greeks whom he has spent much of his career rejecting.

We also need to consider some wider implications of Rorty's philosophical opinions. I wish I could say for certain that Rorty would not condone atrocities and injustices, but sometimes I really am unsure: 'at most times', he writes, 'it sounds crazy to describe the degradation and extirpation of helpless minorities as a purification of the moral and spiritual life of Europe. But at certain periods and places — under the Inquisition, during the Wars of Religion, under the Nazis — it did not'.[555] These views may not have sounded crazy to the persecutors of these atrocities, but they would certainly have sounded crazy to those who were being persecuted, something that Rorty appears to ignore.

Rorty's *instrumentalism*, which basically reduces all truth to nothing but usefulness, would make it impossible for him to both remain consistent with his philosophical opinions and say that genocide is objectively wrong. An instrumentalist could reply that it would not be useful for society to murder millions of its own people. Hitler, however, would disagree. If Rorty is correct that truth is nothing other than what serves some purpose, and if your purpose is to attain a 'pure' German race, and you believe that it is useful to exterminate every Jewish person to attain this goal then, on Rorty's account, you would be entirely justified in doing so.[556] If this is enlightened thought, then I will gladly remain ignorant.[557]

Rorty also expresses the anti-representationalist belief that we should not attempt the impossible and seek something beyond ourselves: 'we should not look for skyhooks, but only for toeholds'.[558] A toehold, however, is something to rest your weight upon, a foundation, but Rorty's whole project is to do away with foundations, so he must do away with toeholds, leaving him with nothing to support his philosophical program. Those who follow Rorty share a common desire; they want to never be wrong. If there really is no objective truth about anything, then no one can ever really be wrong about anything. But as Clark aptly notes, 'the price of never being wrong, of course, is never being right'.[559] It is precisely because realists admit to the possibility of being wrong that they can engage in fruitful argumentation in order to clarify further problematic assumptions.[560]

Finally, on Rorty's view, if we need only seek out certain frameworks of conversation that are useful for certain purposes, then it would be most useful for us to converse as much as possible about Platonic realism, since this position has been so important to various eminent pioneering theoretical physicists.[561] We may also be motivated to understand the rational arguments in the Platonic tradition that aim to lead us to the non-discursive insight that the nature of reality is One, which can lead to a whole life transformation. Whatever our motivation, we can easily find many useful reasons for studying the Platonic tradition, which, on Rorty's account, is all that is required to make Platonism 'true'. It has certainly been useful for developments in both science and religion.

In the end, Rorty believes that 'the anti-essentialist description will make us happier in the long run, because in the long run it will make us freer'.[562] Well, his philosophical opinions do not make me feel the least bit happy, nor do they provide me with any sense of freedom, and that, according to his own criteria, is all I need to say in order to justifiably reject his philosophical opinions.[563]

Where are my keys?

When we come to believe that reason cannot provide us with the certainty or consolation that we desire, we tend to blame reason instead of taking responsibility and facing our own shortcomings. We have believed one argument and then its opposite so many times, or we have failed too often in providing sound reasoning for what we know to be true, that we no longer have any faith in reason, or see any reason to have faith. What we need is the philosophical courage to seek the truth, as exemplified by Socrates and many others throughout history.[564] Many of our ethical dilemmas involve very difficult grey areas that require careful analysis, but only if we search for the guiding light of truth beckoning to us from beyond the darkness of our ignorance is there any hope of attaining right understanding and action. Antirealists, however, keep us ignorant by trying to convince us that this light is an illusion. Having an antirealist as your guide is like looking for your keys at midnight in the middle of a jungle while blindfolded with your hands and feet bound, all the while believing that your keys do not even exist.

Everybody, whether your parents or your preacher, your doctor or your plumber, your lover or your enemy, is trying to convince you of some version of how they see reality. What I am offering, or reinforcing, are some of the tools required to see and understand reality for yourself, whereby you can judge more objectively the accuracy of what I say, what others say, and, most importantly, what you yourself say. A strict adherence to antirealism removes these vital tools, making it far more difficult to discern reality from appearances.

But is there ever a time when you may want to conflate reality with appearances? Yes, there is; when you need to deceive your enemies. We need to conceal the reality of our situation and plans by creating deceptive appearances. Conversely, our enemies will be doing the same thing to us, and so we need to be able to see through their deceptive appearances and discern the reality of the situation, for only then can we hope to make the best plans of attack and defense. If you follow Rorty in his belief that there are no distinctions between reality and appearances then you can easily be deceived and will soon be defeated. If you want to achieve your goal and defeat an enemy, then do not seek the advice of an antirealist. However, if you want to expedite the fall of your enemy, find a way to indoctrinate them with antirealism.

As Clark writes, 'if anti-realists are (absurdly) right, there is no harm following Plato's example; if they are wrong then something very much like Platonism is correct'.[565]

EIGHT: The Laws of Nature and the Nature of its Laws

Rarely do philosophers or physicists consider to the appropriate depth the nature of the laws of physics. If the laws of physics are not real, then there is no viable way to explain the possibility of science, or even the fact that the universe exists at all. Consequently, we must say that the laws of physics are real. However, if the laws of physics are real, we need to be able to explain what they are, how we can discover them, and how we are able to rely upon any one particular law to predict with extraordinary accuracy what will happen in two or more very different situations involving completely different phenomena. In seeking such an explanation, we will have to admit that the laws of physics transcend the physical universe, yet are responsible for how it operates. This chapter undertakes to explain in further detail how the laws of physics are real.

It is essential to keep in mind throughout this chapter that there is a difference between the laws of physics as we state them and the laws of physics as they are in reality. We may believe that we have discovered an unquestionably true law of physics, but in the future we may discover that this law was either false or only partially true. In such cases, those who have not thought deeply about the issue may carelessly proclaim that all of the laws of physics are false or illusory. Yet we could only know that a previous law was false if we happen to find a replacement law that is true (or more approximately true than the previous one), which presupposes that we can in fact discover true laws. In other words, I would be saying that the laws of physics are false and believing that I can prove this point by saying that I have found a true law of physics that proves the falsity of a previous law. Therefore, only by relying on a true law could I argue that there are no true laws, and so the assumption required to prove my conclusion would actually make my conclusion false.

Whenever an antirealist argues that the laws of physics are unreal, they can only possibly be referring to the laws of physics as we state them. They cannot be referring to the laws of physics as they are in reality; otherwise, they would be saying that the universe does not obey itself. As we push our explanations further, we are going to encounter subtle refinements concerning what it means to say that the laws of physics are real, beginning with examining why one of the most plausible rivals to Platonic realism fails to provide a satisfactory explanation for the reality of the laws of physics.

Aristotelian abstractionism: clever isn't good enough

Aristotle, who had been Plato's student, was also an extraordinary philosopher in his own right. Here, however, we are concerned with the failure of Aristotelian abstractionism to explain the reality of the laws of physics, which begins with an important misconception about Aristotle being a realist.[566] Despite the other ways in which classicists may wish to present Aristotle, he seems not to have been a realist about the laws of physics. For example, he offered a view that the later Platonists referred to as *abstractionism*, whereby 'mathematical objects are mental concepts derived from sensibles'.[567] In other words, abstractionism assumes that by watching the way the physical world operates we can deduce laws to explain our observations, although these so-called 'laws' are nothing more than convenient descriptions that have no genuine reality, which is a view not all that different from modern-day positivism. Aristotle seemed to vacillate between realist and antirealist positions, but here he is a straightforward antirealist about the mathematical laws of physics.

It is not my goal to engage in a historical analysis of Aristotelianism; rather, I am going to clarify and argue against the general claim that the laws of physics are nothing but fictitious accounts that we infer through abstracting from the sensible phenomena.[568] My version of a modernized, ahistorical, more or less Aristotelian abstractionist account of the mathematical laws of physics would go something like the following.[569] We only know the meaning of '1' by looking at '1 thing', for all knowledge comes from sense data of particular physical entities. By picking up one rock and holding it in my hand and then looking at or holding another rock, I can then realize that there are two rocks, which is to say that there are two one-rock things. By induction we begin counting 1 rock + 1 rock + 1 rock, eventually realizing that we can simply drop the noun 'rock' and keep the number symbol to represent any object. Similarly, by induction we could conclude that 1 rock being taken away means that I now have no rock, and so 1 rock − 1 rock = no rock, which could be symbolized as '0'. From these basic mathematical foundations, all derived from sensible experiences of concrete particular objects, we can then construct the mathematics that we presently have today. I realize how overly simplified this account may seem, but it does represent the sort of assumptions underlying any abstractionist account of the laws of physics. Let us continue our tale.

From our experience, we find that throwing small, round, smooth rocks at our enemy allows for better accuracy than large, blockish ones. We then realize by induction that the rounder the rock the better the accuracy, and so we look for the roundest rocks possible, which could also inspire the idea of a perfectly round rock, a mental image we conjure up to compare other rocks to when searching for the best throwing weapons. We may be amazed by the power of these round rocks and so desire to draw pictures of them. By induction we come to see that the various round rocks have similar ratios, such as the distance from the center of the round shape to its edge, and that if we keep the distance from the center to the edge of the round line constant, we will draw something most accurately resembling our mental image of a perfect throwing weapon.

We then have the misfortune one day of meeting an enemy who has somehow devised a machine that launches rocks at us from a great distance. We eventually figure out how to construct the same machine after stealing one and copying it, and by induction we realize how to calculate the amount of force required to throw a stone a certain distance. By further developing our previous pictures into more refined and complex diagrams, including further measurements and inductive calculations, we realize that we can apply our findings to a variety of objects, such as arrows, baseballs, and nuclear missiles. But what about things that move naturally, such as ocean tides, celestial bodies, and apples falling from trees?

We decide to collect data about the distance the ocean water progresses toward and away from land and then calculate the time between each complete procession and recession. We then carefully count the seconds it takes for apples and other similar objects to fall from various heights, perhaps by using a stopwatch while dropping bricks from a tall tower, and by induction we realize that there is a common mathematical equation that applies to the rate of acceleration for each of these falling objects.

While up in our tower, we spend many sleepless nights staring at the mysterious starry sky and meticulously recording our observations with increasingly sophisticated diagrammatical representations, which eventually leads us to inductively determine the mathematical equations that describe the motions of the planets. Then, through further clever induction, we recognize mathematical relationships between the tides, the movements of the heavenly bodies, falling apples, and parabolic curves. One day, we somehow or other inductively arrive at quantum theory.

First of all, if the antirealists are correct that there is no truth, or that opinion is no different than knowledge, then my tale is just as true as any other. The main point, however, is that such a story is, in principle, what is entailed by assuming that all mathematical laws are nothing more than constructed fictions inductively achieved through the study of particular phenomena. These inductive fictions somehow allow us to predict what we are likely to observe in the future by using these idealized mathematical equations as a guide, where the role of the sciences is merely to construct such fictions to enhance our success at predicting observations.

The Platonist is in agreement with the abstractionist insofar as we often take the sensible world as a starting point that leads to mathematical inquiry, but the disagreement occurs when the Platonist asserts that the laws we discover are something beyond fiction, something beyond mere particular physical manifestations. Moreover, the abstractionist antirealist often seems unable or unwilling to appreciate the deeper implications of the process of induction, which is quite a mysterious activity. Recall Russell's point that 'it is quite difficult to think of the right hypothesis, and no technique exists to facilitate this most essential step in scientific progress'. But how exactly can we inductively conclude anything about anything without first having some sort of inspiration, a 'flash of understanding', or a 'direct perception'? All of these instantaneous moments of direct understanding, where we see inherent order underpinning what many others see as a dizzyingly kaleidoscopic mess, involve some mystical element that transcends our limited discursive ability. It certainly transcends any kind of viable explanation that the abstractionists may offer.

So why do physical things behave they way they do? If there is an explanation, a reason, for why things behave the way they do, there has to be some sort of cause, be it physical, metaphysical, or both. Abstractionists may say that there is a 'reason-principle', logos, or some sort of intelligence built into the very nature of a thing, inseparable from the thing itself or that *just is* the thing itself.[570] When a human zygote develops into a human being, or when an acorn develops into an oak tree, it is because the zygote and the acorn both contain within themselves their own reason-principles.[571] But if everything in the universe has its own reason-principle then, among other important questions, we have to ask:

1. how everything in the universe received its reason-principle;
2. how a common reason-principle could be shared by a plethora of particular individuals; and,
3. how all the reason-principles are related.

Abstractionists cannot provide satisfactory answers to any of these questions, not unless they admit the reality of the nonphysical realm, which includes higher metaphysical principles responsible for the reason-principles that are manifested in physical reality. In this case, however, they would be in essential agreement with what I am presenting here, in principle if not in detail. There is simply no way to make sense of how one reason-principle can be physically manifested in more than one instance simultaneously if reason-principles are purely physical. No physical thing can occupy exactly identical space-time co-ordinates simultaneously, and so, if we are to accept the abstractionist account, the reason-principle for each and every acorn must be different. If they are all different, how can the abstractionist explain how they are related? The reason-principle for each acorn would still have to have some aspect that remains common to them all; otherwise they could not explain why acorns develop into oak trees instead of apple trees or zebras.

If the reason-principle of an oak tree is what guides an acorn to become an oak tree and not an apple tree, and yet all oak trees contain the same basic reason-principle, then there is a common reason-principle in all oak trees. Each tree will grow in its own unique way, given the availability of sunlight, water, and other factors, but an oak tree is still an oak tree and not an apple tree or a zebra. The reason-principle of the oak tree is the same, or has some common core that is the same, in all oak trees, while allowing for an indefinite variety of physically manifested oak trees (various shapes, sizes, etc.). This potential for indefinite variety is contained within the parameters of the original reason-principle (in interaction with the parameters of other reason-principles, such as those for water, soil, air, and arborists). The reason-principle for each thing, therefore, must be a universal since, although limited in applicability, it is not reducible to particulars and is common to a whole class of particulars.

Proclus offers this important clarification: 'some universals are prior to their instances, some are in their instances, and some are constituted by virtue of being related to them as their predicates'.[572] In other words, a universal can:

1. have independent existence;
2. be inseparable from each particular; or,
3. refer to a name that merely denotes a particular class.

Let us consider each possibility in reverse order, and start by taking the word 'cat' as an example of the third type of universal. The word 'cat' does not exist beyond our language structure. We impose the name upon a certain group of creatures, recognize a variety of similar features that would satisfy our definition of 'cat', and then live with the ambiguity inherent in our own classification systems concerning cat-like creatures. The same can be said about the challenge of how to classify Pluto and other such entities. The antirealists, such as Wallace, are trapped at this level of understanding and consequently attribute magic to words by believing that changing our definitions, or simply naming things, brings material things into existence and has the power to change them physically. It does not take much effort to recognize the incoherency and irrationality of remaining fixated at this level of reasoning. For example, I am at this moment calling my laptop a glass of freshly squeezed orange juice, but the fact remains that I cannot now drink my laptop, because it is still a laptop; it did not just magically morph into a glass of orange juice. Changing the name of a thing does not change reality, although changing a name can cause me to classify things differently, which is a completely different and relatively trivial point. Uttering the word 'cat', in other words, does not suddenly create the reality of cats.

The second kind of universal is one that is manifested in each particular, which we have already discussed above. On one hand, each reason-principle is manifested only in a particular instance, such as a particular zygote or acorn, and so the reason-principle seems to be inseparable from its associated matter. On the other hand, the reason-principle is not reducible to any particular manifestation since it is repeatable and operates in a multitude of particulars at the same time. It must be a nonphysical principle, and yet some aspect of it is manifested in each physical particular. Some abstractionists get stuck on the third kind of universal with their more extreme antirealist partners, but usually they are caught here on the second kind. If they say that there is a different reason-principle for everything in the universe, they have no way to explain how all acorns become oak trees, which can only happen if they have or partake in the same reason-principle. If they admit that these reason-principles transcend particulars, they are essentially in agreement with Platonism. If they do not admit this, they are materialists of one sort or another, and all reason-principles would therefore be nothing more than fictions.

The first kind of universal, those that are prior to their instances and have independent reality apart from their instances, are beyond the limitations of abstractionism. Since the laws of physics are an example of this sort of universal, abstractionists must deny the reality of the laws of physics, and so for abstractionists the fact that we can make such accurate predictions about the behavior of physical reality must, in the end, amount to either magic or extraordinary luck. After all, why should the universe be such that we are capable of creating equations that just so happen to allow us to predict observations to such an extraordinary degree of accuracy, such as we can achieve in quantum physics, and yet these equations have no basis in reality?

Abstractionists rely upon our ability to inductively (in one way or another) invent fictitious laws that just so happen to be able to predict observations, but they cannot explain how this process of induction actually happens. They also cannot explain how Einstein's particular story about physics works, whereas other stories do not. This is a crucial point. If the laws of physics are mere fantasies, we have to explain why some fantasies work while an indefinite number of other possible fantasies do not. $E=mc^2$ is helpful (even if limited in its applicability), but '$E=mc^2 + 17$' or '$E=mc^2 + 2$ bananas' do not allow me to predict anything in physics.

Consequently, if the laws of physics in our textbooks are not true, or if they are not at least approximately true, and we then conclude that the laws of physics are not real, we have to explain why my random equations do not work while Einstein's does. Even if we say that Einstein just so happened to make a lucky guess, it would still be the case that his guess just so happened to match reality to some appropriate degree, in which case the laws of physics are real. Otherwise, we have to say that Einstein has more magical power than we do, allowing him to make the universe bend to his will and follow his equations. Then we would have to explain this kind of magical power and make it a vital aspect of all our scientific research. Surely it is more reasonable to accept that the laws of physics are real than inject magic into every aspect of science.

The abstractionist is likely to reply that there is no such thing as magic, and that just because not all fantasies work it does not follow that some are not useful. In other words, abstractionists end up falling back on the assumption that some fantasies work because they just so happen to allow us to predict how things will behave, and that things behave the way they do according to their inherent reason-principles. We then need only ask to see one of these so-called reason-principles. If I only know about reality by observing it, as abstractionists claim, then I naturally want to *observe* a reason-principle.

However, I cannot observe a reason-principle, because either they do not exist, or they are real but not physical. If they do not exist, then abstractionism is false. If they are real, then they are either physical or nonphysical. If they are physical, we should be able to observe them, at least in principle with appropriate technology. But we cannot observe them, so either we end up saying that they are real and nonphysical, which the Platonist could grant, or the reason-principle *just is* the thing itself, in which case we no longer need to talk about reason-principles at all and should instead just speak of the things themselves. By doing so, however, we fall deeper into the trap of materialism and would have no viable way of explaining how, for example, a particular universal law of physics applies to a plethora of different phenomena simultaneously. All of the other problems with materialism that we have discussed so far would also return to haunt the abstractionist.

If there are no independent universals, then there are also no genuine laws of physics, which will mean that matter behaves the way it does because that is just the way it behaves. We either want a further explanation for why matter behaves the way it does, or we reject any such explanation. If we reject any such explanation, we have to admit that we have no idea why anything does what it does other than saying that it just so happens to do so. But why do apples fall? On this account, apples fall because apples fall, and that's that; there is nothing more to say on the matter. Perhaps they fall because of their 'desire' to be closer to the earth. And why not admit just such a story if the laws of physics are mere fantasies? It is fortunate, however, that at least some scientists wanted to know why apples really fall, which eventually led to our ability to fly to the moon. It is thoroughly anti-scientific to deny that there are deeper explanations for why phenomena behave the way they do, so let us persevere in seeking underlying explanations for these phenomena.

In Plato's *Timaeus*, for example, we are presented with a 'likely tale' about the creation of the universe, rather than a completely literal, detailed scientific account.[573] The *demiurge* (the 'craftsman' or 'maker' or 'one who fashions'[574]) is not the One, or the highest God, but it was the demiurge that brought order to the universe by taking over all that was visible, which was initially in disorder. For the universe to unfold in an ordered way, it would need to follow its own reason-principle, which would be the one lawful structure expressed as the eternal mathematical law underlying all of physical reality. This eternal law of the physical universe is itself modeled on higher metaphysical principles, such as pure symmetry, which itself implies absolute beauty and, ultimately, the One itself.[575] It is toward this higher eternal model, such as pure symmetry, that the demiurge looks when fashioning the one eternal mathematical law for the universe. By the very nature of this eternal law, with its inherent power and order, it is compelled to give birth to the physical universe, which is the birth of time and space, and the universe has been unfolding accordingly ever since.

For those who may feel uncomfortable with this sort of pure metaphysical reasoning, as if it were an unnecessary departure from proper philosophy or scientific reasoning, I offer the following reminders. To cite just two previous examples, Einstein believed in 'Reason' permeating the cosmos, and Pauli believed in a 'cosmic order independent of our choice and distinct from the world of phenomena'. If you still believe it is unreasonable to talk of cosmic order and pure reason permeating the cosmos, bear in mind that in doing so you are saying that Einstein and Pauli were unreasonable. But instead of simply proclaiming that reason and order permeate the cosmos, we are here trying to make sense of what such claims actually mean, and to provide a rational account for why they are true.

If absolute or pure chance is the final explanation for the origins and sustainability of the universe, there is no good reason to assume that the laws of physics will remain constant for any length of time, and it would be only by an absolute miracle that the laws of physics remain constant so that we can do science (and so that anything can exist at all). If absolute chance is the foundation of all reality, there would be no laws of physics in the first place, and neither would there be a universe. Faith in an intelligible First Cause is far more rational than believing that by pure randomness or absolute chance the universe is actually ordered, unified, consistent, and absolutely not random or a perpetual succession of miracles.

If we deny that the universe is rational and ordered, there is no hope for predicting observations other than by appealing to perpetual miracles. Even Dawkins admits that the notion of chance is 'just a word expressing ignorance. It means 'determined by some as yet unknown, or unspecified, means".[576] Dawkins seems to realize that if chance really infected science at its core, science would be impossible. What he refuses to do is to ask seriously what is implied by the fact that the universe is ordered and unified, which would require entering into the sort of metaphysical territory we are presently exploring. Dawkins' atheistic fundamentalism prevents him from considering questions that make him uncomfortable in a similar way that religious fundamentalism prevents its adherents from taking seriously the uncomfortable questions that Dawkins asks them. Dawkins is happy to put others on the witness stand, but runs away when he himself is subpoenaed.

The authors of an astronomy textbook admit that there appears to be 'a high degree of order within our solar system', and that 'the overall architecture of our solar system is too neat, and the ages of its members too uniform, to be the result of random chaotic events'.[577] Unless the universe really is ordered, which is to imply that it has a lawful structure, it would be disordered and random. If the universe truly were disordered and random, we could never do science or make any plausible predictions about anything. In fact, an absolutely disordered universe with no ordered unity whatsoever could never even have come into existence.

If the universe really were absolutely disordered and random, then the only way for the universe to actually behave in an ordered fashion would be if, at every instant, the entire universe miraculously defied its naturally disordered state and by absolute chance happened to behave in a rational, ordered way. In such a case, we would be justified in saying that every instant would be a miracle, since it truly would be miraculous that, just by chance, the universe at every instant did not behave by chance, despite the foundation of the universe itself being pure chance. Atheists abhor the idea of a miraculous creation, and yet, unless they admit to an ordered universe, they must rely upon an *infinity* of miracles at every moment throughout the entire universe. By applying Occam's Razor, which demands that we prefer theories or hypotheses that have the fewest assumptions, we should accept one miracle over an infinite number of miracles.

Abstractionism provides a clever way of avoiding deeper metaphysical territory, but such cleverness is not enough to provide a genuine explanation of the reality of the laws of physics, or even of how we can do physics in the first place. The higher we ascend into abstract, unifying reason-principles, the more we approach greater unifying productive power and simplicity. The entire physical universe obeys its own reason-principle, which immediately implies that there is something nonphysical that transcends the universe. In other words, if the universe is the totality of all physicality and the universe is guided by a reason-principle, the ideal reason-principle could not itself be physical, for it would then be subject to generation and change along with the universe itself. If the reason-principle could change, all of physics would be impossible. Fortunately, the reason-principle of the universe remains what it is eternally and, therefore, to the degree that abstractionism can in any way be made intelligible, it approaches Platonic realism.

Ducks and rabbits (and the Mona Lisa)

In order to address the fundamental questions that concern us in these last two chapters, we will need to ignore the subtle differences between laws, theories, and equations. For example, there is the law of conservation of energy, quantum theory (which has competing interpretations comprised of various and sometimes apparently conflicting components), and equations such as the

Heisenberg uncertainty relations. It is easy to get lost in the details of analyzing any specific law, theory, or equation, but what I aim to do here is provide an abstract analysis of their fundamental nature regarding whether they have eternal and unchanging ontological reality or are simply human constructs. I will argue that we actually discover laws, theories, and equations, which implies that they must have prior existence or being. For simplicity, I will refer to all of the above as laws, for in an important sense they all are. If they are true (within their applicable domains), they cannot be broken and so deserve to be called laws. (For a condensed argument supporting the reality of nonphysical mathematical laws of physics, see Appendix C.)

Physicists themselves are not consistent in clarifying exactly what they mean by a law. For example, the conservation laws are more akin to metaphysical principles. It is not possible even in principle to verify empirically whether or not the law of conservation of energy is valid throughout the entire universe. The only way we could provide such empirical proof would be if someone were to measure every bit of matter/energy in the entire universe at precisely the same moment, a measurement which would have to include the energy of the measuring apparatus and the experimenter, as well as the act of understanding the data in the mind of this experimenter.

Physicists must also assume the prior reality of the principles of symmetry, abstractness, simplicity, and absolute unity. Although such notions are metaphysical in nature, they are the driving force and fundamental assumptions of the mathematical laws that must obey them. We can argue about what exactly the differences are between a mathematical or metaphysical law and a physical theory, even though there appears to be more similarity than dissimilarity between them. If they are all mere fictions, however, arguing about their distinguishing details seems to be irrelevant. Therefore, I argue that they all must have ontological priority over the physical universe, which is the only option capable of making sense of the success of physics.

If phenomena appear to violate some physical law, L_1, there must be a deeper, more abstract and pervasive law, L_2, that accommodates L_1 and explains the behavior of the phenomena that L_1 could not. For example, classical physics is not wrong, as is sometimes falsely assumed; it is just that quantum physics applies to the subatomic realm beyond the limit of applicability of classical physics. The nature of the sciences is to seek theories that are deeper and more unifying and so quantum theory, too, must eventually be superseded. Based on what is already known, we will be able to innovate for quite some time yet, but sooner or later our innovative capacity will run relatively dry until we are able to advance far beyond what we can presently even imagine. Even if we can develop subareas of physics that appear to have nothing to do with quantum or classical physics, we would not be able to maximize our understanding or potential for technological applications unless we first were able to discover the law that reveals their underlying unity.

The common antirealist challenge that the laws of physics are mere fictions can be expanded to say that, therefore, there can in principle be an infinite number of equally good laws that accomplish the same job of accurately predicting the exact same observations. This problem can be made clear through the well-known duck/rabbit illusion, whereby the picture looks like a duck from one perspective and a rabbit from another.

Thomas Kuhn, a physicist who had become a historian and philosopher of science, argued that 'what were ducks in the scientist's world before the revolution are rabbits afterwards'.[578] The antirealist claims that because there are two equally valid ways of perceiving the duck/rabbit picture, there cannot be one correct way of seeing it. Therefore there is *no* correct way, and realism fails. This argument is also the essence of the challenge of underdetermination, and my reply applies equally to both. The (apparent) problem of *underdetermination* results from recognizing two or more competing theories that appear to explain the phenomena equally well, but where each theory contains different or even incompatible assumptions, or postulates different entities or processes.

My reply to Kuhn and those who believe that underdetermination overthrows realism is quite simple, but sufficient.[579] Just because two or more theories may appear to explain observations equally well, it does not follow that therefore no theory can ever be true. It also does not follow that there is no truth about anything whatsoever. The duck/rabbit picture is not the *Mona Lisa*. There are objective limitations on what the lines of this painting can literally represent. For anyone who still disagrees, they can imagine driving their car through a stop sign and smashing into another vehicle. Later, in court, they may tell the judge that they are not guilty because, in *their* reality, the stop sign looked exactly like a purple alien signaling for them to drive forward as fast as possible. After all, the stop sign would have an infinite number of equally true representations, so I am sure that any judge would understand and find you not guilty.

Concerning underdetermination specifically, just because two or more theories may explain a particular set of observations with equal accuracy, it does not follow that *any* theory whatsoever, no matter how ridiculous, would be equally true. Therefore, there must be *real* restraints on what could possibly count as a plausible theory. These real restraints are imposed upon us by the way things actually are in (physical and nonphysical aspects of) reality. Therefore, it is far more plausible to say that one theory or another more closely represents the way things actually are, instead of agreeing with the antirealist that no theory matches anything because there is no reality about anything, or that any theory whatsoever is as equally valid or applicable as any other to explain the same set of observations. If there are no true theories, then any theory should work just as well as any other, and any criteria or lack of criteria should provide just as good a foundation for any ridiculous theory whatsoever. This, however, is not how science works, and that is because there are real external constraints placed upon us by reality.[580]

The big bang computer (or, nature is one smart cookie)

The laws of physics in our textbooks are, in the sense in which Einstein meant it, free inventions of thought.[581] This may sound like I am playing straight into the hands of the antirealists, but I am doing no such thing. Saying that the laws of physics are free inventions of thought is very different than saying that we actually invent these laws that we then impose upon physical reality. Physicists, like other artists, need to be creative. However, to the degree that such invented laws actually match a reflection of the eternal mathematical law, the physicists have hit upon the truth, and their apparent inventions are seen for what they really are—they are *discoveries*. Our minds must be free to make discoveries, which we can subsequently implement in creative ways, although even this stage of creativity is only possible so long as our creations are possible in reality.

Tim Addey, a specialist on the Platonic tradition, makes the same basic point in a slightly different way: 'the first action of the human is *discovery*, not *creativity*. Once the mind discovers principles it then naturally starts to use them in the subsequent creative process – but it must be emphasised that we cannot be truly creative unless we have contacted some *already existing* abstraction'.[582] We need to find creative ways to make an actual discovery; upon making this discovery, we can then employ it creatively. For example, after we discover some actual chemical property, we can then create a new medicine that did not physically exist before. Yet we need to be creative in order to discover this law, as no one before us had been able to figure it out, and we must then utilize a different sort of creative expression to figure out what to do with this discovery (hopefully in a way that benefits rather than harms).

We are free to create in myriad new ways from already real possibilities and already existing things, but our creations are constrained by the laws of physics.

EIGHT: The Laws of Nature and the Nature of its Laws

The degree to which our creative impulse reflects reality is the degree to which we have discovered something true, and the degree to which our discovery is closer to the ultimate cause of all reality is the degree to which we are attaining deeper mystical apprehension. To the degree that a creative insight matches, corresponds to, or approximates a genuine feature of physical or abstract reality, we have made a genuine discovery. Our creativity is also constrained by the lawful structure of the universe and, contrary to the popular idea in our culture that we should have no limits in anything, the fact is that the limiting power of the laws of physics is what actually allows for and makes possible our creativity. If there were no such ordered laws, nothing could ever come into existence in the first place.

If a law of physics was only applicable in one particular instance involving, say, two highly specified variables at some particular moment, then that law would never be applicable to any other phenomena at any other time. If all the laws of physics were like this, we would need a different law of physics for every single moment and every single event or experimental arrangement. Physics would never be able to develop in the way that it has if we were not able to discover more general, unifying laws that apply to a wide variety of phenomena under varying conditions. In other words, the more universal the law, the more powerful it is, and yet the more universal, the less the law is tied to any particular physical phenomena. The more fundamental, powerful, and unifying a law of physics, the more it is detached from physical reality and the closer it resembles the one eternal mathematical law. Even if the universe ceased to exist, this eternal law would remain. The following basic diagram should be helpful in visualizing what I have here been explaining.

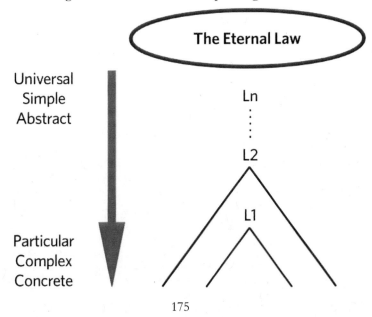

Whatever is universal, simple, and abstract (ideal, intangible, and eternal, rather than a mere abstract image in a particular mind) is metaphysically prior to what is particular, complex, and concrete. However, higher metaphysical principles are beyond and metaphysically prior to the lawful structure. The more powerful the law of physics, the more simple, abstract, and universal it will be, pulling it away from the particular, complex, and concrete, which means it is closer to the one eternal mathematical law. We may have two laws of physics that seem to be incompatible, but that does not make the universe irrational or imply that these laws are fictions. Rather, it means that we have yet to discover how these two different laws are unified. When we do discover the law that clarifies the underlying unity between them, we will have discovered a more powerful law that explains what the previous two laws explained and more. It will be more universal, and so will also be simpler and more abstract, and thereby more closely approach the eternal mathematical law.

According to many astronomers, within 10^{-43} seconds of the Big Bang all the fundamental forces of nature—gravity, electromagnetism, the strong force, and the weak force—were indistinguishable. As noted in Chapter One, an astronomy textbook states that 'the four forces are said to have been *unified* at this early time—there was, in effect, only one force of nature'. Despite the obvious Platonic language, the authors, strictly in their role as astronomers, do not really understand what they mean by this at the deeper level, for they cannot explain how the forces were unified, nor how they (apparently) separated. They also are not yet equipped to deal with the pressing issue of how the one unified force came to be what it is and what its relationship is to the physical universe. They cannot explain the ontological status of this one unified law, or of the subsequent laws, and they certainly cannot explain the very notion of unity itself, which, in the end, is an explanation that is required to make sense of how anything could be unified in the first place. These remarks are not meant as criticisms of the physicists and astronomers, because their goals are generally not aimed at clarifying the metaphysical foundations of their discipline. Conversely, my goal is not to chart the stars.

One could try to argue that the laws of physics are not constant but instead evolve or change over time, even if these changes could only be perceived by a creature existing over the course of a billion years. Somewhat surprisingly, Smolin tries to make this sort of argument by claiming that the laws of physics evolve according to 'cosmological natural selection'. He briefly notes the 'obvious' objection that 'even if a law turns out to evolve in time, there is always a deeper, unchanging law behind that evolution'.[583] Even more surprisingly, however, he does not adequately respond to this objection, and it is with good reason that he avoids doing so, for it would be devastating to his argument.

From the outset, it should be noted that his desire to subject the laws of physics to biological evolution has the whole process backwards, and is ruled out by the assumptions of modern biophysical chemistry.[584] While in some contexts it may be possible to find a way of extending natural selection beyond the biological realm, nevertheless, physics is more fundamental than biology. Claiming that the laws of physics follow biology is akin to the proverbial tail wagging the dog.

Smolin might have been able to avoid his confused argument about the changing laws of physics if he had first read Charles Babbage, the nineteenth-century polymath who is generally considered to have been the 'father of the computer'.[585] Babbage noted that the apparent changes in the order of the sequence of numbers being generated by his Calculating Engine were due to deeper laws implicit from the commencement of the sequence. 'The first engine', Babbage writes, 'must be susceptible of having embodied in its mechanical structure, that more general law of which all the observed laws were but isolated portions'.[586] In other words, a calculating engine may produce ten million consecutive numbers in a series that follows a precise law, a law that we have been able to discover and that explains the output we have seen so far.

With this understanding, we are then able to make precise predictions about which numbers will come up next in the sequence. However, after ten million and one outputs, the law suddenly seems to change, and a new sequence of numbers unfolds. We would be surprised at this new output that contravened our hitherto known law, especially given the inductive confirmation of ten million cases displaying the same order and apparently following the same law. We would be very wrong, however, to conclude that the law had actually changed, for what we took to be the original law was only an approximation, a limiting case, of the more general law.

The calculating engine is not suddenly going to start spitting out numbers at random and forget its initial program, the program that it has no choice but to obey. The original law includes within itself, right from the beginning, the apparent change in sequence at the ten million and first case, and so the law that we had originally attributed to the sequence of numbers was good enough for ten million cases, which means that it was a good approximation. This approximate law was only a partial aspect of the greater initial law that is overseeing the whole of the calculating process. We may discover further approximate laws, and may even in principle be forever incapable of grasping the initial law as it really is, but this initial law is not only real, it is even more real than all the approximate laws, which are really reflections of the one initial general law.

Babbage noted that we may never be able to discover completely this initial law, even though we can discover a series of laws, 'each simple in itself', that 'successively spring into existence, at distances almost too great for human conception'. This initial law, however, 'is itself of the simplest kind'. The apparent changes in the lesser laws 'are in fact only the necessary consequences of some far higher law'. This computing process, Babbage further notes, is analogous to 'several of the phenomena of nature'. For example, 'the laws of animal life which regulate the caterpillar, seem totally distinct from those which, in the subsequent stage of its existence, govern the butterfly'.[587]

Babbage is espousing the same sort of fundamental Platonic realist principles that I have been defending in relation to the mathematical laws of physics, which are necessary consequences of, or reflections of, the eternal mathematical law. This highest law does not change, and it contains within itself in potentiality all the necessary consequences that unfold from it.[588] With Babbage's example in mind, let us return to Smolin, who has only two options in this case: either the 'evolution' of the laws of physics is governed by some further law (or laws), or it is not. If it is not, then it is only by absolute chance that the laws evolve in the extraordinarily precise way that they do, which is no scientific explanation at all. More importantly, we also could never know when these laws would change again, nor could we explain why they do not change from moment to moment. It looks like Smolin will need to rely on perpetual miracles to keep the universe ordered in such a case.

Evolutionary theory itself depends upon the stability of the very laws of physics that Smolin says change according to evolutionary principles. To avoid such problems, there must be one higher law that is responsible for all of the apparent changes in the lower laws, a higher law that itself cannot change. There must be unifying order to account for the systematic harmony of the laws of physics. Every law of physics that we discover will end up having limited applicability, and must ultimately be superseded or embraced by some as yet unknown law, but that process must finally rest in one ultimately simple, abstract, universal law, which itself cannot change.

Let us suppose the sequence of numbers being output from a computing device to be the following: 1, 2, 3...50, 51, 52...$n_{\to\infty}$ (where '$n_{\to\infty}$' means 'continue to infinity'). For this sequence of numbers, we are going to say that the obvious law is to 'begin with 1, then add 1 continuously to each resulting number'. However, there is no logically necessary reason to make this assumption about the law that governs the sequence. I could just as easily say that the law is the following: begin with 0, then add 10,000,000, then subtract 10,000,001, then add 1, then add 6, then subtract 5, then repeat this sequence indefinitely to each resulting number that is produced. This is what the application of this law would look like:

$0 + 10{,}000{,}000 - 10{,}000{,}001 + 1 + 6 - 5 = 1$

$1 + 10{,}000{,}000 - 10{,}000{,}001 + 1 + 6 - 5 = 2$

$2 + 10{,}000{,}000 - 10{,}000{,}001 + 1 + 6 - 5 = 3$

With this law, I can generate the number sequence of 1, 2, 3...50, 51, 52...$n_{\to\infty}$, but it is far more complicated than simply adding 1 continuously. This more complicated law, however, is just a sample. We can in principle generate an infinite number of possible laws that perform the same function as adding 1 continuously. Some of these laws could be so complex that they take an enormous amount of time to generate each succeeding number. In fact, given that we have infinite numbers and infinite possible arrangements of those numbers, we could begin to formulate a law so complex that it would take an infinite amount of time just to write it, let alone begin to execute it.

The key point here is that there is no logically necessary reason to prefer the simpler law to a more complex one, so long as they both generate the same number sequence. We may say that it would be a waste of time to concoct such complicated laws just to accomplish what can easily be done by saying, 'add 1 to the previous number' and, indeed, it would be a waste of time. But saving time is not a sufficient logical reason to prefer the simpler law.

Newton, however, had his reasons for saying that 'nature is pleased with simplicity'.[589] He writes that 'truth is ever to be found in simplicity, and not in the multiplicity and confusion of things....It is the perfection of all God's works that they are done with the greatest simplicity'.[590] Einstein's pragmatic reasoning leads us to the same conclusion: 'our experience hitherto justifies us in believing that nature is the realisation of the simplest conceivable mathematical ideas'.[591] The fundamental laws that enable us to understand and predict the behavior of physical phenomena become simpler and more universal as they approach the one ultimate law.

Heisenberg makes a similar point: 'Nature, in the last resort, is uniformly ordered, [and] all phenomena ultimately take place according to nature's unitary laws. So ultimately it had to be possible to discover the common underlying structure within the different branches of physics'.[592] Even though Heisenberg and Einstein had various disagreements about the philosophical interpretation of quantum theory, they shared the same underlying metaphysical assumptions about unity, simplicity, and the reality of the objective laws of nature, all of which are essential aspects of Platonism. Such assumptions were at the foundation of classical physics, and they must also underpin any future physics.

However you slice it (it's still one pie)

It is true that the ideal, abstract, unifying laws of physics are more real than the physical realm. But it is wrong to assume that, therefore, physics can tell us nothing about physical reality. We must step down, so to speak, from the dizzying intellectual heights of the abstract and discover the more detailed aspects of the particular phenomena that are derivative of the abstract laws. Electrons, or something very much resembling them, are real. But an electron, like everything else that is physical, is a reflection of the higher abstract level of nonphysical reality, specifically the eternal mathematical law. However, the ideal realm of the eternal law, which is the foundation of the physical universe, is not the highest metaphysical realm. Before venturing into this deeper metaphysical territory in the final chapter, we must first consider an important alternative way of thinking about the ontological status of the laws of physics.

It is relatively easy to assume that successful laws of physics are successful precisely because they accurately represent some aspect of reality as it really is. For example, $E=mc^2$ may not be true in all situations, but within its limited domain of applicability this equation works because it corresponds to the relevant aspects of reality. Although this realist claim may be true, realists must also admit that we may be wrong. Therefore, we shall consider how this realist claim could be wrong while realism remains true.

It could be the case that there are an indefinite number of ways of expressing the laws of physics, where each possible way depends upon such things as our perceptual and cognitive abilities, and even our goals and personal creative capacities. Therefore, alien life forms in distant regions of the universe may have discovered what appear to be wholly different laws of physics than what we presently have, and their part of the universe may even operate according to laws of physics that appear to be incompatible with our own. In that case, it would be an objective fact that different regions of space really do operate according to a different set of physical laws, but let us consider a more interesting reason for why Platonic realism is still presupposed even if there are different laws of physics operating in different regions of the universe.

The universe as a whole is one vast system, where all of its parts are integrated in some sort of systematic, ordered way. Even if there are infinite dimensions, or infinite so-called universes, they are all still in some sort of unity, which makes the entire system one. If the whole universe is one ordered system, then even if different parts of it have different functions, as my legs have a different function than my brain, they would still be systematically related. The more closely we investigate the details of my legs, the less they seem to have in common with my brain. But the more holistically we consider my legs and my brain, the more easily we can see how they are interrelated and operating according to the same underlying biological, chemical, and physical laws. There is, as Pauli said, a cosmic order.

Even if creatures from two different species in vastly spanned regions of the universe were to discover laws of physics that appear to have nothing in common, those laws would still be expressible in some sort of mathematical structure and would be unified by some underlying law. In a similar way, if two human scientists offer completely different theories that explain and predict the behavior of the same phenomena with equal precision, then one of two possibilities would apply: one of the theories is mathematically simpler and is thus a more accurate reflection of the actual law, or they are mathematically equivalent even if this is not yet known to be the case. (Recall how Heisenberg's matrix mechanics and Schrödinger's wave mechanics were eventually shown to be mathematically equivalent.) Let us consider an analogy using the division of a circle in order to help us understand this point.

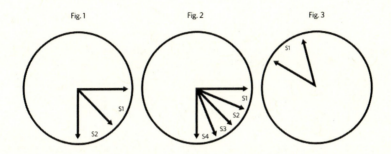

Fig. 1 Fig. 2 Fig. 3

In Figure 1 and Figure 2, we have divided the circle into a one-quarter space and a three-quarter space. Within the one-quarter space, we have made one further division in Figure 1, and three further divisions in Figure 2. (It should be noted that we could have made an indefinite number of such subdivisions in each figure.) Imagine there are different physical attributes associated with each section of the circle, and that all three figures represent the exact same circle. Now imagine that you are living inside Figure 1 (S1) and that I am living inside Figure 2 (S1). Some of what we experience will be nearly identical, since we are the same kinds of creatures and since some of the physical attributes of each of our respective sections will be identical. However, some of what you experience in your space is beyond my boundaries of possible experience, since your section is larger and will include some physical attributes that are different from my section. Therefore, if you had a mathematical law that explained all that happens in your space, it would look different than a mathematical law that explained all that happens in my space. Nevertheless, my law must ultimately be derivable from your law, because your law would be more general and would embrace all that my law could explain and more.

Now imagine that an alien (let's call him Joe) lives in Figure 3 (S1). Joe would have a very different experience than us, since his section of space is in a completely different region of the circle and would therefore have totally different physical attributes associated with it. As a result, the law that he discovers to explain all that happens in his space may look entirely different than the laws that you or I discover for our respective spaces. Nevertheless, no matter how many subspaces we may arbitrarily create within any circle, each one will be mathematically interconnected with every other subspace. For example, underlying the laws of any particular subspace are more general laws that apply to the circle as a whole, such as $A = \pi r^2$ (where 'A' equals the area of a circle, 'r' is the radius, and 'π' is the numerical value of the ratio of the circumference of a circle to its diameter, or approximately 3.1415). This general equation for the area of a circle holds true no matter how many times and in how many different ways we may create subspaces within the circle. Any and all possible subspaces of the circle partake in, and are only possible because of, these more general laws.

In our example, we have been assuming that in each subspace there are living creatures and ecosystems and so forth. As we move our concerns away from such physical differences and toward the underlying mathematical laws that underpin and make possible the variety of physical phenomena, we then discover the underlying commonality between each subspace. Although Joe is, in a sense, on the other side of the universe, and given the constraints imposed upon his subspace by the walls dividing his subspace from the rest of the universe of the circle, the law he discovers will fall under one of the following three categories when compared with the law of any other particular subspace: either it will be equivalent, or it will be more general and so embrace the other law, or it will be more partial and so be embraced by the other law. In all cases, and regardless of whichever symbols each law uses in each subspace, they will either be mathematically equivalent, or they will be mathematically related in such a way that they could be derived from a more general law or one will be derivable from the other.

This metaphor is a way of understanding how there could be completely different laws of physics operating in different regions of the universe, where every possible law in every possible region of the universe must ultimately be mathematically interrelated and derivable from more general laws. Since in the end we must arrive at the most simple, universal, and unifying mathematical law that is the foundation for all the laws and for the entire physical universe, this process ultimately arrives at the one eternal mathematical law. By understanding this metaphor, we can allow for the fact that the laws of physics are, in the sense conceived of by Einstein, products of the free invention of thought, and yet they correspond to their subspaces of reality throughout the universe, as well as cohering with all the other laws of the other subspaces throughout the universe.

The different subspaces, however, are not different realties. They are only different aspects or parts of physical reality, just as there are different aspects or parts of one individual human. What is most real is the underlying mathematical structure, because this is where we find objective stability, whereas the physical aspects of reality, whose physical motions must obey this underlying mathematical structure, are constantly in motion. Physical things are objective in the sense that they exist whether or not we perceive them, but they do not have absolute stability because they are in a state of continual change of one sort or another.

EIGHT: The Laws of Nature and the Nature of its Laws

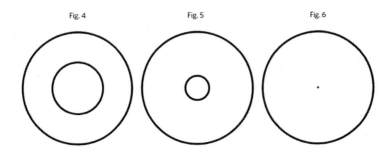

Fig. 4   Fig. 5   Fig. 6

Let us now extend this analogy to consider the absolute center of a circle. Without a center, there obviously would be no circle. A circle also needs a boundary, which we call a circumference, and the circumference, 'C', of any circle is 'π' multiplied by the diameter, 'd', giving us $C=\pi d$. The diameter is a line that cuts the circle in half by going from one edge of the outer boundary to another and passing through the center. Therefore, if there were no center, we could not have a diameter or a circumference, and so we could not have a circle. What we want to know is what, exactly, is the *absolute* center of a circle? Any line that we may draw through the center will have some thickness, and so the center part of the circle will also have a thickness that matches this line. We can always make the line of the diameter thinner and, likewise, we can always make the central part of the circle smaller, as shown in Figures 4, 5, and 6.

The point in the center, however, approaches zero as it gets infinitely smaller. But we must remember that we are seeking the *absolute* center, and not a relatively satisfactory approximation of the absolute center. The absolute center would be that which could no longer mathematically become any smaller, which is to become zero. Zero is that which is approached by the infinitely small, and so zero is the absolute foundation of that which is approaching it. As long as there is still the possibility of getting smaller and smaller as we approach zero, we are still only attaining an increasingly accurate, but still *relative*, center of the circle. We can only attain the absolute center once we have jumped out of the endless series of the infinitesimal and directly into zero.

But zero is nothing; it is no-thing. It is not at all physical. If it were physical, then it could be divided, and then it would not be absolute. If the center is nothing, however, then there is no physical center, and if there is no physical center, then there is no physical circumference, and so there would be no physical boundary. So long as there is no boundary, there is no limit, and with no limiting boundary, which is the circumference, there could be no manifested circle. The absolute center of the circle, in this analogy, would be eternal and nonphysical, and yet it is the foundation of the physically manifested circle. This analogy can help us to understand how there can be one eternal mathematical law at the foundation of physical reality, and yet this law is not itself physical and can never change. It is the foundation of the finite, physical, and changing. It is eternal, just like the absolute center of a circle.

By employing this analogy, which could be extended and developed much further, we have another way of accommodating the various philosophical positions with respect to the laws of physics. We can accommodate correspondence and coherence theories, as well as the problems associated with underdetermination. We can even make antirealists happy to a certain extent by admitting that the laws of physics as we conceive of them could actually have been devised in an indefinite number of ways, and yet they are still real reflections of the ultimately singular underlying law that underpins all of physical reality.

Admittedly, it may be much easier just to accept that whatever physical laws we discover are true representations of limited aspects of reality. However, this alternative perspective can more easily accommodate its rivals while remaining forever realist. (Appendix D contains a speculative rebuttal that could be offered by antirealists. Nevertheless, even under the most extreme case of technological possibility, antirealism still fails.)

NINE: Unity and Ultimate Reality

Throughout this final chapter, we shall explore the following key points:

1. The disunity stance fails to provide plausible foundations for the success and possibility of physics.

2. Physicists seek abstract, simple, unifying laws, which presupposes the one eternal mathematical law. This in turn presupposes higher metaphysical principles, such as symmetry, beauty, and absolute unity or the One, the ultimate nature of reality.

3. The wholeness presupposed by quantum physics, where the observer and the observed (or the measurer and the measured) cannot logically or metaphysically be separated, presupposes the fundamental role of the observer, which was also implied in classical physics.

4. The fact that we can have knowledge of both nonphysical and physical aspects of reality implies that something about our nature is similar to both realms, allowing us to take seriously the notion of soul.

Disunity and disengagement from reality

Philosopher Steve Clarke (not to be confused with Stephen R. L. Clark) makes a strikingly false comment about the lack of unity in the world: 'for a long time scientists have shown us how to make do without the presumption that the world is unified. It is time for philosophers to follow their lead'.[593] We must immediately ask Clarke: exactly w*hich* scientists have shown us how to get by without unity? Certainly not Einstein, Heisenberg, Newton, Planck, or any of the numerous other trailblazers of modern physics. As Clarke is a follower of Cartwright, it will be helpful to give a brief consideration of her views that are most directly related to the assumption of disunity in the sciences.[594]

Cartwright has misunderstood the power of the Platonic dialectic, where we must seek the similarities between different things, as well as the differences between similar things.[595] She focuses only on half of the dialectic, as her *disunification* metaphysics amounts to saying nothing more than that different things are different. I can understand Cartwright's position, in the sense that we can easily see differences everywhere we look. The problem is that she cannot see the other half of reality. She cannot apprehend the underlying unity of the multifarious phenomena. Perhaps Cartwright would benefit from an experience similar to the one Heisenberg had while listening to the violinist on the rooftop.

If there is no unity, then all things are not only different; they are *completely* different. In this case, there would be no way to do science, as scientific explanation presupposes that there is some sort of unity between phenomena, between theories, and between theories and phenomena. But in order to explain this unity, we must first do a lot of metaphysical work, which Cartwright apparently is unwilling to do, as she claims that there is no need to explain how scientific theories are successful: 'theories are successful where they are successful, and that's that'.[596] This is certainly not a very good scientific or philosophical explanation. It is, however, a great blueprint for intellectual laziness.

If Cartwright is correct, then there could be no unity whatsoever between anything. She could not even communicate her denial of the reality of unity, because the very act of communication requires some sort of common understanding, which requires some sort of unity. She would also have a very difficult time trying to explain her own body's existence, which is the most immediately obvious example of the falsity of disunification. If her arms and legs and every other part of her body were not bound together in a unity, into one particular body, then she would not have even been born. The universe itself is the greatest example of an infinite number of different phenomena all bound together into a unified whole making up the entire physical cosmos. To deny the reality of unity is to deny the possibility of science, the universe, and all of reality.

Cartwright asks, 'how unified is our knowledge? Look at any catalogue for a science or engineering school. The curriculum is divided into tiny, separate subjects that irk the interdisciplinist. Our knowledge of nature, nature as we best see it, is highly compartmentalized. Why think nature itself is unified?'[597] We should follow Cartwright's advice and look not only at a science or engineering catalogue but, more importantly, at some of the textbooks that are used to teach these various disciplines. Here is a title that could never exist if Cartwright is correct about rejecting unity in the sciences: *The Physical Basis of Biochemistry: The Foundations of Molecular Biophysics*.[598] The combination of biology, chemistry, and physics in this title entails a very profound unity between these scientific disciplines.

In *Principles of Modern Chemistry*, the authors do not shy away from saying that chemistry has 'been profoundly affected by the new concepts [in quantum theory], to such an extent that physics and chemistry can no longer be considered separate disciplines'.[599] Even the very technical study of the mechanics of materials requires one to first take the prerequisite courses in 'mechanics, physics, and mathematics together with the basic concepts of the theory of elasticity and the properties of engineering materials'.[600] A degree in computer science offers another good example, where typical requisite courses include linear algebra, calculus, and discrete mathematics, while a textbook for discrete mathematics lays the foundation for the course by beginning with truth tables, which are a vital part of logic, itself a branch of philosophy.[601]

Finally, let us consider a basic example from the specialized field of cellular biology. The electron microscope has 'triggered a revolution in the exploration of cell structure and function, and ultimately, in the way we think about cells'.[602] But physics has not just helped biologists by providing new exploratory technologies; it is also embedded in essential aspects of the discipline, such as understanding 'how light energy is converted to chemical energy within a chloroplast or bacterium'.[603] If you are going to explain how light interacts with a chloroplast, you will first need to know something about photons, which leads straight into physics. It is true that each discipline has numerous idiosyncrasies, specializations, and niches, as well as different methodologies and goals, but it is equally true that there is significant underlying unity within and across the disciplines. How Cartwright could possibly refuse to see this most obvious fact is quite a mystery.

Clarke claims that Cartwright 'is concerned to describe the process of scientific explanation as it is practised rather than as it might ideally become'.[604] However, it is not clear *which* scientific practice he is referring to here. Does he mean the practice of Bohr's fanatical behavior or the metaphysical debates he had with Einstein; Fröhlich's experience of leaving his body and becoming the particle; Planck's observation that we cannot prove, empirically or logically, something as apparently obvious as the existence of an external world, and that we must have 'Faith' that it really does exist?

The arguable fact that the majority of scientists practice highly technical work in a laboratory setting does not at all negate or devalue the theoretical and pioneering aspects of science. We should have extraordinary respect for experimentalists; I myself have spent considerable time learning from such specialists in various disciplines, especially in physics, but also in biology and computer science. But we also cannot ignore the ways in which theoretical pioneering physicists approach their own research. Perhaps Cartwright could benefit from shadowing Rowlands or Penrose in their work for a few days. If nothing else, she would doubtless discover that the ways in which they practice science have nothing to do with her opinions about the discipline.

Clarke and Cartwright confuse the necessity of an underlying and pervasive unity throughout the universe with our over-eagerness to assume that we have found exactly what that unity is within our limited scientific theories. It is true that we must remain open about what the underlying order is, or how best to express it, but we cannot *deny* order or unity without immediately running into self-contradiction. By denying the importance of unity, Cartwright and Clarke have offered a very distorted view of physics and the sciences, leaving them very much out of touch with reality.[605]

Systematic unity

While I do not expect unanimous agreement any time soon about what exactly constitutes Platonism, Gerson has recently provided a thorough evaluation of the relevant texts and has offered persuasive reasons for his claims concerning the essence of this tradition. I will not repeat his arguments here, but will consider in detail two of the seven essential aspects of Platonism, which are based upon the interpretation of Aristotle and Plato by the Neoplatonists, whom Gerson believes were essentially correct.

Platonism must actually predate Plato because it points to the eternal, which is timeless and therefore always is what it is at any point in time before or after Plato.[606] In more human terms, Socrates, Parmenides, and the Pythagoreans all had a strong impact upon Plato, and although the Neoplatonists believed that Plato stated this philosophical tradition in one of the most sublime ways, the writings of Plotinus and Proclus, among many others in this and other spiritual/philosophical traditions, equal Plato's genius in their own unique ways.

The first two essential aspects that are foundational to all of Platonism are that:

1. the universe has a systematic unity; and,
2. the systematic unity is an explanatory hierarchy.[607]

Cosmic unity (or cosmic order) must bind together everything in the universe. The explanatory hierarchy says that what is simple is ontologically or conceptually prior to what is complex, which is connected to the other notion that what is intelligible is prior to what is sensible. In other words, the laws of physics are simple in comparison to the complexity of the physical phenomena that must obey them. These laws are accessible to us in the mind only, as they do not have any tangible reality, which is to say that they must be intelligible. Therefore, since the laws must have conceptual priority over the physical universe that obeys these laws, the intelligible must have conceptual priority over the physical universe that we perceive with our senses. Consequently, if we want to understand the complexity of the physical universe that we perceive with our senses, we must understand some relevant portion of the cosmic unity that operates behind the veil of everyday appearances. One way to understand a portion of this cosmic order is to understand the simple, intelligible laws that the sensible, complex universe must obey.

These mathematical laws are simple, not in the sense that they are easy for us to comprehend, but in that they are unchanging and unifying, and are described relatively succinctly. The greater the number of laws that are required by us in order to understand some phenomenon, the farther away we are from having a deeper, more powerful understanding of it. To put it another way, the greater our understanding of some phenomenon, the fewer laws we need to explain it. Our understanding increases when we discover simpler, more unifying laws. As a result, the ultimate explanation must reside in the simplest of laws. Scientifically speaking, the simplest law would be the one eternal mathematical law. Yet this law would itself require further explanation, which is to enter into pure metaphysical territory. At the end of this investigation, we must ultimately arrive at absolute simplicity and unity, which would be the One, or God.

It does not make any difference to my arguments if a historian or classicist disagrees with Gerson's exposition of these essential or foundational aspects of Platonism. I would still argue for the eternal truth of these metaphysical principles regardless of whether or not there is any historical precedence and, in any case, I could just stipulate that I am assuming such metaphysical principles to be the foundation of Platonic realism. In other words, if one were to argue, or somehow or other even prove, that these fundamental principles have nothing to do with Platonism, I would drop the term 'Platonism' and keep these principles. I would have no choice, for these principles are accurate representations of reality and are essential aspects of the metaphysical foundation of physics. Nevertheless, it is significant that Gerson has argued compellingly for such a historical understanding of Platonism. (We should remember that we could drop the term 'Platonism' at any time, for it is only a label used for convenience.)

If there were no such thing as unity, physics would not be possible. Nothing at all would be able to exist, as Proclus argued in the first proposition of his *Elements of Theology*.[608] If something does not partake in unity, it cannot be unified and so can never become a whole, a 'something'. Only by being unified, by becoming a 'one something'—whether an electron, an apple, a universe, or a mind—can anything ever exist or be. It is not possible to object to this reasoning, because any objection presupposes that there is something unified in one's objection; otherwise, it would not be an intelligible response. In other words, if an objection has no unity, no rational way of bringing different words, concepts, and thoughts into a coherent whole, then this objection cannot possibly make sense. Indeed, a person could not even exist in the first place unless all of their different parts were bound together into a unity.

Physicist John C. Taylor has no hesitation in proclaiming the fundamental importance of unity in physics. An important example concerns the hidden unity between electricity and magnetism which, although different, are interconnected to such an extent that they are considered 'two aspects of a unified whole' known as electromagnetism. Taylor continues: 'this pattern of unification is fairly typical. Every time such a unification is achieved, the number of 'laws of nature' is reduced, so that nature looks not only more unified but also, in some sense, simpler. More and more apparently diverse phenomena are explained by fewer and fewer underlying principles'.[609] This description further clarifies how physics presupposes the metaphysical assumptions of Platonism, and how thoroughly out of touch with physics Cartwright and Clarke are with their talk of disunity.

Physicists must continually seek to uncover such hidden unities, and they have always made deep progress when they discover how seemingly disparate phenomena are interconnected and unified. Physicists seek to understand the laws of nature, which provide greater explanatory and predictive power the more simplified and unified they are, which is to say the more hierarchically prior they are to the complexity of physical diversity. The ultimate goal, therefore, is to seek the simplest and most unified mathematical law. This eternal mathematical law, however, is not the simplest metaphysical principle. The logical corollary is thus to seek the absolute simplest and most unified principle of all reality, which is responsible even for the eternal mathematical law. This principle is the absolute One.

The ancient Platonic philosophers emphasized the discovery of metaphysical unity underlying all of physical reality with the aim of personal transformation in our quest to know ourselves and 'become like God' so far as possible. But physicists, in their capacity as physicists, are generally more concerned with discovering and understanding the physical manifestation of this underlying unity expressed according to unifying mathematical laws, while they are less apt than the Platonic philosophers to seek the metaphysical unity that makes the physical laws and the physical universe possible in the first place. But the Platonists also encouraged and provided the metaphysical foundation for the development of science, and many physicists have been keenly interested in metaphysics and mysticism.

Recall Heisenberg's point that 'the search for the "one," for the ultimate source of all understanding, has doubtless played a similar role in the origin of both religion and science'. It is not that Platonism per se is identical with modern physics, but rather that modern physics rests or relies upon the essential Platonic assumptions of hierarchical unity. It is in the Platonic tradition that we find the most detailed and rational defense of these foundational beliefs, which are usually not given a proper defense by physicists but are instead simply assumed to be true, often unconsciously or without any penetrating reflection.

What is most interesting with respect to Taylor's work is that he has outlined some of the essential aspects of Platonic metaphysics purely through studying the ways in which physics has developed. Norton, too, shows how Einstein's 'confidence in the supreme heuristic power of mathematical simplicity' had little to do with philosophy and, instead, everything to do with 'the practical conclusion of a physicist who sought to adopt quite pragmatically whatever approach would lead to successful physics'.[610] In other words, Einstein was looking to adopt the approach to physics that worked the best, which just so happened to be the approach that assumed Platonic realism at its foundation.

The reason this approach 'works' is because it is the way reality is, and this approach must continue to be the guiding principle in future research, *especially* as the discipline becomes increasingly technical.

Harris also notes that Einstein and his colleague Leopold Infeld believed that the simpler our picture of the world, 'the more facts it embraces and the more strongly it reflects the harmony of the universe. The criteria they recognize and seek are those of comprehensiveness and coherent unity'.[611] Planck agreed: 'the idea of potential is superior to that of force, partly because it simplifies the laws of physics, and also because the significance of the idea of potential has a far greater scope than that of force'.[612]

Planck's point about the simpler principle having greater scope, and thereby being superior to that which reaches out to fewer phenomena, was given a profound metaphysical defense by Proclus nearly fifteen centuries earlier: 'the cause of more numerous effects is therefore superior in its being to that which produces fewer'.[613] Moreover, since that which is the more powerful cause must have greater unity, it more closely resembles the absolute simplicity and unity of the One, the ultimate originator of all.[614] Another implication of these views is that there must be *degrees* of truth and reality. The closer a principle is to the One, the greater is its reality (or being) and truth.[615]

Rowlands provides a clear overview of what a unified theory would look like, where the mathematics of such an ultimate theory of physics, besides being abstract and simple, must also be derived from symmetry principles.[616] He also argues that there must be no model-dependent structures of any kind, which is a direct refutation of materialism, and that 'physics works because it has successfully, and uniquely, avoided characterizing nature'.[617] He is correct in an important sense. The deep power of physics is due to its abstract, universally applicable nature that, by virtue of being abstract and universal, cannot be physical. The simpler, more abstract, and more unifying a law is, the more powerful and fundamental it is, and the closer it is to the eternal mathematical law. Fundamental physics cannot characterize any particular aspect of nature; otherwise, it would be tied to particulars and lose its universal and unifying power. The foundation of physics, therefore, must remain abstract and simple.

However, this rejection of materialism can lead physicists to forget that a lack of characterization of nature is not sufficient to make physics work. For example, many theories may fail to characterize nature yet remain nothing other than silly fabrications, such as my new theory that A=BC. Not only does this theory not characterize nature, it also does not tell me anything about anything, and so it is not a theory that anybody should take seriously. Accurate predictions following from fundamental physical theories make it more likely that the theory is true or approximately true, and once the theory is applied in a particular case it does (to the degree that it is true) characterize a limited aspect of physical reality.

Nevertheless, Rowlands is correct that the more fundamental a theory of physics, the more universal, simple, abstract, and unifying it must be, which means that the theory itself can never be tied to any particular aspect of physical reality. As Rowlands writes, 'the truth is that simple facts are not concrete and concrete facts are not simple'.[618] In other words, a simple law of physics is not concrete, nor is it physical, and a concrete fact, such as a brick wall, is actually quite complex.

Harris states clearly what is required of philosophy if it is to be relevant to the sciences. It should be a 'metaphysic holistic in type, and a logic of order, system and hierarchical structure. A pluralism devoid of any overarching principle of unity would be entirely out of keeping with the scientific trends.'[619] Harris is correct, except for two important caveats. First, while it is true that the sciences imply a hierarchical structure that requires an overarching principle of unity, we must remember that this metaphysics has been recognized and developed for over two and a half millennia by the Platonists. Second, as previously mentioned, Harris sometimes seems to be arguing that this metaphysical view is only *implied* by the sciences instead of also being *presupposed* by them, which leaves open the possibility that further developments temporarily seeming to revert to a materialistic basis could then appear to overthrow this metaphysics. (I say 'temporarily seeming to revert' to materialism because in reality this is not possible, no matter how popular a materialistic interpretation may become in the future.) As I have been arguing, however, not only is it true that modern physics implies Platonic metaphysics but, more significantly, physics has had to *presuppose* its truth in order to make fundamental progress in its understanding and developments. Perhaps Harris would also agree. One thing is for certain; the disunity metaphysics of Clarke and Cartwright could not be further from the theoretical foundations of physics.

Soul in science and science in soul

It is quite astonishing that we have been able to pretend that we—the observers, experimenters, and theoreticians—can be excluded from the scientific enterprise, while simultaneously believing that we are being objective and giving as full an account as possible of whatever aspect of reality we are investigating. It is true, nonetheless, that we can still produce theoretical and practical feats while ignoring ourselves (or pretending to be able to do so), as if we had no role to play and were merely mindless automatons following some program (a program that is often too easily assumed to have no programmer). But as soon as we begin to analyze rationally what is really happening in any experiment, we cannot help but conclude that we are center stage in the entire scientific enterprise. To the degree that we ignore this fact, we are not being logically or rationally consistent.

As philosopher Jonathon Scott Lee notes, Plotinus also recognized this problem more than seventeen centuries ago. Even though Plotinus would insist that an ultimate explanation of things requires an appeal to soul and the intelligible realm, nonetheless, physics can generally get by without any explicit consideration of these higher metaphysical aspects of reality.[620] This is fine, so long as we do not falsely believe that the limit of scientific explanation is the limit of reality.

Developments in quantum theory, however, have made talk of the soul seem more plausible. In fact, as shall be made clear here, the central role of the observer in the scientific enterprise has always been implicit in physics. The sciences of the future, at least certain branches, must become more and more overtly conscious of the reality of soul, and develop in ways that approximate more closely the relationship between soul and physicality. Research in consciousness is a step in the right direction; however, soul is not reducible to functional consciousness.[621]

Is all this talk of soul complete and utter nonsense? Perhaps. But at least I am in good nonsensical company with Schrödinger, Eddington, and other physicists. Physicist Shimon Malin also writes that 'scientific evidence is based on human experiences; the human mind is the ultimate measuring apparatus. Yet the nature of the Subject of Cognizance is never raised as a scientific issue! This is like using a telescope to investigate the heavens and never bothering to inquire what it is'.[622] Malin is not completely correct in one strict sense, as some physicists have raised the role of the observer as a scientific issue. However, he is correct in general, as many scientists do dismiss this question as being anti-scientific and therefore meaningless.

But considering the role of the observer is neither meaningless nor anti-scientific. Harris notes that although scientists have stressed the importance of observation, 'no account was taken, or could be given, of the place of the observer in the scientific panorama. The observer was set apart from the scene that was being surveyed, a scene viewed from the outside, as it were, through a telescope or a microscope, or at least through the portals of the senses'.[623] But if empirical evidence is essential to the sciences, then observation is also essential, which places the observer—the one who perceives, interprets, and understands the empirical evidence—in the spotlight.

If we look at the stars through a telescope, we need to know how the instrument functions in order to be sure that we can rely upon it as a means for scientific discovery. Yet even though the physicist is the final 'instrument' for all experimentation and theorizing, we still generally believe that it is not necessary to know anything about that particular person, least of all their fundamental nature as a conscious being. This ignorance is tantamount to saying that we have no idea what a telescope is or how it works, and therefore we have no way of knowing whether or not it is even functional, yet we unquestioningly rely upon it to give us important empirical data. Scientists would not accept this situation with respect to a physical instrument, but they are happy to accept it with respect to conscious observers, such as themselves. Logically, this makes no sense whatsoever. This situation is especially odd since, as already noted, much of the sciences are explicitly about understanding ourselves, and yet we continue to exclude ourselves from the process.

Schrödinger offers an interesting perspective on why it is so difficult to pinpoint the observer in our scientific investigations. He writes that 'the reason why our sentient, percipient and thinking ego is met nowhere within our scientific world picture can easily be indicated in seven words: because it is itself that world picture. It is identical with the whole and therefore cannot be contained in it as a part of it'.[624] This view and his earlier noted remarks about consciousness being universal and singular cast aside the chains of materialist dogma.

It is impossible for any purely material body, such as a rock, to be self-cognizant, meaning it is impossible for it to reflect upon itself. Only nonphysical entities are capable of self-reversion or direct self-awareness. If a rock (or any other physical entity) can be aware of itself, there would necessarily be some nonphysical aspect involved. Direct self-awareness (where 'self', in this case, is referring to the immaterial soul) involves the self reverting back or reflecting upon itself and becoming directly aware of itself. In other words, the self moves away from its self while remaining its self, and then reverts back upon itself. A physical object is not capable of such a feat, and if a future machine is ever capable of genuine self-awareness in the sense being discussed here, there would have to be some nonphysical aspect involved.[625]

As Proclus argues, 'it is not in the nature, then, of any body to revert upon itself so that the whole is reverted upon the whole. Thus if there is anything which is capable of reverting upon itself, it is incorporeal and without parts'.[626] This act of the knower knowing the known, the self becoming the direct object of the knowledge by the self, entails that, as Proclus puts it, 'knower and known are here one, and its cognition has itself as object'.[627] Therefore, 'every soul is an incorporeal substance and separable from body'. Since 'the soul has knowledge of principles superior to itself, it is capable *a fortiori* of knowing itself, deriving self-knowledge from its knowledge of the causes prior to it'. Consequently, the soul is capable of reverting upon itself and of having direct knowledge of itself, which entails that it is incorporeal and must be separable from the body, since 'what is inseparable from body is incapable of reversion upon itself'.[628] In other words, if we can attain genuine self-awareness, that part of us that has attained such awareness would be separable from the physical body, a metaphysical view that is not only Platonic but also consistent with Buddhist teachings.

Despite the assumption of wholeness in quantum theory, which has often been presumed to unite the knower and the known, or the observer and the observed, we have not yet attained a sophisticated understanding of what this is supposed to mean. Bohr, for example, claimed that 'the unambiguous account of proper quantum phenomena must, in principle, include a description of all relevant features of the experimental arrangement',[629] which is the only way we can achieve an 'objective description'.[630] But it seems an impossible task to give a completely *unambiguous* description of *all* relevant features. And who decides what is relevant or ambiguous? If anything is relevant, surely the observer who discovers or postulates or uses a theory, designs the experiments, and reads and interprets the results is *most* relevant. Take away the experimenter and there would be no experiment.

But we do not need to evoke the holistic nature of quantum theory to prove this point about the importance of the observer. We need only look at the role of the observer in relativity and Newtonian physics, as well as the fact that it is a logical inconsistency to exclude ourselves from what we are studying, since we are the central aspect of the interconnected relations between all the relevant phenomena. As Harris states, relativity depends upon an act of thought, 'the choice of a frame of reference'. Since the motion of an instrument or other body cannot be determined absolutely, a conscious being must make a logically arbitrary choice that constructs boundaries for a frame of reference. Thought is necessary.[631]

However, Harris then remarks that, therefore, 'physical fact is and can only be what thinking makes it'. The act of thought does divide up our perceptions, whether in the moment of making a measurement in physics or even by simply looking at a tree. But my act of choosing a frame of reference in that particular instance, which is like taking a snapshot of the unbroken wholeness suggested by Bohm, does not *create* reality. It merely determines which theory-independent facts or objective physical truths will become accessible to me. Reality itself remains what it is, which is why experimental results are repeatable (at least to extraordinarily accurate approximate degrees).[632]

Moreover, as Rowlands clarifies, 'the mathematical laws of physics are general differential equations, which have to be reinterpreted ('solved', using different boundary conditions) every time a measurement is taken'.[633] It is inevitable that some conscious being is required to solve the equation (or create a program to do so) and to impose boundary conditions in order to accomplish any sort of practical knowledge or disclose various facts in physics. All that quantum physics has accomplished in this respect is to more dramatically emphasize the extraordinarily high level of abstraction, intricacy of subatomic experiments, and unifying essence of the relations between the experimenter and the experiment (or the observer and the observed). It has now become more obvious and much harder (even logically impossible) to deny the essential role of the conscious choices and awareness of the observer.[634] My choices do not create reality, but they are still an intimate part of the scientific process that provides a pathway to unveiling aspects of objective reality.

Bohr and Heisenberg, perhaps out of the fear of being accused of destroying the objective character of physics, tried to overlook the obvious implication that the consciousness of the observer must be taken into account, and tried to say that only measurements taken by some sort of appropriate instrument were necessary.[635] They are cheating here, in that they are simultaneously denying what they are requiring. First of all, a measurement is not necessarily the same thing as an observation, since I can observe my desk, for example, without physically measuring it. But if my act of observing the data has no effect whatsoever upon the data, we have hereby admitted that physical reality is what it is regardless of whether or not I, or any sentient creature, perceives it. This means that even if I do not know what the data are, the data will still be whatever they are, which is to say that my knowledge is not the limit of reality. This means that physical reality is objective. The whole antirealist game that has been built upon quantum theory emphatically denies this obvious conclusion.

The denial of objective reality leads to the conclusion that reference to the consciousness of an observer is essential, since according to this view reality depends upon the human mind. Bohr and Heisenberg denied this conclusion, as they were not antirealists. We already know that antirealism fails, but the relevant reasoning of Heisenberg and Bohr denying the importance of conscious awareness also fails. We therefore find ourselves in a situation where antirealism fails, while it also leads to a true conclusion regarding the important role of consciousness or the mind of the observer. However, while the reasoning of Bohr and Heisenberg fails, once it is corrected it also leads to the conclusion that the role of consciousness or the mind of the observer is of paramount importance. In other words, by being ruthless realists we cannot avoid admitting the importance of the human mind.

Let us return to Baggott's point that, according to the Copenhagen interpretation, 'it is not meaningful to regard a quantum particle as having *any* intrinsic properties independent of some measuring instrument'. If we take this assumption seriously, coupled with the assumption that quantum mechanical laws apply throughout the universe, then nothing could have intrinsic properties until measured. Consequently, a measuring device and the result that it has recorded while making a measurement would also have no intrinsic properties until a measurement is made of both the apparatus and the recorded results. This means that *we* are required to make the final measurement of both the apparatus and the resulting data. However, our way of measuring is not just to observe something; we have to know, recognize, or understand what it is that we are observing; otherwise, we do not have any knowledge of what the data is supposed to mean.

These acts of cognition would then also need to be measured by some other entity or device in order for these cognitive acts to become actual. Harris also writes that 'what the measuring instrument registers, however, has meaning as a value only when (and as) read by the investigator. Hence observer and observed are united in the experimental result, which, according to Eugene Wigner, occurs only with the observer's consciousness of it. Thus matter and mind can no longer be treated as separate'.[636]

Schrödinger makes the following apt remark, upon which Malin later builds: 'for our organs of sense, after all, are a kind of instrument'.[637] This is obviously true, for we cannot have any scientific knowledge, whether theoretical or experimental, unless there are conscious minds capable of attaining such knowledge. However, the antirealist falsely extrapolates from these observations to the conclusion that the human mind *creates* all of reality, which I have already argued against in detail.

The realist, however, cannot ignore the fact that we—the humans who discover physical laws, devise theories, develop instruments, design experiments, and read and interpret the resulting data—are one of the most relevant aspects of the entire scientific process. Nor is it logically or scientifically warranted to cast aside the role of consciousness. According to Bohm and physicist Geoffrey Chew, consciousness and matter are part of an undivided whole.[638] Baggott also reminds us that von Neumann, who provided 'an unassailable mathematical foundation' for quantum theory and an approach to the interpretation of quantum measurements that 'shaped virtually all subsequent thinking on the subject',[639] carried the Copenhagen interpretation to its logical conclusion by recognizing the necessity of the role of consciousness.[640]

Von Neumann's logical conclusion must be extended even further to recognize that some immaterial consciousness or mind is necessary for the existence of the physical universe.[641] This is not just wishful thinking; it follows logically from the standard assumptions of quantum theory. Eddington also notes that 'those who in the search for truth start from consciousness as a seat of self-knowledge with interests and responsibilities not confined to the material plane, are just as much facing the hard facts of experience as those who start from consciousness as a device for reading the indications of spectroscopes and micrometers'.[642]

In other words, those who accept the fact that consciousness is not material, which we can know from our own direct sense of self-awareness, are being just as scientific as those who think consciousness does nothing but act as a measuring device. Eddington even goes so far as to say that 'in comparing the certainty of things spiritual and things temporal, let us not forget this—mind is the first and most direct thing in our experience; all else is remote inference'.[643] There are certainly many physicists who would like to exclude all talk of consciousness, mind, or soul, or at least reduce them all to the physical level of the brain, but there are also several pioneering physicists who have taken these concepts seriously.

Even though there is no logically consistent way to avoid including ourselves in our scientific descriptions, it does not therefore follow that we must give a full (or even any) account of the observer when conducting every, or even most, scientific studies. In a similar way, we need not write a treatise on the epistemological justification of the possibility of knowledge before attempting to answer any and all questions about any given topic. Nevertheless, there is no logically consistent alternative to philosopher Renee Weber's point that it is really mysticism, rather than science, that logically pursues the Grand Unified Theory, since the mystic demands that we include the questioner within the answer. Weber writes that 'although the scientist wants to unify everything in the one ultimate equation, he does not want to unify consistently, since he wants to leave himself outside of the equation. Of course, with the advent of quantum mechanics, that is far less possible than it was in classical physics'.[644]

I am aware of the controversial territory we are entering into in this last chapter, which is one reason I have stressed the logical, physical, and metaphysical importance of the role of the observer—us conscious, intelligent entities. Many of us, however, are still likely to hesitate at any suggestion that *soul* really has some role to play. But why should we fear the word 'soul', which is also at the root of 'psychology' (*psyche*—ψυχη—means soul), when a pioneering physicist like Schrödinger was not afraid of the word? In the first chapter we noted that Schrödinger agreed with Jung that 'all science...is a function of the soul, in which all knowledge is rooted'. The soul is at a higher level of reality than the eternal mathematical law, and therefore it cannot be contained within it. It is the universal soul that contains the eternal mathematical law.[645]

Although we should try to do science as far as possible even if we cannot scientifically describe the soul, it is a mistake to claim that any reference to the soul is *anti*-scientific and, therefore, nonsensical. The nature of the soul may be an issue that many, or even most, scientific experiments and theories need not address in any direct way, but this issue still underlies all scientific inquiry. In fact, some aspects of our future sciences will have to engage directly with this very issue if we want to develop our sciences in greater depth, allowing for significantly expanded power and applicability. Recall Schrödinger's point that the most important task of science is to help answer the question 'who are we?' Think about how much of our scientific research is directly concerned with who we are, what our bodies are, where we come from, and where we might be going or what we might become. We engage in science for us—who else? Why, then, do we grow uncomfortable when forced to admit that which is so patently obvious?

NINE: Unity and Ultimate Reality

The phrase 'the observer' really refers to the self or soul. However, just like the notions of 'mass' and 'energy', terms such as 'consciousness', 'self', and 'soul' are notoriously difficult to define unambiguously.[646] For our purposes here we can consider soul to be the essential nature or essence of a person that will not perish with the body. As Gerson writes, 'all Platonists accepted the view that in some sense the person was the soul and the soul was immortal'.[647] As Socrates says, a person is either nothing, or if something, 'he's nothing other than his soul'.[648] Malin also notes positively that by 'soul' Plotinus means the 'self'; that is, 'he speaks about you and me'.[649] Even Heisenberg uses the word 'soul' to refer to 'the central order, to the inner core of a being whose outer manifestations may be highly diverse and pass our understanding'.[650] Through our own being we are able to apprehend the central order or cosmic unity because we are part of the same cosmic order, which permeates our very being, the soul that we are.

Consider the following simple argument:

1. Mathematical laws of physics are themselves not physical and do not change.
2. We are able to discover mathematical laws of physics.
3. We are able to perceive physical reality.
4. Given (1-3), we are able to perceive (in the widest sense of the term, including direct understanding, non-discursive insight, and regular physiological perception) both nonphysical, unchanging mathematical laws, and the changing, physical universe.
5. It is generally much easier for us to understand and communicate with a chimpanzee than with a shrimp, for the basic reason that we are more similar to the former than the latter.
6. Given the general applicability of (5), it is most likely that we are less able to understand, apprehend, or be in relationship with whatever is most dissimilar to us.
7. Given (6), since humans have the potential to comprehend the mathematical laws of physics and the physical world, there must be something about humans that is similar to both the nonphysical and the physical.
8. Conclusion: there must be something about humans that is akin to both the timeless immaterial aspect of reality and the time-bound physical aspect of reality.[651]

We can discover nonphysical, objective, timeless truths even though we are usually focused on the physical, relative, and ever-changing aspects of reality. Norris is a powerful ally for any realist; however, even he has been led to believe that it would be a relief for the realist once the 'Platonist burden' is dropped, once we are free from 'the impossible task of explaining how our minds achieve quasi-perceptual contact with a range of purely abstract [mathematical] entities'.[652] Norris is correct that such an explanation is extraordinarily difficult. Even Proclus, who argues that the mathematical entities and relations imitate the true excellence of the highest metaphysical ideas, such as pure beauty and intellect, does not tell us exactly where the mathematical realm is located.[653]

How can anyone say *exactly* where the mathematical realm exists? It is not as though this realm is located in a physical place. Given his assumptions about Platonism and Aristotelianism, Norris' conclusion about the impossible epistemological task of coming into contact with this mathematical realm is well founded. However, as I have already shown, *Aristotle* was the antirealist about mathematics, while it was *Plato* who was the realist. Antirealism would make science rest on a fictional foundation, a view we have already thoroughly rejected. But if we are realists about the mathematical laws of physics, which we all must be in one way or another, then we have to accept that these laws are real. If they are real, they must have some sort of reality that is of a higher order than the ever-changing physical realm. If we can discover these laws, then there must be something about us that is somehow or other similar to that nonphysical realm.

Let us here consider the relevant views of Gödel, the great logician and companion of Einstein. In order not to offend the analytic philosophers and positivists any more than necessary, Gödel often seemed to try to downplay his Platonism. This did not stop Russell from calling him an 'unadulterated Platonist' after engaging with him in private discussions. Even more significantly, Gödel's fourteen major philosophical beliefs have been ignored by almost all academics, which is another symptom of the fear in academia of admitting to the reality of the spiritual realm. It would be a difficult task to find a set of beliefs more Platonic than those espoused by Gödel. I will here provide only six:

1. The world is rational;
2. There are other worlds and rational beings of a different and higher kind;
3. The world in which we live is not the only one in which we shall live or have lived;
4. Materialism is false;
5. Concepts have an objective existence;
6. There is a scientific (exact) philosophy and theology, which deals with concepts of the highest abstractness; and this is also most highly fruitful for science.[654]

One of the greatest logicians in history believed in some version of reincarnation and that there are other rational beings who are of a higher kind than us. He also believed that there is a scientific (exact) philosophy for dealing with the highest abstract concepts, such as we find in Platonism. Hao Wang, a philosopher specializing in logic and mathematics from whom we learn about Gödel's fourteen beliefs, is honest about the fact that he has 'only a very partial understanding of many of the subtleties' of Gödel's thoughts.[655] Wang is correct that Gödel's fundamental beliefs go 'far beyond' what Wittgenstein would have allowed, but that can hardly be seen negatively.[656] Wang also admits that he does not know how Gödel's 'beliefs are interconnected, or how they might be convincingly supported', and he thinks that it is clear 'that we can neither prove nor refute them, although they are certainly of interest in widening the range of possibilities we can envisage'.[657] Wang also summarizes some of the things Gödel had said to him in the 1970's, such as the following: 'mathematics has a form of perfection. In mathematics one attains knowledge once and for all. We may expect that the conceptual world is perfect and, furthermore, that objective reality is beautiful, good, and perfect'.[658]

One of the main goals of this book has been to show the interconnectedness of such beliefs, which can slice through our false beliefs and turn our attention toward higher aspects of reality. They can also cause some people considerable frustration. The difficulty in trying to talk about nonphysical reality in a way that appeases the materialist is similar to the current inability of physicists to express the subatomic realm in terms that satisfy our everyday experience of the macro world. But it is even worse in this case, because the materialist simply denies that there is any such thing as nonphysical reality from the outset, despite the overwhelming evidence and plausible rational arguments to the contrary. They will say, 'we must listen to science', but when some of the greatest pioneers in physics tell us that materialism is false, they refuse to listen.

It is as if the rattlesnake asks the hawk to explain how they catch their prey, then restricts the hawk to speaking only within the rattlesnake's conceptual and experiential limitations. Every time the hawk begins to describe stalking prey from the air, the rattlesnake protests that this way of speaking is illegitimate or irrational, since nothing can hunt from the air. The hawk is then forced to use metaphors, analogies, and so forth to describe what it is like to fly, which the rattlesnake dismisses as not being logically rigorous and going far beyond what is possible to experience. The hawk replies that all hawks experience hunting from the air every day, at which point the rattlesnake becomes upset and slithers away. The rattlesnake thinks that the hawk is absurd, but the hawk simply shrugs its wings, takes flight, and continues to catch prey in ways beyond the conceptual grasp of the rattlesnake.

Breathing eternity

Let us conclude this final chapter with one last venture deep into the territory of pure metaphysics. It will be a very condensed exploration for which we have been preparing since the first page of this book, one that is also preparing us for a future journey, should we have the desire and motivation to undertake it. If it really is true that there is not only an eternal mathematical law but also higher metaphysical principles such as symmetry, justice, truth, beauty, the good, and the One, then there is also an eternal moral law. We are then compelled to change our lives to live in harmony with this higher law. Genuine freedom comes to us as our lives are transformed to more closely reflect the eternal moral law, a topic that will have to wait for a future publication.

Below is a fuller version of the metaphysical hierarchal structure discussed in Chapter Two. There are further subtle distinctions and clarifications to be made, but this basic structure is suitable for our purposes:

> The One / Absolute Good   *God / Father*
> Intellect / Pure Mind (or Consciousness) / Absolute Beauty   *Wisdom/Sophia*
> Soul   *Time / Spirit*
> Nature / The One Eternal Mathematical Law   *Son / Logos*
> Body / The Physical Universe

Those theoretical physicists who are searching for the most unifying, simple, abstract mathematical law that underlies the cosmos are, as Heisenberg recognized, ultimately being driven to seek the One, the ultimate source of all. The eternal mathematical law is itself a downward projection from the universal soul, which gazes upward, receiving from and reflecting upon pure intellect. The universal soul is the midpoint between pure intellect and the physical universe. It transcends the universe, is not reducible to it, and yet animates the whole of it.[659] We could say that nature is moved by universal soul and in turn moves or brings form to matter. Nature could be understood as the reason-principles of the universe, and it is at this metaphysical level that the Aristotelian abstractionist halts their inquiry (although Aristotle seems to have stopped at the level of intellect). Stopping at the level of nature is at least a better place to take intellectual rest than being content with the false belief that matter is the highest principle of reality.

Whittaker notes that for Plato, 'the objects of mathematics and the faculty of understanding (διάνοια) [*dianoia*] that deals with them come between dialectic and its objects above, and sense-perception and its objects below'. He continues by saying that for Proclus, 'the mathematical forms and the reasonings upon them…result from the productive activity of the soul, but not without relation to a prior intellectual norm, conformity to which is the criterion of their truth'.[660] What Whittaker is basically saying is that dialectical reasoning (pure metaphysics guided in the right direction) leads beyond the mathematical laws of physics and points us to the ideas that, so to speak, reside in intellect, and ultimately to the metaphysical edge of the One itself.⁶⁶¹ It is necessary at this point to rely increasingly upon metaphors, such as referring to the metaphysical edge of the One, because discursive reason cannot penetrate the One.⁶⁶²

Proclus, much like Jung and Schrödinger fifteen centuries later, argued that the mathematical forms are rooted in the productive activity of the soul. The reason these laws are rational, intelligible, and powerful is that they themselves are products of the higher principles, such as symmetry (or harmony) and beauty, which are at the level of intellect. Each individual soul can be regarded as a reflection or aspect of the universal soul, which transcends and yet embraces all souls. Proclus provided some of the most powerful rational arguments ever produced to show that each limited, particular soul is 'indestructible and imperishable'⁶⁶³ and has 'an eternal existence but a temporal activity'.⁶⁶⁴ Because of our limitations, we cannot discursively apprehend more than a portion of the cosmic order that permeates the cosmos. We should now be better able to appreciate the depth of Einstein's words: 'I am satisfied with the mystery of the eternity of life and with the awareness and a glimpse of the marvelous structure of the existing world, together with the devoted striving to comprehend a portion, be it ever so tiny, of the Reason that manifests itself in nature'.⁶⁶⁵

At this point it is worth recalling Penrose's belief that 'Beauty and Truth are intertwined, the beauty of a physical theory acting as a guide to its correctness in relation to the Physical World'. He also believes that we can find 'absolute Truth' in mathematics, ⁶⁶⁶ which implies that there is also absolute beauty. The logos or 'Reason' that becomes manifest in nature is not pure intellect, but the first activity of universal soul as it receives the reflection of absolute beauty and truth, which is the shining radiance of intellect. It is in this way that the logos is born, so to speak, and then projected into the formation of the reason-principle of the universe. The reason-principle of the universe is lower than the universal soul, as it is a projection or reflection of what the universal soul receives from intellect. But the reason-principle of the universe is, in a sense, a metaphysical step higher than the eternal mathematical law, which can be made clearer through the following example. When we throw a baseball accurately, we are somehow or other doing amazing mathematical calculations unconsciously, or we are directly intuiting the necessary motions to accomplish our goal, which, in either case, is really to say that we are tapping into a tiny portion of the reason-principle of the physical universe.

Every physical action, from dancing to having sex, fighting, or breathing, is an expression of the eternal mathematical law. But we do not need to consciously perform the necessary mathematical calculations in order to breathe or to dance. We do them 'naturally' because the soul that we are transcends the reason-principle and the eternal mathematical law of the universe. We are able to intuit the reason-principle of the universe even when we lack the metaphysical ability to speak about it directly, or the scientific knowledge required to speak intelligibly about some particular aspect of it. However, if we are going to understand the mathematical laws such that we can then employ them directly for our creative purposes in the physical realm, we need a different sort of intellectual and intuitive capacity to make these laws manifest to our minds.

What matters more for our purposes here is that we can recognize and understand that there are higher metaphysical principles beyond the eternal mathematical law, and that we, as particular souls, have the potential to attain a glimpse of them, as Einstein had suggested. We can then manifest our new understanding in unique ways, not just in the sciences, but also in the arts, in business, and in every other aspect of our daily lives. In case we start to feel that we have carelessly allowed ourselves to be carried away by nothing more than metaphysical fantasy, we should remember that we find ourselves in good scientific company when engaging in these profound and challenging metaphysical reflections. Eddington offers a good reminder; he writes that 'the idea of a universal Mind or Logos would be, I think, a fairly plausible inference from the present state of scientific theory; at least it is in harmony with it'.[667]

In the same way that the universal soul is a midpoint, so, too, are individual souls midpoints with various capacities.[668] Mathematical objects and their relations are also a midpoint between the ideas in intellect and the physical universe. Such reasoning leads us to conclude that the mathematical laws of physics, which are further reflections of the one lawful structure or reason-principle of the universe, are contained within the universal soul, in a similar way as we contain a thought within our minds. Consequently, individual souls have the potential to comprehend and express various aspects of the universal reason-principle or the eternal mathematical law.

Those who are inclined to discover such truths regarding the mathematical laws of the physical universe must not only seek clues by looking below to see how the physical world actually operates, but must also look within themselves to discover the laws that are responsible for such physical operations. This idea has been a guiding belief for many of our greatest physicists; indeed, the most profound theoretical discoveries of fundamental physics have not occurred in laboratories, but *in the mind* of the contemplative theoretician. Galileo, for example, believed that the 'fundamental laws of motion and of rest are laws of a mathematical nature. We find and discover them not in Nature, but in ourselves, in our mind, in our memory, as Plato long ago has taught us'.[669] This belief makes sense within the metaphysical hierarchical structure of reality given above, from the One down to the Body.

What we find in our soul in relation to the mathematical laws of physics is only a genuine discovery (and not just our own whimsical fancy) when it correlates to the objective truth. In other words, personal conceptions are validated by their correspondence to the objective reality of intellect. This process applies not only to physics but also to all of science, the arts, and in every other aspect of our lives to varying degrees.

I realize how improbable it may seem to a 'no-nonsense materialist' to assert that soul is real, that the laws of physics are nonphysical and metaphysically prior to the universe, that we need a certain kind of faith in the rationality of the cosmos to underpin science, reason, and logic, that direct perception, insight, or even mystical experience all have a role to play in the foundation of the sciences, and that the ultimate nature of reality is the One. But these sorts of claims are precisely the ones that follow from aiming for the most comprehensive, systematic, rational argumentation to defend the view that physics is not a mere fiction. I will not have convinced everyone (and perhaps only those who did not really need convincing in the first place) of my strong claim that Platonism is true. My weaker claim that Platonism is a viable alternative to the competition will be harder to dispute. What is beyond doubt is that there *are* plausible rational arguments in defense of Platonism, and that we can no longer deny the importance of this tradition for the foundations of modern physics.

New frontiers

Every physical and metaphysical aspect of reality is driven by pure power.[670] The metaphysical hierarchy is unchanging in the sense that universal soul will not suddenly upstage intellect, no more than my leg will suddenly usurp every possible function of my brain. However, we are able to ascend and descend according to our capacities and merits, which very much depends upon our level of commitment and the intensity of our desire to become like God so far as we are able. We can, as individual souls, turn our attention to the universal soul, and thus to pure intellect and, to the degree that we are able, even to the One itself. By turning our attention upward, which requires reflecting inwardly, even as we are challenged in all our outer activities and relationships, we have the potential to achieve far greater self-knowledge, which is a key ingredient of genuine personal power.

We all have different capacities, abilities, and so forth, and we must aim to discover and unfold them in the most beautiful ways we can. While we focus on the higher aspects of reality, we cannot forget our embodiment, and that even matter owes existence to the One, and so matter, too, is intrinsically good in its own way. By turning our attention to the higher metaphysical principles, we will be in a better position to develop the sciences *and* produce a just, harmonious way of life. We must not forget, however, the importance of those powerful moments of trans-rational intuition, the flash of insight or direct understanding, and even divine mania.[671] We must also remember that we need to explore the darkest depths of our psyche as part of our purification. We cannot simply pretend that everything is rosy when there is also so much ugliness surrounding us and found within us.

We are often too easily distracted by the vicissitudes of daily life, such as when we are late for class, get a flat tire, or experience conflict with a family member or colleague. How much more easily do we forget the higher metaphysical aspects of reality when we find ourselves in some of the darker moments that can befall any of us, or even when taking in the news about all of the wars, murders, and other devastations in our world? This is one reason Socrates advised us to engage in meaningful dialogue to some degree every day, to keep our attention focused on the higher aspects of reality while remaining firmly grounded in practical life.

While there are many spiritual paths and ways of life that we may choose to explore, we are all bound by the same objectively real laws. Our technological power is only possible because we have discovered, and found unique ways to express, an extremely small portion of these laws. In a similar way, profound inner power and the highest form of personal freedom become more available to us as we discover our own unique way to live in accordance with the higher metaphysical laws. By understanding how certain ancient philosophical ideas are foundational to modern physics, we are better able to understand and appreciate objective truth and reality.

If we choose to enter further into the metaphysical realm, we can gain access to higher forms of power. What we do with this power, however, requires a different sort of exploration.

APPENDIX A

(From Chapter Four)

A full treatment of the nature of the uncertainty principle would require a far more detailed analysis than what I am providing in this Appendix, but it will have to suffice since such exploration on this particular topic, however important and relevant, is not the main goal of this book. As already noted, Barbour outlines three possibilities for interpreting the uncertainty principle, stating that the uncertainty is either:

1. due to temporary human ignorance;
2. inherent to experimental or conceptual limitations; or,
3. in nature itself.[1]

Bohr apparently gave up on the uncertainty principle, or the so-called disturbance argument, as a defense against Einstein's attacks, for it implied, as Baggott contends, an 'almost classical realist conception of the measurement interaction',[2] which would have played into Einstein's demands for a reasonable theory.[3] However, a 'classical realist' conception really just means a materialist metaphysics. Pauli also wrote to physicist Max Born in 1954 and told him that Einstein '*disputes* that he uses as criterion for the admissibility of a theory the question: Is it rigorously deterministic?'.[4] In fact, Pauli adds that 'Einstein's point of departure is 'realistic' rather than 'deterministic'".[5] We can already begin to see some of the important philosophical confusion surrounding this issue, and if we expect to gain an appropriate understanding of the uncertainty principle, we are going to have to explore difficult metaphysical territory.

I will here give a summary of Barbour's explanations of these three possibilities and critically assess his views. Einstein, Planck, Bohm, and de Broglie, for example, held to the first view.[6] According to this view, the quantum system must be objectively determined even though subjectively we may never be able to predict with absolute certainty any particular outcome. To give a macro example, tossing a coin could in principle be predicted with complete accuracy if we could know all the relevant variables. Yet because such knowledge is extraordinarily difficult to attain, the result of the coin toss appears to be random. However, I would add that a robotic arm could toss a coin with

---

[1] Barbour, 1966, pp. 298-299.
[2] Baggott, 2004, p. 186,
[3] Baggott, 2004, p. 133.
[4] Pauli quoted in Mermin, 1985, p. 40.
[5] Pauli quoted in Mermin, 1985, p. 40.
[6] For example, see Planck, 1931, pp. 47-48.

pre-specified momentum in a controlled environment that could allow us to predict the outcome with great precision and, in principle, without error.

Barbour also discusses Bohm's hidden-variables alternative and states that 'most scientists are *dubious about such proposals*. In the absence of any clear experimental evidence, the defense of determinism rests largely on philosophical grounds'.[7] Norris has written extensively on precisely this sort of dogmatism and lack of philosophical clarity resulting in a prejudiced opposition to any alternative, such as Bohm's. What Barbour misses here is that *any* interpretation of quantum theory (or any alternative theory) necessarily rests on philosophical assumptions, and so resting on philosophical grounds is no criticism of hidden-variables theories or of any theory whatsoever. Moreover, the idea of a measuring device interfering with the system it measures is not new. Goethe was already raising the issue of 'the disturbing influence of the instruments on the 'natural' phenomena',[8] which was supposedly one of the most significant revolutionary points of quantum theory. Newton also intimated indeterminacy, as is shown by a comment recorded by David Gregory: 'A ray of light has paroxysms of reflection and refraction and indeterminate ones at that'.[9] Schrödinger also remarks that the contention that subject and object are inextricably interwoven 'is almost as old as science itself'.[10] Many of the so-called revolutionary aspects of quantum theory are not so revolutionary after all;[11] it is just that we could more easily ignore these ideas before quantum theory.

According to Barbour, Bohr was a chief proponent of the second view, where the uncertainty is not just due to temporary human ignorance, but is a fundamental limit on human knowledge. There are two versions of this view: the experimental and the conceptual. The *experimental version* says that the uncertainty is due to the disturbance of the physical system through interaction with an instrument or observer, although Barbour is quick to point out that he does not endorse any reference to the consciousness of the experimenter.[12] However, this view cannot account for uncertainties when nothing has disturbed the system, such as radioactive decay.[13] 'The unpredictability of the atomic realm, then, appears to be a distinctive feature of quantum mechanics—from whose postulates the Heisenberg Principle can be derived without reference to disturbances introduced by the observer'.[14] The *conceptual version*,

---

[7] Barbour, 1966, p. 300 (original emphasis).
[8] Pauli in Jung and Pauli, 1955, p. 206.
[9] Quoted in Rowlands, 1992, p. 224.
[10] Schrödinger, 1952, p. 52.
[11] On this point in general, see Rowlands, 1992.
[12] Barbour, 1966, p. 301.
[13] Barbour, 1966, p. 302.
[14] Barbour, 1966, p. 302.

however, concerns our human epistemological limitations.[15] On this view, we can never know physical reality in itself and 'the ontological question of the character of the world is ignored or dismissed as meaningless',[16] which is why it is rightly assumed to be positivistic in this sense.

The third position, which Barbour approves, is endorsed by many physicists and 'holds that *indeterminacy is an objective feature of nature*, and not a limitation of man's knowledge'.[17] Barbour follows what is generally considered to be Heisenberg' claim that the uncertainty comes about only when the observer makes a choice as to how and when to make a measurement, which results in actualizing one of multiple possibilities. However, Gibbins also notes that sometimes Heisenberg 'writes as if it is only our *knowledge* of the world that is indeterminate, not the concepts which we take to apply to physical systems big and small'.[18] Instead of jumping into what Heisenberg's views may or may not have been, let us first of all examine the problematic notion of indeterminism in quantum theory.

Since it is believed to be the case that energy levels for an electron are not continuous, there must be a cut-off point between each level, which would mean that there must be definite and determinate energy levels (even if they are a bit fuzzy around the edges). If there were no definite and determinate levels, then each level would necessarily blend with the one closest to it, which would allow a continuous transition from one energy level to the next (assuming that we could even demarcate energy levels in such a case). But this admission would defeat the backbone of quantum theory that initially seemed to separate it from classical physics; namely, that energy comes in discrete packets, or quanta. An electron may have an indeterminate motion within its defined energy level, which is to say that, given that the initial conditions remain exactly the same, the same electron may well have a different position or momentum within the same energy level at the same moment in time. In other words, the conditions of the electron could have been different and so would not have been strictly determined in the first place. However, there are three problems with such a view.

First, macro results will always be (more or less) the same, even when applying quantum laws. For example, if X breaks in my computer, Y will be the result, and the same result would occur in your computer (if it is the same model). Science would not be possible if the universe did not follow the law of cause and effect.

---

[15] Barbour, 1966, p. 302.
[16] Barbour, 1966, pp. 302-303.
[17] Barbour, 1966, pp. 303 (original emphasis).
[18] Gibbins, 1989, p. 53 (original emphasis).

Second, even assuming that indeterminism is true in some way at the quantum level, there is not yet any proven or genuinely plausible explanation for the transition from the micro level to the macro level. Heisenberg cites Bohr as saying that 'we don't know where to draw the line between small and large objects'.[19] However, despite this epistemological limitation, Norris notes that Bohr maintained the ontological belief that there must be 'some cut-off point on the micro-to-macro scale at which quantum effects such as superposition or wave/particle dualism gave way to determinate (classically decidable) states such as position [and] momentum'.[20] But if there is really some cut-off point between the micro and the macro, then it must be a real feature of reality that is objectively the case despite our (current) inability to know it, which is to admit to realism.

It makes no sense, however, to maintain that there is a *real* discontinuity between the micro and macro levels of physical reality, not if that means there is an unbridgeable disunity between them, because in actual physical reality there cannot be such a dichotomy. In actuality, the macro is made of the micro. Macro technology allows us to penetrate into the micro realm, and our theoretical understanding of the micro realm in turn allows us to build macro technology, which is only possible if the micro and macro are somehow connected, if they are in some way always unified. Nevertheless, if any quantum effects are possible at the macro level, we seem unable to witness them. Even if we are in fact seeing genuine quantum effects at the macro level without consciously being aware of that fact, it is still the case that we are not yet very good at recognizing them or taking advantage of them, which is not necessarily to say that we will not be able to do so in the future.

Third, there is no way in principle ever to provide a definitive empirical test for the notion of either indeterminism *or* determinism. We can only give philosophical arguments. The only way we could unambiguously prove empirically that either metaphysical view was true would be to conduct experiment A, and then later reset all the initial conditions of experiment B to match those of experiment A. However, quantum theory entails that the entire universe is holistically interconnected, where we cannot logically or scientifically separate any object from any other object. The best we can hope for in conducting experiments is to create approximations that are as accurate as possible, but those approximations always have an element of subjectivity and, therefore, do not guarantee certainty. Thus, we could only repeat all the logically and physically interconnected initial experimental conditions by resetting those conditions for the entire universe, which is not possible, unless we could go back in time and rewind the whole universe in its entirety (which, apparently, we

---

[19] Heisenberg, 1972, p. 107.
[20] Norris, 2002-A, p. 41.

cannot do). Even if Gödel was correct that we could travel back in time, we still could not reset all the initial conditions of the entire universe exactly as they were for experiment A.

Even if we could reset every initial condition for the entire universe, we would still be no better off, because if literally every aspect of the entire universe were rewound to exactly the same conditions as they were during experiment A, then experiment B would never be possible, because every time we repeated experiment A we would—obviously—be repeating experiment A and not performing experiment B. In other words, even if we go back in time to the moment of experiment A, we do not find ourselves at the moment in time of the future experiment B. To put it another way, if we reset all the initial conditions of experiment A, we would necessarily be back at the same moment in time when we conducted experiment A, and not at the moment in time when we would be conducting experiment B. Moreover, just because experiment A produced effect X the first time around, it does not logically follow that the same result will necessarily occur next time, even if all the original conditions for the entire universe were reset exactly as they were previously. This is because there is no *deductively* logical way to guarantee a conclusion in the future based upon the past. But in order to make such an *inductive* conclusion, we would need to have faith that the laws of physics would remain constant in both experiment A and experiment B (which would actually just be experiment A, anyway). We would also only be able to conclude inductively that the original experiment A and the succeeding experiment B (which would actually be the original experiment A) would have the same effect X if we assume not only that our faith in the laws of physics is justified, but that determinism is true in the first place.

However, if we were to repeat the experiments in the way described above, we would not even be able to know that we had done so, for we would ourselves be reset back to the original conditions and so could not know what the outcome was of the original experiment A when conducting the 'second' experiment A, which means that we would be unable to make any comparisons. But if I, the conscious, aware perceiver, am nonphysical, then in principle I would be able to transcend and not be affected by the resetting of all the original conditions of the entire universe.[21] Only in such a case could I be aware

---

[21] Deterministic laws pertaining to physical objects do not preclude free will. The notion of free will hinges on very deep metaphysical assumptions about the nature of the self and its relation to physical reality. It is worth noting that evoking quantum theory to support free will merely gives in to the materialism that quantum theory apparently overthrows and, in any case, if one rejects consciousness as merely being an epiphenomenon of neurological activity (instead of being the immaterial foundation of all physicality) then there is no necessary logical contradiction between maintaining free will for sentient creatures such as humans and holding on

that I was conducting experiment B through resetting all the initial conditions. Some physicists do believe that consciousness is not physical and that it is still required in order to make a quantum measurement, so let us suppose that assumption to be true. In that case, my original state of mind, or my original state of conscious awareness, during the original experiment A would then be a relevant initial condition. Thus, if we were to test determinism versus indeterminism by repeating experiment A to see whether or not the effect would be the same, we would have to reset *all* the original conditions, including my own state of mind, and so, again, I would not even know that I was repeating experiment A.

This exploration is now getting quite confusing, and it could quickly get more confusing still, for any counterargument will easily be met by pushing the exploration still further. This sort of applied metaphysical reasoning is anything but a pointless word game or bunch of nonsense. Only the intellectually lazy or incapable put forth such objections. In practice we are forced to settle for approximations and logically arbitrary (though practically reasonable and relevant) experimental boundary conditions, but we then must admit that we do not have definitive empirical or logical proof of either determinism or indeterminism. Adopting either determinism or indeterminism, then, is a matter of considering empirical data and then providing rational arguments for one view or the other (or some other option), which is to engage in applied metaphysical reasoning. Thus, metaphysical arguments take on preeminent value when considering such fundamental scientific notions as determinism and indeterminism, and there is no purely empirical or observational justification for accepting or rejecting one over the other.

---

to strict causality in the physical universe. Planck would agree: 'all studies dealing with the behaviour of the human mind are equally compelled to assume the existence of strict causality…human free will is perfectly compatible with the universal rule of strict causality' (Planck, 1931, pp. 84-85).

Apparently the Dalai Lama has also 'confessed to having difficulties with the philosophical implications of quantum physics, especially the role of chance and causality in nature. As the idea of determinism is central to Buddhism, the existence of purely random acts might call into question Buddhist doctrine' (quoted by Schiermeier, 2005). If the physical universe and even our thoughts, intentions, and nonphysical stream of awareness (to use the Buddhist phrase) do not obey strict causality, then the notion of karma becomes impossible to defend. I could kill a million people just for fun and yet, if there is no strict causality, still end up being reborn as a self-realized Buddha in the next life. There would be no karma, no perennial action of justice, which is why the Dalai Lama would reject indeterminism. Yet I am free to choose my actions within objective limitations. It is just that after I have chosen, my body must then obey the laws of physics, and I am compelled to obey the law of karma.

However, if determinism were not true, our ability to provide such accurate predictions of empirical observations would seem quite miraculous. Barbour, however, claims that his third position—that the uncertainty is in nature itself—negates *absolute causation* (or determinism) and is better called 'weak causality', since 'the probabilities at one instant are precisely and unambiguously determined by the wave-functions at earlier instants'.[22] But, in that case, there are determined and definitive limitations on what actions a particle can take at any given moment. What Barbour wants to say is that there is still inherent unpredictability due to inherent randomness within these defined parameters; a particle may take path X at this moment when it could have taken path Y or several other possible paths instead, which is to say that the result could have been otherwise. In strict determinism, or strict causality, however, the path a particle takes at any moment could not have been otherwise, given all the initial conditions. Yet we still do not know empirically that a particle could have taken another path, for all the reasons we have just discussed. The question of whether or not the result of an experiment at the quantum level could have been otherwise is very much a metaphysical question.

Since the machine or the physical body of the sentient creature making the measurement or observation is also susceptible to the same quantum laws, these machines and physical bodies will also have been actualized by some other sentient creature or a machine, which in turn must have been susceptible to the same laws. Clearly, then, the universe itself must ultimately be susceptible to the same laws, and so would have to have been actualized by something outside the universe. We are now entering theological territory where, for example, we either say that there was no beginning to the actualization of potentiality but simply an infinitely extended stream, or we say that there was a first cause, a beginning that must have been self-actualized in some sense. There is no escaping these sorts of questions, for they are logical consequences of applying this interpretation of quantum theory to the universe itself. One may reject this interpretation, but then equally difficult metaphysical questions shall be lurking around the corner.

Recall the three views concerning the interpretation of the meaning of the uncertainty principle, whereby the uncertainty: (1) is due to temporary human ignorance; (2) is inherent to experimental or conceptual limitations; or (3) is in nature itself. Epistemological limitations, whether due to experimental error, physical interference with the system, conceptual difficulties, or simple human ignorance, more or less amount to the same thing, and so options (1) and (2) can reduce to a more fundamental position that simply says that the uncertainty is not in nature itself, as opposed to option (3). However, the three different viewpoints are not necessarily mutually exclusive options, because we could still

---

[22] Barbour, 1966, p. 305 (original emphasis).

have our own epistemological limitations (such as in (1) and (2)), and yet the uncertainty could still be inherent in nature itself. Nevertheless, it is wholly unwarranted to claim that it has been empirically and unambiguously decided that the uncertainty is in nature itself. Such a claim involves presupposed metaphysical assumptions and applied metaphysical reasoning, although it is also on the edge of pure metaphysics. Nevertheless, nonlocal effects can appear to offer such support for indeterminism at first glance.[23] While nonlocality is, indeed, a strange phenomenon, it still does not change the fact that merely calling a nonlocal process 'random' does not actually make it so. In other words, we do not *know* that any decay process is random, for there very well may be some sort of determining possibility of which we have not yet conceived and, in any case, we do not really know what we mean by the word 'random'.

If we say that a 'random' process means a process that is unpredictable, we are encountering the same sort of problem as trying to make sense of the uncertainty principle. We do not know if the unpredictability is due to our ignorance of all the relevant variables, or if it is truly unpredictable because, even if we have infinite knowledge, we still could not predict the outcome. If it is the first option, then it is not really a random process but simply an unknown or unknowable process. If it is the second option, then even God apparently could not know the outcome and, in fact, we could not explain how it is that science is possible because all events would truly be unpredictable due to their inherent randomness. One could respond that the randomness is only at the micro level and does not happen at the macro level, but there is no clear distinction between the micro and the macro and, even if there were, there is no satisfactory explanation for why the randomness disappears at the macro level other than saying that, if it didn't, the universe would not be here.

Claiming that the uncertainty is in nature does not actually make it so, no more than calling a process random actually makes it so, and any possible evidence must be based upon an interpretation of the data, which entails using applied metaphysical reasoning. Even if I wanted to accept that the uncertainty is in nature, it is still wholly unclear as to what is meant by the concept of 'nature'. Nature could refer to physical reality or to the nonphysical laws, or it may include both. These laws may describe probabilities, and those probabilities are supposed to reflect what is actually happening in the physical world, but the laws themselves do not change, and so the laws of physics cannot be uncertain.

---

[23] An example of a nonlocal effect would be the apparently 'random' decay of one particle instantly requiring the decay of its entangled anti-particle. The relative distance between the two entangled particles (such as particles called B-mesons) as they are moving through space is such that they could not send signals to each other without going faster than the speed of light, and so there appears to be information passing between them that is outside the bounds of space and time. (Thanks to physicist Mike Houlden for this clarification.)

Therefore, the question of whether or not uncertainty is in nature hinges on the question of what we mean by 'nature', and the ambiguity of this essential term creates much unnecessary confusion.

If by 'nature' we are referring to physical reality, then it may be true that there is inherent uncertainty in nature, in the trivial sense that all physical reality is constantly changing and is therefore in some sense epistemologically uncertain. But physical nature cannot be ontologically uncertain if uncertainty implies random chance, for random chance would never have allowed the discipline of physics to develop in the first place. The only genuinely possible uncertainty comes from our own epistemological limitations about both physical and nonphysical aspects of reality, which is not really saying anything more than that we are not God and so we cannot know everything, and thus there are some things (probably an infinite number of things) about which we are uncertain.

Bohr accepts a sequence of cause and effect in the atomic realm that conforms with 'elementary demands of causality', although he claims that we still must abandon 'ideal determinism', which seems to mean strict determinism.[24] But there are various difficulties here. First, it is unclear how he can admit that there is a sequence of cause and effect while simultaneously denying ideal determinism, which is a denial that every single effect has some cause and that the effect could not have been otherwise. He seems to want to say that the subatomic particles (or waves) follow a precise cause-effect sequence, but that the outcome could have been different. However, if the particles follow a precise cause-effect sequence, then the outcome could *not* have been otherwise. If the outcome could have been otherwise, then at some point in the cause-effect chain something must have been altered. If the alteration occurred according to some higher order of causal law that we presently cannot comprehend or access, then there is still strict determinism. If there is no such higher causal law, then the break in the casual chain at some point is going to come down to a random event. In such a case, either the random event occurs at the same moment or in the same part of the causal sequence every time the experiment is performed, or the random event must itself occur at some random moment or place in the causal sequence. If there is only one random event that occurs at the same place or moment in the causal sequence every time the experiment is performed, then that seems to indicate a higher order of causality overseeing the apparently random event which, if there is such a higher order of causality, would indicate that the assumed random event is not random at all, but is actually susceptible to some hidden order.

If there is no higher order of causality that dictates when a pivotal moment of randomness in the causal sequence is supposed to occur, there is no reason

---

[24] Bohr, 1963, pp. 4-5.

whatsoever to expect the randomness to be limited to any particular moment or place in the causal chain. There would likewise be no reason to presume that random events would not happen at each and every moment in the causal sequence. We would then be left with no causal sequence at all, because all events would be random at each and every moment. And so, if there were any randomness whatsoever, there would have to be total randomness, or else the randomness would only appear to be random while in fact following higher causal laws that may well be enacted through one or more hidden variables (though not necessarily mechanical or material variables). Consequently, Bohr cannot both say that the atomic realm conforms with 'elementary demands of causality' while simultaneously abandoning determinism. And it is redundant to say that there is such a thing as strict or ideal determinism, for determinism is, by its very nature, uncompromisingly strict.

Even if indeterminism is true in any way at all at the micro level, it is still the case that this indeterminism is confined to a very narrow range of possibilities; a subatomic particle is not suddenly going to turn into a dancing pink elephant. The specific behavior of any particular subatomic particle apparently cannot be known prior to the result of the experiment, which may just be due to our lack of knowledge about how the micro level actually operates. In any case, the ontological claim that nature is uncertain is *not* definitively supported by physical experiments. Experiments alone give no conclusive support for determinism, either. Even if the uncertainty is in nature itself, it would therefore be an objective feature of nature. If it is an objective feature, then it is mind- and instrument-independent, which means that realism is true. The uncertainty, however, cannot mean randomness and must represent an objective epistemological limitation within the applicability of quantum theory. Nature is not random, that much is for certain.

We should now have a better understanding of some of the inherent conceptual difficulties surrounding the uncertainty principle, and we should also be in a better position to understand why, as Einstein believed, we are not scientifically justified in drawing the ontological conclusion that there is no determinate quantum action.[25] As Harris notes, if we accept the third option, that the uncertainty is in nature itself, then the assumed probabilistic nature of physics must permeate all of science. 'Thus, reality, so far as we have any indication of its nature, seems to be wholly random and unaccountable in its underlying activity'.[26] Harris further notes that those who think in this way generally try to make sense of the apparent order of the world by saying that in science *we* impose this order.

---

[25] For example, see Yourgrau, 2005, pp. 107-108.
[26] Harris, 2000, p. 158.

We cannot impose order unless we ourselves are already ordered, but if there is no order in the universe then, since we are part of the universe, we would have no order either. If we had no order, we would not even exist, for nothing about us would be arranged in any ordered way whatsoever. However, even if we could exist without order, we could not then impose order onto something else for, being unordered, we would not know how to do so. There is a relatively trivial sense in which we can impose order upon physical reality, such as gathering a bunch of stones and placing them in such a way so as to make a bridge across a stream. In such a case, we are analogously acting like Plato's demiurge (the working god) in the *Timaeus*, bringing order out of chaos or imposing order onto chaos. In constructing the physical world through imposing lawful structures, the demiurge looked to an ideal model that already had order, symmetry, and beauty. Therefore, even though we can in a limited sense impose order upon our surroundings, it is only because order or 'Reason' or the logos has ontological priority over the physical world.

One may reply that order is merely apparent, for all that occurs is statistical averaging. However, if the statistical laws continuously allow us to make accurate predictions, then there is clearly an ordering principle behind these statistical laws. Without a presupposition of underlying order, statistical laws would be meaningless or impossible. If events are truly random, we cannot make any useful prediction at all and there cannot be any sense of order in science or in our experience. If we can make accurate predictions (which indeed we can), then this presupposes order.

Harris notes that in actuality the statistical laws of modern physics 'imply a certainty that the modern physicist recognizes'.[27] He also reminds us that not only Planck and Einstein but also Penrose have thought that uncertainty cannot be an actual property of physical reality,[28] and I have here provided a sample of independent arguments in support of such views. Physical reality is not ontologically uncertain in the sense of being random, and the nonphysical mathematical laws of physics are not uncertain either. The fact that quantum theory yields such powerfully accurate predictions contradicts the assumption that there is really uncertainty in nature itself, especially when considered together with my above arguments. If the uncertainty is merely an epistemological limitation, then we need not have any serious concern, for essentially everything in our daily lives is uncertain from our limited perspectives.

There is a further irony to consider here. Penrose tells us that 'Euclidian geometry is accurate to smaller than the width of a hydrogen atom over a meter's range'; Newtonian mechanics is accurate to about one part in 10^7;

---

[27] Harris, 2000, p. 31.
[28] Harris, 2000, p. 31.

Maxwell's electrodynamics in conjunction with quantum mechanics corresponds to a range of scales of 10^{35} or more; Einstein's relativity is accurate to about one part in 10^{14}; and quantum field theory (which is the combination of quantum mechanics with Maxwell's electrodynamics and Einstein's Special Theory of Relativity) has results accurate to about one part in 10^{11}.[29] Quantum mechanics can also explain such phenomena as the stability of atoms, spectral lines, chemical forces, blackbody radiation, the reliability of inheritance, lasers, superconductors, and superfluids.[30] The supposed uncertainty in quantum mechanics yields extraordinarily precise predictive results, far beyond what classical physics, the supposed harbinger of materialism and determinism, was able to offer. This fact alone should be enough of an indication as to where the uncertainty really is; it is in those who are uncertain about the uncertainty.

---

[29] Penrose, 1997, pp. 50-51.
[30] Penrose, 1997, pp. 54-55.

APPENDIX B

(From Chapter Seven)

Bas van Fraassen offers a slightly different type of antirealism than Rorty's version, which he calls 'constructive empiricism'. Among other significant problems, I will show that his empiricism is not only inadequate for physics, but would also undermine his Catholicism. To avoid this second danger, van Fraassen fragments his belief systems, completely separating reason and faith, which would not be of so much concern if he did not also virulently attack metaphysics, claim that science is the paradigm of rationality in our world, and refuse to submit his religious beliefs to his own empiricist criteria, which he uses to attack realism.

Constructive empiricism basically means that scientific theories need only be empirically adequate, and while science may provide truths about aspects of the universe that are observable, it cannot give us truth about the unobservable.[31] While van Fraassen's antirealism shares certain similarities with Rorty's, his attacks are more sharply focused, probably because of his specialization in the philosophy of science, and although he has published various technical works in this field, they will not be my main concern. If a building's foundation is so weak that it is about to collapse, there is little point in arguing about what color to paint the penthouse walls. In other words, I will bring to light some generally neglected problems at the foundation of the whole of van Fraassen's philosophical stance, particularly the fact that he keeps his scientific, philosophical, and religious views fragmented in order to protect himself from rational criticism.

To put it simply, in order to be philosophically coherent, van Fraassen either has to give up his constructive empiricism or give up his Catholicism. Catholicism claims that God is the ultimate cause of the entire cosmos, which would mean that God is also the ultimate scientific explanation for why anything is at all and why phenomena behave the way they do, and so any claim about God is necessarily relevant to the foundation of science. As Pope Paul VI remarks, 'the highest norm of human life is the divine law—eternal, objective and universal—whereby God orders, directs and governs the entire universe and all the ways of the human community by a plan conceived in wisdom and love'.[32] In any case, constructive empiricism fails of its own accord without having to enter into any exploration of religion.

---

[31] For an overview of constructive empiricism, see Mohler and Monton, 2008. For arguments against it, see Norris, 1997, and Ladyman, 2004.

[32] Pope Paul VI, 1965. Pope Benedict XVI also states that Augustine's

van Fraassen's philosophical position is not easy to summarize, and any summary will not be fully (empirically) adequate. Nonetheless, while recognizing that there is much more to his overall position than I will discuss here, we can still proceed to deal with some central underlying problems. Listed below is the relevant information that I shall draw upon in the subsequent analysis.

1. van Fraassen is a Catholic,[33] and he is an analytic philosopher,[34] but he is not a Catholic philosopher.[35] He also stays 'far from the "science and religion" debates' and does 'not engage in philosophical theology'.[36] He is against intellectual or rational theology, such as Catholic Thomist philosophy, and he 'cannot see philosophy as being able to be more—at most—than a voice in the wilderness, clearing the way for the Lord'.[37] He also believes that 'the philosopher *qua* philosopher is not contributing to science', since 'separating out scientific questions entangled in the philosophical puzzles is for the philosopher only brush-clearing, removal of impediments'.[38] 'Philosophy itself is a value- and attitude-driven enterprise; philosophy is in false consciousness when it sees itself otherwise'.[39]

2. van Fraassen claims that 'the God of the philosophers is dead' because this God is 'a creature of metaphysics—that type of metaphysics—and metaphysics is dead'.[40] By metaphysics, which is 'merely idle word play'[41] and involved in 'trivial pursuits',[42] he means 'the speculative art',[43] a position 'reaching far beyond the ken of even possible experience, on what there is,

---

thirst for truth was radical and therefore led him to drift away from the Catholic faith. Yet his radicalism was such that he could not be satisfied with philosophies that did not go to the truth itself, that did not go to God and to a God who was not only the ultimate cosmological hypothesis but the true God, the God who gives life and enters into our lives (Pope Benedict XVI, 2008).

[33] van Fraassen, 2007, p. p. 177.
[34] van Fraassen, 2002, pp. xvii, 30.
[35] van Fraassen, 2007, p. p. 177.
[36] van Fraassen, 2002, p. xiv.
[37] van Fraassen, 2007, p. 179; also, van Fraassen, 2003, pp. 11-12. To put it another way: 'according to constructive empiricism, the aim is only to construct models in which the observable phenomena can be embedded. (Slogan: the aim is empirical adequacy)' (van Fraassen, 2002, p. 198-199).
[38] van Fraassen, 1991, p. 481.
[39] van Fraassen, 2002, p. 17.
[40] van Fraassen, 2002, p. 1; also, 'metaphysics is dead' (van Fraassen, 2002, p. 4).
[41] van Fraassen, 2002, p. 27.
[42] van Fraassen, 2002, 30.
[43] van Fraassen, 2002, p. 31.

or on what the world is really like'.[44] At the same time, he also believes that 'there is one special case of personhood: God. The God of Abraham, Isaac, and Jacob is a person'.[45] He further claims that 'an encounter with the divine is a personal encounter. As we human persons do, so God too manifests himself to us only through the familiar materials among which we live; how else? There is no similar localizing constraint; God's work goes on everywhere and every-when, throughout history'.[46] We persons are 'indeed beings of flesh and blood. This is a simple truism'.[47]

3. van Fraassen claims that scientific realism implies that 'if the model is to count as entirely successful, then every element of the model must correspond to something real. Thus elements purporting to represent unobservable structure must have corresponding elements in reality as well'.[48] The 'correct statement of scientific realism' is that 'science aims to give us, in its theories, a literally true story of what the world is like; and acceptance of a scientific theory involves the belief that it is true'.[49]

4. van Fraassen is an empiricist ('or at least [he tries] to be'),[50] and by empiricism he means 'the philosophical position that experience is our source of information about the world, and our only source'.[51] 'To be an empiricist is to withhold belief in anything that goes beyond the actual, observable phenomena, and to recognize no objective modality in nature'.[52] The empiricist claims that there is no 'extra-scientific foundation for

---

[44] van Fraassen, 1991, p. 17. He says that he does not reject all metaphysics but that the reversion to a 'seventeenth-century style of metaphysics' is 'disastrous' (van Fraassen, 2002, p. xviii). However, it is very difficult to discern exactly what kind of metaphysics he would actually approve. (Apparently only his own kind.)

[45] van Fraassen, 2002, p. 192.

[46] van Fraassen, 2002, p. 193.

[47] van Fraassen, 2002, p. 192.

[48] van Fraassen, 2002, pp. 198-199.

[49] van Fraassen, 2003, p. 8. This quote, beginning with 'science aims', was originally in italics.

[50] van Fraassen, 2002, p. 195.

[51] van Fraassen, 1989, p. 8.

[52] van Fraassen, 2003, p. 202-203. He continues:

To develop an empiricist account of science is to depict it as involving a search for truth only about the empirical world, about what is actual and observable. Since scientific activity is an enormously rich and complex cultural phenomenon, this account of science must be accompanied by auxiliary theories about scientific explanation, conceptual commitment, modal language, and much else. But it must involve throughout a resolute rejection of the demand for an explanation of the regularities in the observable course of nature, by means of truths concerning a reality beyond what is actual and observable, as a demand which plays no role in the scientific enterprise (van Fraassen, 2003, p. 202-203).

science'.⁵³ van Fraassen's anti-metaphysics, anti-realist position, which he calls constructive empiricism, 'says that the aim of science is not truth as such but only *empirical adequacy*, that is, truth with respect to the observable phenomena....but the criterion of success is not truth in every respect, but only truth with respect to what is actual and observable'.⁵⁴ 'Observation is perception, and perception is something that is possible for us without instruments', and although 'we can detect the presence of things and the occurrence of events by means of instruments', that does not generally count as observation.⁵⁵ Consequently, 'the instruments used in science can be understood as not revealing what exists behind the observable phenomena, but as creating new observable phenomena to be saved'.⁵⁶ However, he also claims that each phenomenon to be saved, such as a rainbow, is actually different for each individual. It is not that we have different perceptions from different points in space-time, but rather that we are not actually seeing the same rainbow, for the rainbow I see is in a different place than the rainbow you see, and so it must be a different rainbow. Thus, 'nature creates public hallucinations'.⁵⁷ He further claims that 'we can still distinguish science from *scientism*, a view in which science, which allows us so admirably to find our way around in the world, is elevated (?) to the status of metaphysics.... Scientism is also essentially negative; it denies reality to what it does not countenance'. He also thinks that 'the success of current scientific theories is no miracle. It is not even surprising to the scientific (Darwinist) mind. For any scientific theory is born into a life of fierce competition, a jungle red in tooth and claw. Only the successful theories survive—the ones which *in fact* latched on to actual regularities in nature'.⁵⁸

5. van Fraassen claims that, according to empiricism, we should be deferential to and express admiration for 'science as a paradigm of rational inquiry', but that we should not have such an attitude in matters of opinion; instead, we should hold 'a certain epistemic detachment with respect to the content of current or even ideal science'.⁵⁹ He claims that 'as an empiricist' you 'see the

---

⁵³ van Fraassen, 1991, p. 481.
⁵⁴ van Fraassen, 1989, pp. 192-193.
⁵⁵ van Fraassen, 2001, p. 154.
⁵⁶ van Fraassen, 2001, pp. 154-55.
⁵⁷ van Fraassen, 2001, p. 156.
⁵⁸ van Fraassen, 2003, 40.
⁵⁹ van Fraassen, 2002, fn 28, p. 236. Also: 'obviously I applaud all of empirical science; and people can wear many hats; but I don't applaud philosophical activity that has the form of scientific theorizing but is not at all part of genuine empirical science' (van Fraassen, 2002, fn 26, p. 242). How, then, he can applaud his own work is certainly puzzling.

empirical sciences as a paradigm of rationality in a largely irrational and anti-rational world',[60] but it will not profit us to gain the whole world through scientific knowledge if we 'lose our own soul',[61] since the objectifying inquiry, the 'sine qua non of the development of modern science', is not applicable to matters of opinion, such as in religion. We should not be tempted to 'think that by our reason we can supply humankind with a morality, meaning, or cognitive access to the ultimate question of what there is'.[62] However, 'in our beliefs and other epistemic attitudes' he thinks that 'we are as free as birds in the air. There are no rules of right reason, rationality is only bridled irrationality'.[63] But he also claims that 'reason rightly pursued will place the secular thinker into a position where a religious world-view becomes a genuine option'.[64] However, he also states that 'there is no point at which faith can be anything but foolishness in secular thinking'. He is 'not referring to doctrine here, least of all to creationism, transubstantiation, miracles, or resurrection, or any of the other quasi-philosophical, quasi-scientific beliefs which the secular think they can attribute to us in the sense in which they understand them'. What he means is that religious people 'see themselves and the world we live in differently, see them in a different light, and to the secular this can never be other than baseless *Schwaemzerei*'.[65] Even being or becoming an empiricist 'will then be similar or analogous to a conversion to a cause, a religion, an ideology', and he thinks that 'all the great philosophical movements have really been of this sort, at heart, even if different in purport'. What he wants is for us to make such a conversion 'without false purport'.[66] Moreover, he claims that 'a leap of faith' is required to accept belief in scientism (as he has defined it), but that does not necessarily make a person irrational. However, he says that we should still be able 'to live in the light of day, where our decisions are acknowledged and avowed as our own, and not disguised as the compulsion of reason'.[67]

6. van Fraassen believes that the stakes are great in the sciences, so we had better get things right, and so 'for science it is a real wager' while 'for ontology it could merely be pretend'.[68] The metaphysician may gain or lose a false belief, which van Fraassen thinks is not a very significant occurrence.

---

[60] van Fraassen, 2002, p. 195.
[61] van Fraassen, 2002, p. 195.
[62] van Fraassen, 2007, pp. 180-181.
[63] van Fraassen, 2001, p. 168.
[64] van Fraassen, 2007, p. 179.
[65] van Fraassen, 2007, p. 179.
[66] van Fraassen, 2002, p. 61.
[67] van Fraassen, 1991, p. 17.
[68] van Fraassen, 2002, p 15.

Against the metaphysician, he says that 'if simplicity, strength, coherence, and all those explanatory virtues are to be placed in the balance against truth, may we please be shown the balance, the gauge, the units, the scale? As long as we have nothing like that, the defense of ontology as scientific points only to the form its theory and theory choice take. But nonsense can come in the same form as wisdom. Form is nowhere near enough'.[69] However, he also says that 'an encounter with God does not involve solving a theoretical equation or answering a factual query: its searing question is an existential demand we face in fear and trembling. As with a human person, the encounter coincides with a call to decision: possible stances toward ourselves and to our world come to the fore and ask for choice. The choice is momentous and sometimes, in some ways, inescapable, for it pertains to our ultimate concerns'.[70]

Some of van Fraassen's incompatible beliefs are already evident in the above summary, and although I will address some of the relevant aspects of his position, there will still be much that will not be relevant to discuss here. From the outset, it is important to recognize that if van Fraassen really is an antirealist, if he really denies mind-independent reality, then he cannot also believe in the mind-independent reality of God.[71] He will, therefore, have to be some sort of a qualified realist, one who admits that the ultimate nature of reality is mind-independent, which is to concede an essential aspect of Platonic realism.[72] van Fraassen tries to avoid addressing these sorts of metaphysical problems by rejecting metaphysics and claiming that his faith is not susceptible to reasoned argument. His rejection of the vital interplay between reason and faith, however, is at odds with the teaching of Pope Benedict XVI, who states that 'these two dimensions, faith and reason, should not be separated or placed in opposition; rather, they must always go hand in hand'.[73]

In an important sense, faith is the foundation of reason since, as I have already argued, reason itself requires some trans-rational foundation. But saying that faith is the foundation of reason is significantly different than saying that faith and reason have no relationship, as van Fraassen claims. The First Vatican Council, however, claims that the divine mysteries 'far surpass the created

---

[69] van Fraassen, 2002, p. 16.

[70] van Fraassen, 2002, p. 193.

[71] First Vatican Council—1869-1870, Chapter 1: 'since he [God] is one, singular, completely simple and unchangeable spiritual substance, he must be declared to be in reality and in essence, distinct from the world' (Tanner, 1990, p. 805).

[72] van Fraassen's position, on a charitable interpretation, seems to be a mix between Luther's and Kant's, but I will not pursue this conjecture here (even though it is not beyond the ken of all possible experience). See, for example, Bergvall (1992, pp. 117-118) and Schnädelbach (1998, pp. 10-11).

[73] Pope Benedict XVI, 2008.

understanding'; it also states that reason and faith can never be at odds with one another and that 'they mutually support each other'.[74] Mystical knowing really *is* possible, and such knowledge or direct non-discursive experience cannot be reduced to logical or rational analysis. But we must be able to rationally unfold this mystical knowledge so far as possible, and we need to be able to integrate it with all facets of existence so far as we are able. Indeed, the Church Councils may have been grounded in mystical knowledge, but it took them a very long time and great discursive, scholarly, rational effort to understand, integrate, and unfold this knowledge. If reason and faith cannot be in disagreement, then van Fraassen's philosophy of science and his religious beliefs also should not be in disagreement. For example, he would need to show us how his empiricism could offer an explanation of the compatibility of the laws of physics and transubstantiation.[75]

I am sympathetic to an important implicit aspect of van Fraassen's view. If one has had a deep religious, spiritual, or mystical experience, it is quite difficult to articulate it rationally in such a way that is likely to fully convince someone who has not had such an experience and refuses even to acknowledge the possibility of having one. There are those (the materialists, for example) who will always aim to reduce all experience to neurochemistry or ultimately to physics while refusing to examine the foundations of physics, which points to the nonphysical aspects of reality. However, just because fundamentalist materialists refuse to countenance the possibility of genuine religious experience, it does not therefore follow that we should never attempt to offer rational arguments to support and clarify our experiences and beliefs about reality. The problem is not that 'reason rests on faith', as Clark has put it, but that van Fraassen wants to *separate* his faith from reason, which would generally

---

[74] See the First Vatican Council—1869-1870, Chapters 2-4. Tanner, 1990, pp. 806-809.

[75] As van Fraassen himself mentions transubstantiation, the topic is worth a bit more consideration. As stated by the Council of Trent—1545-1563 on the subject of transubstantiation:

> But since Christ our redeemer said that it was truly his own body which he was offering under the form of bread, therefore there has always been complete conviction in the church of God—and so this holy council now declares it once again—that, by the consecration of the bread and wine, there takes place the change of the whole substance of the bread into the substance of the body of Christ our Lord, and the whole substance of the wine into the substance of his blood. And the holy catholic church has suitably and properly called this change transubstantiation (Session 13, Chapter 4; Tanner, 1990, p. 695).

The church response to anyone who denies transubstantiation has been to say, 'let him be anathema' (Session 13, *Cannons on the most holy sacrament of the eucharist*, Tanner, 1990, p. 695).

be considered to be contrary to the teaching of the Catholic church. An atheist or any non-Catholic may well argue with me by saying that faith and reason have nothing to do with one another, but I would direct such people to my discussion on the importance of scientific faith found in Chapter Five. As for the Catholic who embraces van Fraassen's constructive empiricism, they will have to take their argument to the Pope.[76]

It is true that our discursive reasoning capacity will always fall short of ultimate truth. It is also true that we can use reasoned arguments in apparent support of atheism, pantheism, and essentially any position whatsoever, and so a conversion of sorts seems to be required for us to hold any position. This conversion can be trans-rational or pre-rational, and although they may look the same on the surface, they are actually at opposite ends of the spectrum. What is most obviously problematic here for van Fraassen is that he claims that 'there are no rules of right reason', while also telling us that 'reason rightly pursued' will help a secular person better appreciate a religious worldview. Clearly, however, if there are no rules of right reason, then it is impossible to pursue reason rightly.

In order to fend off rational criticism, van Fraassen is forced to fragment his religious beliefs completely from his philosophical and scientific beliefs. In fact, he would simply prefer to say that he has a philosophical 'attitude'. Philosophy, however, is not simply preparing a way for the Lord, nor is it merely brush-clearing for the scientist, and it is definitely something more than adopting an attitude. If philosophy is just an attitude, then my attitude is that Platonic realism is true, and I do not need to say anything more. My faith in the truth of this attitude is enough, and no rational criticism is permitted. Such a 'stance' is easy to adopt, and I can see why it would be seductive for van Fraassen to maintain it with respect to his incompatible positions. Nevertheless, when he tries to make his religious beliefs immune to rational criticism, he ends up being no different than any other religious fundamentalist.

Contrary to common misrepresentation, Platonism helps us to unveil the hidden unity of the entire cosmos. Our scientific and spiritual worldviews can be understood as not only being compatible, but also as being so interconnected that we can clearly see how each rationally implies the other. However, an intellectual understanding of this rational integration is not generally enough to bring about full emotional assent or total personal transformation. The mystical moment of direct understanding, in any case, is also vital in order to unleash so far as possible the dynamic and purifying power of this understanding, and faith is needed to sustain us in the process until we have verified for ourselves the ultimate truth. This ultimate truth, however, can never be held in totality or fully

---

[76] On Papal Infallibility see Vatican, 1993.

actualized by any individual or group, for the absolute, by its very nature, cannot be absolutely reduced to anything other than itself.

van Fraassen also believes that successful theories are the ones that have '*in fact* latched on to actual regularities in nature'. But he has to explain the nature of these actual regularities, or else he must ignore them and admit that he does not really know what he is talking about. In any case, he wants to forbid any serious explanation of the actual regularities in nature, because such an explanation will inevitably involve metaphysics. His way out of this dilemma is simply to evoke his antirealism and say that ontology is only pretend, so there is nothing to say about the regularities other than the fact that they are regular—such is the poverty of empiricist attempts at explanation. In other words, he chooses not to bother explaining why or how there are regularities in nature, for if he did try to explain them he would necessarily become involved with pure metaphysics and have to talk about the 'Reason' or 'cosmic order' that permeates the cosmos and is responsible for these regularities (which are actually laws). This metaphysical investigation would then involve him in bringing together reason and faith with his philosophy of science, something that he adamantly refuses to do.

One would reasonably assume that a philosopher would want to provide, or would be compelled by the nature of philosophy to provide, some sort of rational justification for their beliefs, especially those beliefs that they holds about the ultimate nature of reality. van Fraassen, however, will not provide arguments in support of his Catholicism, and when someone claims that their faith is impervious to reason or rational criticism, which is essentially to say that it is beyond question, then that person is a fundamentalist. Fundamentalists, such as van Fraassen, certainly do not then have the right to demand a rational explanation from others regarding their beliefs.

van Fraassen also believes that 'the success of current scientific theories is no miracle', which may seem to be an odd statement coming from someone who believes in miracles. van Fraassen would say that anybody who is not a Catholic would have no hope of understanding what he means by miracles. van Fraassen, however, is not a scientist, yet he feels quite confident in making numerous pronouncements about the sciences. It is also surprising that he draws a Darwinist analogy for successful scientific theories, which is a similar sort of error made by Smolin, as discussed in Chapter Eight.

van Fraassen also makes a similar error as Rogers when he claims that using scientific instruments to aid our perceptual capacities does not uncover what is already there, but actually *creates* something new. Moreover, if only unaided perception counts as observation, and science only deals with such observations of things that are physically actual, then a vast amount of the most important aspects of modern physics could not be counted as part of genuine science, but would instead be relegated to mere 'trivial' metaphysics. Einstein, Penrose, and other pioneering theorists would not be very happy to hear that most of their scientific work was trivial and non-scientific. In fact, if we take constructive empiricism seriously, then mathematics could not be a part of science because nothing about mathematics is observable, whether with unaided or aided perception.

Let us suppose that van Fraassen is somehow or other correct. Anyone who wears glasses (or contact lenses) would be quite upset, since only through scientifically aided perceptual enhancement can they clearly see certain objects that would otherwise be visually inaccessible. The objects that they see would themselves not be part of the scientific process, since van Fraassen tells us that they must actually be different objects than the ones I see, and that only observations made by the unaided senses are admissible under his constructive empiricism. Strictly speaking, anybody who wears glasses could never be a scientist.

Moreover, van Fraassen does not have any consistent way of saying how one person is able to contribute to scientific knowledge through unaided perception over another person who also has unaided perception but does not see quite as well as the first person, especially since, according to him, nobody ever sees the same objects anyway. He is risking falling into speciesism and is ignoring the rapid developments of robotics and the various offshoots of biology. If we can create a 'human liver chimeric mouse',[77] for example, then why not a human with a hawk eye or digital camera eye? It seems that only humans with eyes like van Fraassen's can make unaided observations that count as part of the sciences.

van Fraassen has also completely ignored the fact that so many pioneering theoretical physicists have argued that they need metaphysics, which I have already discussed at length. Metaphysics is not dead, and everyone holds some sort of presupposed metaphysics, as I have already shown. If I believe my computer exists when I do not perceive it, then that belief presupposes realism. If I think that I can distinguish my cat from an apple, then I believe that there are wholes that are distinguishable from other wholes, where each whole is itself an integrated unity of many parts, and each of these parts is also a whole. I am assuming, therefore, that there are wholes and parts and unifying forces, which I am somehow able to perceive and interpret or understand (whether consciously or not). This assumption also presupposes that I am existing and that I can trust my senses and abilities to reason about my senses; all of these presuppositions (and more) are implied as soon as I think that my cat is different from an apple.

Try as he may, van Fraassen cannot deny that he engages in metaphysics, for he believes that he can use reason (rightly pursed or not) to make claims about reality. Under the strictures of his constructive empiricism, however, he is not

---

[77] Karl-Dimiter Bissig, et al, 2010, p. 1.

permitted to do metaphysics. His theological claim that 'God's work goes on everywhere and every-when, throughout history' would clearly require a pure metaphysical defense, for such a claim is beyond scientific explanation and even beyond the boundaries of applied metaphysics. It is certainly beyond the ken of all possible experience, whether through our regular or technologically aided senses.

APPENDIX C

(From Chapter Eight)

Let us here consider a more detailed argument for the nonphysical reality of the mathematical laws of physics. We could greatly expand this argument into an even more exhaustively detailed analysis, but what I have provided will suffice as an example of the sort of reasoning process we can use to help us understand the rationality behind accepting the reality of the nonphysical laws of physics. My opponents may make certain objections, such as disputing my claim in premise (1) (see below) that valid mathematical structures are either real, nonphysical, and unchanging, or they are fictions; however, I have already argued for this bifurcation. There are possible middle positions, but when each of them is pushed to their limits they will collapse either one way or the other.

Consequently, either the laws of physics are real or they are not. If they are real, then we are either materialists or abstractionists, whom I have already argued against, or we are Platonic realists. If they are unreal, then we are either positivists or antirealists, whom I have also already argued against. In any case, not only can we meet these objections from our opponents, but we could easily turn the tables by critiquing *their* assumptions, premises, and logical methodologies, as well as asking *them* for much deeper clarification of their own positions. The main point to understand here is that there are, in fact, rational arguments for the reality of the nonphysical mathematical laws of physics. Below we will see a more formally structured example.

Starting Assumption (SA):

Appropriate aspects of mathematics are applied successfully in physics (and other sciences) to control and predict physical reality to the extent of applicability for each mathematical law.

1. Given (SA), then genuine mathematical structures either:
   a. have substantial being that is beyond the physical realm (they always are just what they are without change); or,
   b. are constructed fictions (whether in the materialist, abstractionist, positivist, or antirealist sense).
2. Assuming (1b), then either:
   a. the sciences are based on pure fiction and need not match reality; or,
   b. some fictions match reality and others do not.
3. Assuming (2a), we should be able to write any scientific fiction we want in order to change reality at will, which would then make science impossible.

APPENDIX C

4. Science is not impossible; therefore, we cannot write any scientific fiction we want in order to change reality at will; therefore, (2a) must be false.
5. If (2a) is false then (2b) must be true, and so some fictions match reality while others do not.
6. Given (SA) and (5), mathematical laws either:
   a. change; or,
   b. never change.
7. If (6a), then the mathematical laws would not have the predictive power required for physics to be successful, because the laws would be changing and we would not be able to predict those changes, and so (SA) would be false.
8. (SA) is not false, so (6b) must be correct.
9. (6b) reduces logically to (1a).
10. Conclusion: (1a) is correct, which is to say that genuine mathematical structures have substantial being that is beyond the physical realm: they always are just what they are without change.
11. Corollary: it is reasonable to be a mathematical Platonist.

APPENDIX D

(From Chapter Eight)

One of the main arguments against antirealism goes something like the following. If there is no objective reality, I should be able to do or create anything I want at every moment. Since I am unable to do or create anything I want at every moment, then there must be some sort of objective limitations that inhibit me. If there are any objective limitations, then there is something about reality that is objectively real, which means that antirealism is false.

Some antirealists will not want to accept this wholesale antirealism and will try to qualify their position in one way or another. What must be understood, however, is that there cannot be *any* qualification *whatsoever*. If there is even one qualification, then there will be some sort of limitation, and all such limitations must be either subjective or objective. If they are subjective, then they are conventional or arbitrary and there is no need to accept them, and so there is also no need to accept antirealism. If they are objective, then antirealism is false. So, either antirealism is objectively false, or there is no reason to accept it.

But if the only limitations are subjective, then they could in principle be broken, which means that in reality there would be no limits. In that case, any qualified antirealism would necessarily imply wholesale antirealism, as outlined above. Since wholesale antirealism is false, then any position that necessarily implies wholesale antirealism must also be false. Therefore, any qualified antirealism must either imply an objective limitation, which proves antirealism to be false, or it has a subjective limitation, which necessarily implies wholesale antirealism, which is false, meaning any qualified antirealism must also be false. In any case, it is not really necessary to try to prove antirealism to be false, since antirealists deny that there are any objective rational standards that could be used to adjudicate the truth or falsity of anything in the first place.

It is important, however, to provide examples of various rational paths out of the insidious antirealist trap for those who want to escape or avoid succumbing to it. To that end, let us consider two final analogous examples. What I would like to do is offer the antirealist exactly what they want: a world where they can do or create whatever they want instantly and without limitations. The first scenario is that our dreams actually propel us, or enable us to consciously enter into, endless alternative realities. The second scenario is a technologically plausible imaginable future where we can create and do whatever we want while in a waking state.

Scenario One: Dreaming Reality

1. When I dream, the part of me that remains aware of my dreams is able to enter into that dream reality. This is an actual reality that is at least as real as what I experience while I am awake.
2. In dream reality, I have no limitations. I can fly, jump backward and forward in time, and create whatever I want instantly.
3. Every other sentient creature capable of dreaming is able to do the same things in their dream reality, so there are endless subjective realities that all exist simultaneously.
4. Therefore, antirealism is true.

The first point is that 'I', the being that is aware in waking and dream realities, requires a thorough analysis. What, exactly, is it that is experiencing these realities? I will set this question aside here. The second point is that even if I could enter an infinity of dream realities and do anything at all without limitation, there would still be other objective limitations in place. For example, I would need to have a body to be able to dream in the first place. However, perhaps a body is not necessary; maybe I could exist as a disembodied being, endlessly traversing multiple intangible alternative realities. We may grant that, but then three other difficulties follow. First, the fact remains that we cannot do in waking reality what we can do when visiting our dream realities. At the very least, I can say that if I am currently in a dream reality when I believe that I am actually in waking reality, it is still the case that I, and whoever else is here with me, cannot act as we do in other dream realities. Therefore, there are objective limitations in waking reality (or the dream reality version of waking reality), which is to say that in some part of the totality of all possible realities, there are objective limitations.

The second point is subtler. If there are endless dream realities, then either they are in some way connected, or there is absolutely no connection or relationship between them. If there is no connection between them, then all the people who seem to inhabit my dream reality are actually not real, as there would be a real objective limitation that prohibits any crossing over between different dream realties. Moreover, in that case, once I ended up in a particular dream reality I could never leave it, because there would be no way to jump into another dream reality since they are not connected and there is no way to move between them. In that case, every time I think I have entered another dream world, I would be objectively mistaken.

However, if these dream realities are connected, then we could potentially move between them, but then we have to admit that all the realities are bound together into a unity, which would be an objective fact. This objective fact would actually be a limitation, since it would prohibit endless dispersion and instead maintain cohesion. However, it is this cohesion that allows for the interconnections between each dream reality. Moreover, even if there were no unity between the dream realities, it would still be the case that each separate dream reality would have to be a self-contained unity, keeping itself separate from the infinity of other dream realities, but this act of containment would also be an objective limitation.

The third point is that something cannot be what it is and what it is not at the exact same moment in the exact same dream reality. To prove me wrong, we need only draw a circular square, or drive a car that is an apple—not an apple-shaped car, but an actual apple. We can be sure of one thing; if there were no objective limitations in this waking reality (or dream version of waking reality), most of us would be living very different lives than we presently do. We can be sure that there are objective limitations of one sort or another, even if this dreaming scenario is true.

Scenario Two: Projecting Reality

1. I can connect a small device to the back of my head that reads all my thoughts and instantly translates them into dynamic holographic images.
2. A far more sophisticated and expensive model enables these holographic images to become actual physical replications.
3. Every intelligent creature that has the more sophisticated replicator is also able to do the same as me, so everyone in principle can instantly create any physical reality they want.
4. Therefore, antirealism is true.

If this scenario is not possible in reality, then there is an objective limitation, so antirealism is false. But even if it could be possible, there are still objective limitations. For example, you could not make a herd of fifty adult elephants suddenly appear on your fingernail. If we say that we need only replicate an enormous fingernail for ourselves, then we have to remember that the objects that we are replicating (manifesting) into our regular waking reality have to take up space, which means that there are going to be limits on what size objects we can replicate. You cannot even replicate two small objects on the exact some spot at the exact same moment, as no two (macro) physical objects can occupy the exact same space at the exact same time (at least in this waking reality). Thus, there would be objective limitations. Moreover, you could only replicate things from your thoughts if there were a machine that enabled you to do this, and in the absence of such a machine, you would not be able to do so. Therefore, there are objective limitations. This technology, of course, would only be possible if we were to discover the objective laws of physics that would enable us to create it in the first place.

If those people who claim to be able to make objects materialize really have such power, why can't they instantly materialize very large objects, like a fleet of airplanes, or the moon? These sorts of possibilities could be explored in more detail, but it remains the case that antirealism must always fail in the end.

NOTES

Notes for chapter one

[1] Penrose, 2004, pp. 1028-1029. See also pp. 17-23.

[2] Nobody wants to hold false beliefs, and yet we all do. One of the main founders of the contemporary philosophical counseling movement, Pierre Grimes, has spent decades helping people rationally uncover, understand, and transcend (or dissolve) a certain class of sick beliefs (the 'pathologos'), which block us from attaining our most significant goals. (See Grimes and Uliana, 1998).

[3] See Benn, 1882, p. 82.

[4] As Socrates stressed in Plato's dialogues, we must follow the argument wherever it leads us through the darkness of the unknown in pursuit of the light of truth, which is also vital for the sciences. (For example, see Harris, 2000, p. 24 and Polanyi, 1964, p. 14.) I am not concerned here with the otherwise interesting and important arguments aimed at making distinctions between the historical Socrates and Plato; see Cohn, 2001, for such a discussion.

[5] Einstein, 1954, p. 11. Whittaker makes the similar remark that 'the emotions in which philosophy and science had their common source was exactly the same in ancient Greece and renascent Europe. Plato and Aristotle, like Descartes and Hobbes, define it as "wonder"' (Whittaker, 1928, p. 8).

[6] Plato, 2001, p. 133, *Apology* 38a.

[7] Bible, 1992. 1 Corinthians 14:34-35

[8] There is a way in which such beliefs can impact the *practice* of science, such as deterring or excluding women from becoming scientists. But that important sociological point is not relevant here. $E=mc^2$ is true, or false, or approximately true, regardless of whether it was discovered by a woman, a man, or an alien from another galaxy.

[9] Freud seems to be paraphrasing and agreeing with Le Bon. (Freud, 1960, p. 16.)

[10] Horstman, 2009, p. 60.

[11] Colet in Blau, 1995, p. 222.

[12] Lee in Blau, 1995, p. 223. UFC champion Georges St-Pierre also understands the importance of pattern making and breaking. As St-Pierre steps into the cage, one of his coaches, Greg Jackson, says that 'human brains are looking for a pattern. Establishing that pattern and then breaking it can be very powerful' (Jackson quoted in O'Brien, 2012).

[13] Einstein, 1954, p. 66.

[14] Einstein, 1954, p. 9.

[15] Penrose, 2004, p. 1029.

[16] Dawkins, 1989, p. 330.

[17] Planck, 1932, p. 214.

[18] Schrödinger quoting Jung in Schrödinger, 1967, p. 129.

[19] Frattaroli, 2001, p. 13.

[20] Frattaroli, 20.01, p. 6.

[21] Frattaroli, 2001, p. 115.

[22] Schrödinger, 1952, p. 51 (original emphasis). MacKenna translates the relevant part of this passage as 'what are We?'. (Plotinus, 1991, p. 453; *Ennead* VI. 4. 14.)
[23] Planck, 1932, p. 217
[24] Pauli in Jung and Pauli, 1955, p. 152.
[25] Einstein, 1954, p. 11.
[26] We should keep in mind that the word 'logos', as Rist notes, 'is perhaps the most difficult term in Greek philosophy' (Rist, 1977, p. 84).
[27] Schrödinger, 1964, p. 94.
[28] Einstein quoted in Blackmore, 1983, p. 396. See also Jaki, 1978, p. 185.
[29] Crease, 2002.
[30] Cartwright, 1983.
[31] It is not really clear if each 'world' is equivalent to a 'universe' or just another world within the totality of one universe. Regardless of whichever interpretation we adopt, we will find similar metaphysical trouble.
[32] As surprising as it may seem, the claim that history can offer us no guidance for the present is a view that is often vociferously championed by various historians and classicists (for example, see Murphy, 2007, pp. 12-14).
[33] See Capra (1991), Zukav (1980), and McFarlane (2002). For an examination of the various 'pitfalls' in too simplistically assuming parallels between Eastern mysticism and modern physics, see Restivo (1978).
[34] Shimon Malin (2001) is an unusual and welcome example of a contemporary physicist arguing for a Western perspective of quantum physics, drawing from Platonism and Whitehead. However, Malin argues against realism, apparently because of his confusion concerning materialism, realism, and Platonism, and he adopts what seems to be more of a version of Berkeleyan Idealism. I discuss these sorts of errors in general and with respect to other authors, but interested readers could profit from reading Malin and gaining an alternative but related view.
[35] For example, although the ancient Egyptian sages distrusted the Greeks, 'whom they charged with being fickle and inconstant', they reluctantly assented to train Pythagoras in their mysteries (Schuré, 1923, p. 21). In fact, 'Pythagoras crossed the whole of the ancient world before giving his message to Greece. He saw Africa and Asia, Memphis and Babylon, along with their methods of initiation and political life' (Schuré, 1923, p. 11). See also Bernal, 1987.
[36] Newton quoted in Stokes, 2010, p. 31.
[37] Chaisson and McMillan, 1999, p. 628 (original emphasis).
[38] Schäfer, 1997.
[39] Private correspondence with Stephen Snobelen. See also Snyder, 2006 and Beck, 1959, p. 21.
[40] Ayto, 1990.
[41] Tipler, 1994, p. xv.
[42] Gibbins, 1989, p.1.
[43] Kahney, 2009, p. 173.

Notes for chapter two

44 Isaacson quoted in Ferenstein, 2011.

45 Heisenberg, 1974, p. 117.

46 For example, central state materialism says that 'mental states are contingently identical with states of the central nervous system', but unlike logical behaviorism it 'does not imply that mental sentences can be translated into physical sentences' (McLaughlin in Audi, 2001, p. 687).

47 K. Ward, 1996, p. 99.

48 Dennett, 1991, p. 33 (original emphasis). Non-reductive materialists partially appear to escape this definition, but mental properties either emerge as something physical or as 'something' nonphysical. If they emerge as something physical then, given basic materialist assumptions, their claim of irreducibility is dubious. If they emerge as something nonphysical, then materialism would necessarily be false.

49 Some philosophers use the words 'physicalism' and 'materialism' interchangeably, while others believe that there is a distinction. For example, physicalism tends to allow for forces such as gravity that may not fit the traditional definition of matter, which may be stated as being 'an inert, senseless substance, in which extension, figure, and motion, do actually subsist' (Berkeley, 1992, p. 79, *Of the Principles of Human Knowledge*, Part 1, paragraph 9). (See also F. W. McConnell in Steinkraus, 1966, pp. 50-51, Daniel Stoljar, 2009, and Blackburn in Bunnin and Tsui-James, 2003, p. 65.)

50 Finally some attention is starting to be paid to such deeply metaphysical physicists. For example, see Atmanspacher et al (2007), who hosted a conference in Switzerland in 2007 on 'Wolfgang Pauli's Philosophical Ideas and Contemporary Science'.

51 See Pappas in Audi, 2001, p. 418.

52 Norris, 2000-A, p. 63.

53 See Zeyl's discussion in his Introduction to his translation of Plato's *Timaeus*, 2000, pp. xxix-xxxii.

54 Norris has already argued successfully against seeking such a compromise when he showed several problems in what are referred to as Response-Dependence approaches, which need not concern us here. See Norris, 2002-A, 2002-B, and 2003.

55 Baggott, 2004, p. 118.

56 Faye notes that although it was commonly assumed that the Copenhagen interpretation was subjectivist and positivist, today 'anyone who has studied Bohr's essays carefully agrees that his view is neither'. However, what Faye has glossed over is that the current consensus remains fixed on the apparent antirealist aspects of his thought to the exclusion of his obvious realism. But, as Faye continues, 'there are, as many have noticed, both typically realist as well as antirealist elements involved in it, and it has affinities to Kant or neo-Kantianism' (Faye, 2008). But Kant had also been influenced by Platonism. For example, in his review of T. K. Seung's *Kant's Platonic Revolution in Moral and Political Philosophy*, J. Ward quotes Seung as saying that 'Platonic ideas 'provide the ultimate goal for all [Kant's] maneuvers; they give the unity and integrity to all his works. The spirit of criticism is only a dutiful

handmaiden to his grand Platonic vision'. (J. Ward, 1996, p. 281). However, Brittan claims that in arguing against Hume, Kant wanted a 'fully realist, or *material*, interpretation of Newtonian physics' (Brittan, 1978, pp. 125-126, original emphasis), which would seem to put Kant at odds with Platonism in this respect, at least if Brittan's interpretation is correct.

[57] Baggott, 2004, Preface (original emphasis).

[58] Feynman, 1985, p. 10. Even Heisenberg sometimes felt that nature might be absurd. After talking with Bohr for hours on end and falling into a state of despair, Heisenberg walked through the park alone and reflected: 'can nature possibly be as absurd as it seemed to us in these atomic experiments?' (Heisenberg, 1958-A, p. 42). To answer his question: no, nature is not absurd.

[59] Dawkins in Twist, 2005. See K. Ward, 1996, esp. p. 11-12, and McGrath, 2004, for a critique of Dawkins with respect to his distortion and misrepresentation of religion and philosophy. See Barr, 2004 for several examples of how Dawkins gets his science wrong. See Midgley, 1979 for philosophical criticisms of his position.

[60] Quoted in Burtt, 1925, p. 64.

[61] Proclus, 1987, pp. 602-603, *Commentary on Plato's* Parmenides, BK VII, 74K.

[62] Gibbins, 1989, p. 2 (original emphasis).

[63] Laszlo, for example, writes that 'Einstein substituted the relativistic universe for Newton's mechanistic clockwork universe' (Laszlo, 2004-B, p. 536).

[64] Rowlands, 2007, p. 57.

[65] For example, see Janiak, 2006.

[66] See Shorey, 1927, p. 181.

[67] Santillana, 1957, pp. 15-16.

[68] Rowlands, 2007, pp. 59-60.

[69] Gibbins, 1989, p. x.

[70] Heisenberg, 1958-A, p. 28.

[71] Gibbins, 1989, p. 55.

[72] For example, see Computational Physics and the American Institute of Physics.

[73] Gibbins, 1989, p. 24.

[74] Although entanglement is usually considered necessary for nonlocality, the two concepts are not identical or reducible to one another. However, entanglement may also sometimes be a local phenomena. For example, see Hewitt-Horsman and Vedral, 2007.

[75] Baggott, 2004, p. 105.

[76] Gibbins, 1989, p. 47.

[77] Faye, 2008.

[78] See Heisenberg, 1974, p. 159 and Einstein, 1954, p. 323.

[79] Afshar et al, 2007 and Rowan University, 2007 (discussing the same experiment). However, this experimental conclusion is heterodox, as the majority of physicists would not likely agree with it.

[80] For example, see Gibbins, 1989, pp. 9-10.

[81] For example, see Yourgrau, 2005, p. 35.

[82] For example, see Powell, 2009.

[83] See Norris, 2002-A, p. 40, and Santos in Selleri, 1988, p. 389. Norris writes that 'Von Neumann's apparent mathematical proof against the possibility of a hidden-

variables theory has been argued to be conceptually flawed; Bell, like Bohm, thought that nonlocality was a small price to pay in comparison with the various conceptual dilemmas imposed by the orthodox theory' (Norris, 2002-A, p. 40). Physicist Emilio Santos also argues that the experiments by Aspect and collaborators have not disproved local hidden-variable theories, concluding that 'the search for hidden-variables theories should be a priority of theoretical physics' (Santos in Selleri, 1988, p. 389). Although one may argue that a hidden-variables theory is not actually an interpretation of quantum theory but is more of an alternative to 'standard' quantum theory, it is still the case that philosophical assumptions play a significant underlying role in our decision to approve or disapprove of hidden-variable possibilities.

[84] See Baggott, 2004, p. 105 and Gibbins, 1989, p. 56.

[85] Gibbins, 1989, p. 48.

[86] Gerson, 2005-A, p. 256.

[87] For example, see Rasmussen (2005), Cleary (1999), Dillon (1977), Gersh (1986), and Jayne (1952).

[88] For example, see Gerson, 2005-A and 2005-B.

[89] Gerson, 2005, pp. 32.

[90] The following sources are a small sample of relevant secondary literature: Siorvanes (1996), Lowry (1980), Rosán (1949), Lloyd in Blumenthal and Lloyd (1982), Whittaker (1928), Bos and Meijer (1992), Blumenthal in Huby and Neal (1989), Beierwaltes in Blumenthal and Markus (1981), and Clark (1994, 1997).

[91] Siorvanes, 1996, p. x.

[92] Corrigan, 2005, p. 235.

[93] On the freedom of thought at Plato's Academy, see Hadot, 2002, pp. 64-65.

[94] Emerson, 2000, p. 421.

[95] Rorty, 1990-A, p. 311.

[96] Rorty, 1979, p. 43.

[97] Benn, 1882, pp. 386-387.

[98] Benn, 1882, pp. 336.

[99] Arieti and Wilson (2003) also offer a distorted misrepresentation and caricature of Neoplatonism. Levinson (1953) offers a detailed and fair representation of some of Plato's key detractors before arguing persuasively against them. See also Greene (1953).

[100] Gerson, 2005-A and 2005-B; Siorvanes, 1996.

[101] The philosophy of mathematics is an interesting and technical field, but such specialists still often overlook the deeper metaphysical issues underpinning their theories. For an overview of the philosophy of mathematics, see Horsten, 2007. I am not going to discuss Balaguer's *Platonism and Anti-Platonism in Mathematics* (1998) beyond a few points given immediately below. Balaguer seems to ignore essential relevant concepts of modern physics while promoting his brand of positivism: 'the central viewpoint that I am trying to motivate in this book is that we don't have any sound arguments for or against Platonism, let *alone* the necessity or impossibility of that view' (Balaguer, 1998, p. 167). Well, there are no sound arguments for or against Balaguer's arguments that there are no sound arguments for or against Platonism in mathematics, and so, on his own terms, there can be no fact of the matter about his arguments. As I have written elsewhere (Spencer, 2007), Balaguer's

claim is in principle no different than someone arm wrestling with themselves and, being unable to win or lose, claiming that there are no facts of the matter about winning or losing. If there is no fact of the matter (whatever that would actually mean), then there would at least be the fact that there is no fact of the matter, which would then *be* the fact of the matter. But if there really is no fact of the matter, then clearly it does not matter which view one holds, and if there are no good or bad arguments for or against mathematical Platonism, we may as well just have faith in whichever view we want with no further arguments required. We must wonder why Balaguer would bother writing a book that concludes that there is no fact of the matter about the topic of his book.

[102] Sriraman, 2004, p. 134. (Sriraman is quoting from Davis and Hersh, *The Mathematical Experience*, 1998, p. 318, which I do not cite.)

[103] See Koyré, 1968-B, pp. 36-37 and Wedberg, 1955, p. 137. Sriraman, 2004, p. 134.

[104] Koyré, 1968-B, pp. 36-37. Despite this antirealist aspect of Aristotelianism, Gerson has argued that Aristotle was still a Platonist in the most fundamental sense (see Gerson 2005-A, pp. 269-276 and Gerson 2005-B).

[105] On theurgy, see Sheppard, 1982.

[106] Siorvanes, 1996, p. 9.

[107] This comment is especially for those interested in Ken Wilber and integral theory. (Other interested readers are directed to the works of Wilber and many other relevant authors, as I am not going to discuss this topic except for the following observation.) While much of Wilber's work is of vital importance for helping us to understand integrated systems, it is equally important to recognize that Platonic metaphysics cannot be reduced to one of Wilber's quadrants. Since the quadrants themselves are in some sort of unity, the whole system presupposes that unity is in fact possible, which ultimately means that we need a way of explaining unity itself so far as we are able, which is one of the main goals of Platonism. In other words, to the degree that integral theory is viable, it *presupposes* Platonic realism as well, which must underpin any integral theory in much the same way as it underpins any theory in physics. This point does not negate or devalue the importance of Wilber or integral theory. However, it does make sure that we are aware that the very viability of integral theory necessarily presupposes the prior reality of unity, which the Platonists specifically sought to clarify rationally so far as humanly possible, thus transcending any sort of classificatory scheme, including their own.

[108] Einstein, 1954, p. 274.

[109] Norton, 2000, p. 135. Also see Yourgrau, 2005, p. 34.

[110] Heisenberg, 1974, p 173. Even the foundations of modern medicine are directly dependent upon Platonism to a significant extent. For example, see Lindberg (1992, p. 125-127), von Staden (1995, p. 62), and Frattaroli (2001, p. 287). In the second century CE, Galen of Pergamum, the 'leading medical authority of antiquity—rivaled only by Hippocrates...[who had] unparalleled influence well into the modern period' (Lindberg, 1992, p. 125), was influenced by Aristotle yet esteemed Plato and Hippocrates as his classical heroes. 'Galen in his principal anatomical work more than once exhorts his reader: "Follow Plato and me"' (von Staden, 1995, p. 62). Galen adopted Plato's tripartite division of the soul to correlate with 'the three basic

physiological functions defined by Erasistratus' (Lindberg, 1992, p. 127). Even Freud's 'model of the tripartite soul (the *It*, the *I*, and the *I that stands above*) resembled Plato's but...was more specifically designed to explain the processes of inner conflict' (Frattaroli, 2001, p. 287).

111 Shorey, 1927, p. 172.

112 On Copernicus being a (Neo)Platonist, see: Burtt (1925, p. 43), Shrimplin-Evangelidis (1990), Siorvanes (1996, p. 36), and Koyré (1968-A). For a brief dismissal of Copernicus being a (Neo)Platonist, see Rosen, 1983.

113 Santillana, 1957, p. 24. Santillana notes further that it was an 'absolute Pythagorean faith which made Copernicus insist that what he presented was not merely an abstract model such as mathematics was supposed to prove, but a true physical system' (Santillana, 1957, pp. 157-158).

114 Whitehead, 1948, pp. 42-43.

115 Whitehead, 1948, pp. 42-43. Also see Whitehead, 1953, pp. 47-48. Burtt offers further clarification: 'the famous Pythagorean doctrine that the world is made of numbers is apt to appear quite unintelligible to moderns till it is recognized that what they meant was *geometrical units, i.e.,* the sort of geometrical atomism that was taken over later by Plato in the *Timaeus*' (Burtt, 1925, p. 30, original emphasis). Also see Burtt, 1925, p. 41. Aristotle, however, had tried to separate Plato and Pythagoras by raising the following distinction: 'Plato says that the numbers are apart from the objects of sense; whereas the Pythagoreans say that things themselves are numbers, and they do not place the "mathematicals" in between them' (Aristotle, 1963, pp. 48-49. *Metaphysics* Bk I.6). But numbers and their relations cannot be reduced to physicality or mere sensations. '1+1=2' is true regardless of what physical things we are applying this relationship to, and abstract mathematical relations hold true regardless of any particular reference to physical reality. Yet the mathematical laws of physics must be grounded (or be capable of being grounded even if we cannot yet prove it) in the actual objects of physical reality, otherwise they are not genuine laws but are just mathematical constructions. Therefore, in one sense, all physical things are just manifestations of numbers in motion according to mathematical laws, and yet these laws are not reducible to particular things, or to anything physical. Aristotle seems not to have grasped the subtlety here.

116 Burtt, 1925, p. 71. See also Santillana, 1957, pp. 226-227.

117 Stokes, 2010, pp. 129-130. Newton was at least a mathematical and scientific Platonist, and Newton's *Principia* is certainly in the spirit of Plato's *Timaeus*. However, as J. E. McGuire and P. M Rattansi argue, 'Newton's Platonism was not entirely the Platonism of More and Cudworth, with their stress on such intermediaries as the Hylarchichal Principle; but it was also a Platonism in the spirit of the early Church Fathers' (McGuire and Rattansi, 1966, p. 124). For a sample of further relevant readings on Newton that offer a variety of views, see Snobelen (2001 and 2006), Westfall (1962), Rogers (1979), and Forsyth (1932).

118 Newton, 1999, p. 940, *General Scholium* (original emphasis).

119 Despite such obvious Platonic commitments, Newton also rejected certain aspects of Platonism that he thought were not congenial to his vision of Anti-Trinitarian Christianity. He rejected Platonists, Kabbalists, and Gnostics for similar reasons, such as for their supposed injection of metaphysics into theology, which he

thought was wrong because Scripture, he believed, was meant to teach us morals, not metaphysics. (For example, see Coudert in Force and Popkin, 1999, pp. 30-31.) Here is where we must part ways with Newton, because if scripture cannot be subjected to metaphysical scrutiny and historical analysis then there seems to be no way to prevent a slide into fundamentalism.

[120] Siorvanes, 1996, p. 36. See also Kepler (1997), Burtt (1925, pp. 58-59), Taylor (2006, p. 6), Koyré (1968-B), and Pauli in Jung and Pauli (1955).

[121] Kepler, 1997, pp. 298-302.

[122] Kepler, 1997, p. 302.

[123] Bergethon, 1998, p. 10.

[124] Laudan, 1969.

[125] See Zhmud, 1998.

[126] See, for example, Lloyd (1968), Tait (1986), and Hammond (1935).

[127] See, for example, Whittaker (1928, p. 13), and Norris (2002-A, p. 137-138).

[128] Whittaker, 1928, p. 40.

[129] Gerson, 2005-A, p. 263. See also Santillana, 1957, pp. 24–25. As Clark notes in a different context, Platonism is an older realism (Clark, 1995). Clarifying the meanings of 'realism', 'idealism', and 'antirealism' is not easy, but this further brief technical exploration should be helpful.

Nadeau and Kafatos argue that the terms 'antirealism' and 'idealism' do not apply to Bohr. They also claim that idealism 'properly applies to the so-called realists who assert the existence of an ideal system with properties that cannot be simultaneously measured', and that Bohr was actually 'brutally realistic in epistemological terms' (Nadeau and Kafatos, 1999, p. 98). Indeed, some scholars would rather refer to Platonism as idealism, while others would prefer to call it idealist-realism. In fact, Corrigan argues that Plotinus 'spells out his own peculiar idealist-realist, but definitely Platonist, version of perception' (Corrigan, 2005, p 67; see Ennead I, 1, 7). Unfortunately, the phrase 'idealist-realist' may falsely suggest a compromise between the two positions, which is especially problematic (if not impossible in principle) if one accepts *modern* idealism and *materialist* realism. However, ancient idealism, as just noted, is really Platonic realism, which says that abstract nonphysical laws, principles, and universals are real and, on my account, it also accepts that matter is objectively real. In this sense, we can say that Platonism is idealist-realist, but it is less confusing to simply say 'Platonic realism' or just 'Platonism'.

The word 'realism' derives from the Latin word for 'thing' (Ayto, 1990), which implies that anything real must be a physical thing However, in philosophy, 'realism' has three general meanings:

'the doctrine that universals or abstract concepts have an objective or absolute existence. The theory that universals have their own reality is sometimes called **Platonic realism** because it was first outlined by Plato's doctrine of 'forms' or ideas'. (Oxford Dictionary of English.)

'the doctrine that matter as the object of perception has real existence and is neither reducible to universal mind or spirit nor dependent on a perceiving agent'. (Oxford Dictionary of English.)

Truth and reality are objective, which follows from the definitions 1 and 2.

'Idealism', however, is usually defined in philosophy as denying 'the existence of material things' (Audi, 2001, p. 563), and can often include the belief that 'objects of knowledge are held to be in some way dependent on the activity of mind' (Oxford Dictionary of English). The etymology of 'idealism' is 'via Latin from Greek idea 'form, pattern', from the base of idein 'to see'' (Oxford Dictionary of English). However, Plato used the word idea 'in the specialized sense 'archetypal form of something...but as far as the modern English noun [the word "idea'] is concerned, its sense 'notion, mental conception' developed (in Greek) via 'look, appearance,' 'image,' and 'mental image' (Ayto, 1990). Thus, 'idealism', can mean the appearance or image, or it can refer to an archetypal form.

On one hand, if we take the word realism in the literal etymological sense, then we would seem to be committed to materialism. However, the denial of realism in general leads to the denial of objective truth and reality, leading to relativism, which would be the end of the very possibility of science and all rational discourse. On the other hand, if we accept the ancient meaning of idealism, then Platonism is idealist, although it is not clear that this ancient idealist position also allowed for the mind independent reality of physical entities, which Platonists certainly did believe were real. So, if we use the word 'idealism' in the modern sense, we end up denying the reality of the material world and can easily slip into full-blown antirealism, and if we use 'realism' in the strict etymological sense we end up endorsing materialism. We can avoid these problems by more clearly stipulating the meaning of such key terms, such as I have done for Platonic realism, which allows for the reality of the physical universe and mind-independent objects, as well as the underlying nonphysical aspects of reality. In other words, 'Platonic realism' or, simply, 'Platonism', is extreme realism that accepts and attempts to explain the corporeal and incorporeal aspects of the totality of reality.

Through private correspondence (April, 2012 – not listed in the Works Cited), Schäfer makes some very interesting points. For example, he explains that the German word "Realität" (reality) is actually inferior to another term, "Wirklichkeit", which comes from the German "wirken", 'meaning affecting you'. In other words, anything that can act on you is real, which would include both physical and nonphysical aspects of reality. Therefore, it seems to me that perhaps the best word to use is neither 'realism' nor 'idealism' but 'wirkenism'. This neologism is not likely to catch on, however, so I will stick with Platonism.

[130] Poli, 2001, p.279.
[131] Wigner, 1960.
[132] Gerson, 2005-B, p. 34. See Plato, 1996, p. 129, *Theaetetus* 176b.
[133] Schäfer, 1997, p. 157.
[134] Schäfer, 1997, p. 157.
[135] There is a new trend called 'experimental philosophy'. For example, see Knobe and Nichols, 2008.
[136] Trusted, 1991, p. xi.
[137] See Walsh, 1963, p. 11.
[138] See Berlin, 1957 and Walsh, 1963, p. 13.
[139] Hacking, 1983, p. 46.

[140] Trusted, 1991, p. 144. Trusted continues: 'even if science and mathematics could be adequately assessed in positivistic terms there would still be metaphysical assumptions underlying the scheme of knowledge expressed entirely in terms of sense experiences' (Trusted, 1991, p. 144).

[141] Burtt, 1925, p. 224.

[142] Taylor, 1936, p. 13.

[143] Harris, 2000, p. 5.

[144] Mattick makes a related point when he argues that Marxism, 'not being a theory of physical materialism and not bound to Newtonian determinism, is not affected by the new physics and microphysics' (Mattick, 1962, p. 358). It does not follow, therefore, that our metaphysical theories should be completely distinct from the sciences, but it does mean that we should be cautious about using the sciences as a foundation for our metaphysics.

[145] Eddington, 1935, p. 276. Also see Eddington, 1929 and 1935, to see how far he engages in metaphysical speculation and ventures into mysticism.

[146] Whitehead, 1953, p. 21.

[147] Proclus, 1970, p. 62, *A Commentary on the First book of Euclid's Elements*, Chapter VIII. 75.

[148] Rorem in Pseudo-Dionysius, 1987, footnote 2, p. 135.

[149] McGinn, 1991, pg. 25.

[150] These first three are from McGinn, 1991, p. xv-xvi.

[151] Daniels, 2005, p. 235. (I have removed the original italicization of the entire quote.)

[152] Benn, 1882, p. 82.

[153] Back cover, Hyland and Rowlands, 2006.

[154] See Fröhlich in Hyland and Rowlands, 2006, p. 329. (Original emphasis.)

[155] Wilber's *Quantum Questions: Mystical Writings of the World's Great Physicists*, 1984, is an edited collection of writings from several pioneering twentieth century physicists. Although Wilber had the correct intuition to go to the original sources, he does not undertake the rigorous metaphysical work required to explain, expand, and critically evaluate and correct these physicists' views. But this was not his purpose, and so my point is not a criticism of his important work. It is unfortunate that so few academics have engaged directly with these vital original writings of such eminent physicists. This is a major oversight that this book can help to rectify.

[156] Baggini and Fosl, 2003, p. 200.

[157] Stenger, 1995, p. 11.

Notes for chapter three

[158] Heil in Audi, 2001, p. 26.

[159] Rorty, 1999.

[160] Sorrell in Sorrell and Rogers, 2005, p. 1.

[161] Solso, 2003, points out the interesting statistic that in 1950 there were about 23 articles published on consciousness, whereas in the year 2000 there were approximately 11,480. While there is now growing interest in the role of consciousness, this vital topic was mostly neglected by philosophers when quantum

theory was first unfolding. Despite important exceptions, the general consensus in philosophy, psychology, and psychiatry is that consciousness, if it exists at all, is merely an epiphenomenon resulting from neurological complexity. However, many pioneering physicists have thought of consciousness as being immaterial and prior to physical reality. The concepts of emergence and epiphenomenalism are highly problematic, but I will set aside the issue here. (Clark, 2002, argues that 'the reductive fantasy is no more coherent than the epiphenomenalist fantasy'.)

[162] See Harris, 1983, p. 37.

[163] Hawking and Mlodinow, 2010, p. 5. Hawking has believed for a long time that philosophers are incapable of really participating in a serious understanding of the sciences, and many years ago he paraphrased Wittgenstein as saying that 'the sole remaining task for philosophy is the study of language'. He then adds, 'what a comedown from the great tradition of philosophy from Aristotle to Kant!' (Hawking, 1988, p. 175). He also writes that 'in the nineteenth and twentieth centuries, science became too technical and mathematical for philosophers, or anyone else except for a few specialists' (Hawking, 1988, p. 174), and his not-so-subtle stab at Plato (and all philosophers after Kant) may well be aimed at Penrose. Clark has an apt rebuttal: 'the weird assumption that only those with "scientific" training can actually think is as obvious a piece of self-serving ideology as that of any ancient priesthood....Misplaced animism may be an intellectual sin, but so is misplaced mathematization' (Clark, 1990, p. 31).

Physicist Lawrence Krauss makes similar errors as Hawking, and he simply defines the terms of his game so as to eliminate his potential opponents. He also claims that philosophy is useless while relying upon arguments to defend his position, which is to do philosophy. Unfortunately, you are not likely to argue well when you deny the importance of arguing well.

We could take any of his key premises and dismantle them and deconstruct them back into 'nothing'. One very brief sample shall suffice. Krauss claims that 'the ultimate arbiter of truth is experiment' (Krauss, 2012, p. xvi), which is merely to hold to the philosophical position known as empiricism. Indeed, every brief critical comment I make below could be expanded and developed much further.

1. Krauss is also holding a naive realist position in assuming that there is any such thing as truth in the first place. I agree with him, of course, that truth is real, but he has not done the work to convince us that he knows what he is talking about when he uses the word. Is truth physical, or can it also be nonphysical? If it is only physical, then he is also holding the philosophical position of materialism, and he would have to address the issue of the nature of the laws of physics that he thinks can be discovered (or pop out of nowhere). If he accepts that truth could be nonphysical, then he is starting to sound like a Platonist.
2. An 'experiment' is not literally capable of arbitrating anything; only we, or other intelligent creatures, can be the arbitrators. That does not reduce truth to our opinions, and it does not mean that we can never be wrong. Reality certainly is what it is regardless of what we happen to believe about it, but there could be no such thing as an experiment without conscious, intelligent creatures who are able to create an experiment and collect, read, and interpret the resulting data. Indeed, there is no such thing as a purely observational statement. All so-called

data must be interpreted, which requires the use of philosophy, as well as philosophical assumptions even to begin using rational thought or constructing experiments in the first place.
3. To rely solely upon experimental data to tell us what is true is necessarily to require relying upon our own senses. Of course, our senses can deceive us; so relying purely on sense data to tell us what is true or false is to open the way to all the relevant skeptical arguments. Krauss cannot guarantee that he has found anything true when he relies solely upon his senses. In fact, he attempts to use arguments coupled with interpreted sense data, but the very act of using arguments is to do philosophy, which he denies has any value. So he cannot therefore value rational thought, which means that either his own thinking is not rational, or he has just made a blatantly irrational contradiction by denying the reality of rational thought while simultaneously relying upon the reality of rational thought.

The underlying assumption that Krauss expects us to accept is that just because he is a scientist, he is therefore a better thinker than anybody else. Well, let us grant that he is correct. Then he will have no possibility of saying that the theistic or spiritual conclusions of other scientists are false, because these other scientists must also be better thinkers than anybody else, according to his own self-serving doctrine. These theistic scientists may very well accept all the same experimental data as Krauss, yet they draw different conclusions from it. In that case, experimental data cannot help us to decide anything because everyone agrees on the data as such. What they disagree about is what the data mean and what the data imply, and the only way to present and adjudicate such discrepancies is to use rational arguments, which is to do philosophy.

[164] Hawking and Mlodinow, 2010, p. 5.
[165] Hawking and Mlodinow, 2010, p. 7.
[166] See Hawking and Mlodinow, 2010, p. 7.
[167] Newman, 2009.
[168] Wilson, 2009.
[169] Hylton in Bell and Cooper, 1990, n. 1, p. 165.
[170] This confusion is simply a part of analytic philosophy's history. As Hylton notes, 'given the length of Russell's active philosophical life, and the multiplicity of positions that he held, few claims about his views can be made without qualification as to time' (Hylton in Bell and Cooper, 1990, n. 1, p. 165).
[171] Rorty, 1979, p. 171.
[172] Dummett, 1996, p. 14.
[173] Hart writes that 'logical positivism, centered around Schlick and especially Carnap, is perhaps the most interesting public movement in analytic philosophy between the two world wars.... [B]eginning between the wars, Carnap was to develop one of the most articulate and influential conceptions of clarity yet to emerge in analytic philosophy' (Hart in Bell and Cooper, 1990, p. 198).
[174] Collins, 1998, p. 703-704.
[175] Collins, 1998, p. 704.
[176] Dummett, 1996, p. ix.

[177] Collins, 1998, p. 705. See also Bell in Bell and Cooper, 1990, p. vi. How analytic philosophy became so dominant so quickly is an important question that I will set aside here. Mandt offered a less than ideal, though perhaps pragmatic, solution to the conflict between the 'pluralists' (essentially a conglomerate of the non-analytic philosophers) and the status quo (the analytics): 'the two camps do not speak to each other or understand each other's language; they lack a shared culture. I suspect that the best immediate solution would be the cultivation of mutual indifference. Share the spoils and let each go his or her separate way' (Mandt, 1986, p. 277). Some analytic philosophers claim that there is no such thing as analytic philosophy, which is merely a self-serving fiction, but in trying to find space for their philosophical practice among the dominant analytics, the 'radical' pluralists clearly go too far (and in effect endorse antirealism) by claiming 'that philosophy does not exist; there are only philosophers' (Mandt, 1986, p. 268). See also Sibelius, 1993, who has argued that 'analytic philosophy has failed within the philosophy of science due to the way the dynamic aspect of scientific theories is traditionally treated' (Sibelius, 1993, p. 558). He claims 'that the methods of analyzing scientific theories used by analytic philosophers fail to recognize the distinctions between kinematics/dynamics and description/explanation in a way which would do justice to their importance in science' (Sibelius, 1993, p. 560).

[178] Carnap, 1995, p 29 (original emphasis). Collins, 1998, p. 722 notes the (apparent) antirealism of some of the early positivistic physicists. While I cannot give a detailed analysis of all the various aspects of positivism and empiricism, I do discuss and clarify the most relevant points with respect to my goals. For a more detailed discussion of positivism, see Hacking, 1983, and Collins, 1998.

[179] See also Spencer, 2007, p. 106, and Adams and Spencer, 2007.

[180] One could argue that something similar to my hypothetical medical example is, in one form or another, actually occurring in the field of medicine.

[181] Heisenberg, 1974, p. 18.

[182] Schrödinger, 1964, p. 37.

[183] Frank, 1941, p. 212.

[184] Jaki, 1978, p. 159. For a critique of Mach's views and how Einstein ended up rejecting him, see Yourgrau, 2005, pp. 34-35, 37, 49.

[185] For example, Baggott writes that Heisenberg took 'a fairly uncompromising positivist stance' (Baggott, 2004, pp. 103-104).

[186] Heisenberg, 1972, pp. 122-123. Unfortunately, Heisenberg's remarks about atoms being 'parts of observational situations' helped lead to an antirealist interpretation. Although such remarks need not end up in antirealism, he also was not being philosophically clear. For example, he would have to explain what an 'observational situation' means. Looking at my car would be an observational situation, and surely it is a physical object, so why can't atoms be physical too? In this respect, it seems as though Mach's influence was still causing philosophical trouble for Heisenberg.

[187] Daniels notes that 'this view [that ancient philosophy was a waste of time] was eventually overcome by Gilbert Ryle who found that Plato's later dialogues were actually 'doing' analytic philosophy and were thus of some interest after all!' (Daniels, 1998). We certainly do not need Ryle to tell us that Plato is worth studying and that

he sought logical clarification of ideas and terms so far as possible. 'The method of making distinctions to solve philosophical problems is ubiquitous. Reading Plato should suffice to convince one that the method goes right back to Socrates' (Mortensen, 2000, p. 342).

[188] Heisenberg, 1958-A, p. 145.
[189] Heisenberg, 1972, p. 213.
[190] Heisenberg, 1958-A, p. 86.
[191] Comte in Hacking, 1983, p. 47 (original emphasis).
[192] Frank, 1941, p. 215.
[193] Collins notes that 'the last notable act of the *Encyclopedia*, and hence of the Vienna Circle's organizational core, was to commission Thomas Kuhn, a physicist turned historian, to write *The Structure of Scientific Revolutions*' (Collins, 1998, p. 730). Hacking notes that Kuhn also seemed to be a realist in many ways and believed that theories should be 'simple in structure and organize facts in an intelligible way' (Hacking, 1983, p. 13).
[194] Collins, 1998, p. 722.
[195] Dummett, 1996, pp. 22.23. Cf. Yourgrau, 2005, p. 35.
[196] Hylton in Bell and Cooper, 1990, p. 138.
[197] Hylton in Bell and Cooper, 1990, p. 145.
[198] See Baillie, 1997, p. 42.
[199] See Sloss, 2000, pp. 166, 171.
[200] Collins, 1998, p. 712. Laszlo also writes that Russell's pessimistic belief that all life and every action, thought, and feeling end up being meaningless, because all things are destined for 'extinction in the vast depth of the solar system', may 'be the chimeras of an obsolete view of the world' (Laszlo, 2004-A, p. 15. See also Russell, 1903).
[201] Audi, 1999, pp. 139-140.
[202] Norris, 2000-B, p. 107.
[203] Sorrell in Sorrell and Rogers, 2005, p. 2.
[204] Sorrell in Sorrell and Rogers, 2005, p. 2. Norris, too, disapproves of the fact that 'one distinctive feature of the work in the broadly 'analytic' tradition is its tendency to treat philosophical issues as if they spring fully formed at each moment and can therefore be addressed with a minimum of reference to episodes in their own formative prehistory' (Norris, 2002-A, p. 18).
[205] Rorty, 1999.
[206] Baillie, 1997, p. 4.
[207] Wittgenstein, 1918, Preface.
[208] Cottingham in Sorell and Rogers, 2005, p. 26.
[209] Cottingham in Sorell and Rogers, 2005, pp. 26-27.
[210] Dummett, 1996, p. 1.
[211] Nichols, 2006, p. 34.
[212] UCD News, 2007.
[213] Putnam, 1994, p. 445.
[214] Wertheimer, 2006.
[215] For example, Philo, 'the famous Jewish philosopher' (c. 20 BCE), was 'as much Greek as Jewish'. 'He believed wholly in the Mosaic scriptures and in one God

whose chief mediator with the world is the Logos or 'Reason' of God'. (Philo, 1941, Book Summary—Front Cover Panel.) In fact, Philo's
> account of the Creation is almost identical with that of Plato; he follows the latter's "Timæus" pretty closely in his exposition of the world as having no beginning and no end; and, like Plato, he places the creative activity as well as the act of creation outside of time, on the Platonic ground that time begins only with the world. [As well,] the influence of Pythagorism appears in the numeral-symbolism, to which Philo frequently recurs. (Jewish Encyclopedia, 1906.)

[216] Putnam, 1994, p. 445.

[217] See Adams and Spencer 2005, 2007 and Hadot, 1995, 2002. Here is the relevant passage from Putnam:
> It would be absurd, however, to make the reactionary move of trying to believe what philosophers who lived two hundred or two thousand years ago believed. As John Dewey would have told us, they lived under wholly different conditions and faced wholly different problems, and such a return is impossible in any case. And even if it were possible to go back, to do so would be to ignore the correct criticisms of the abandoned positions that were made by later generations of philosophers. (Putnam, 1994, p. 445.)

Besides the brief criticisms I have already discussed, it is worth noting what Dewey himself said about Plato: 'nothing could be more helpful to present philosophizing than a "Back to Plato" movement' (Dewey in Adams and Montague, 1930, p. 21). Dewey does not think much of systematic philosophy, self-admittedly because he is dubious about his own ability to reach inclusive systematic unity, but Plato still provides his 'favourite philosophic reading' (Dewey in Adams and Montague, 1930, p. 21). It is also hard to see why we should ignore philosophers from two thousand or even two hundred years ago but still take seriously a philosopher such as Dewey who, at the time Putnam had published his paper, had already been dead for forty-two years (and now for nearly sixty). Perhaps two hundred is the magic number in this equation. On Putnam's account, it would seem that I do not actually have to prove him wrong; I merely need to say that, by his own standards, his views will be rendered false (or absurd) in two hundred years, and so will mine and everyone else's.

[218] Rorty, 1999. I have deleted an obvious typographical error in the original, where he has written 'in being in being'.

[219] See Yourgrau, 2005, p. 28 and Collins, 1998, p. 713. Ayer wanted to eliminate metaphysics (Ayer, 1952), so it is interesting to see how he tried to deal with his near-death experience (see Rosenthal, 2004). We so often ignore the fact that we make judgments about what is or is not possible based on our own experiences, ignoring the fact that our particular experiences are not the limit of possibility.

[220] Russell, 1997, p. 189.

[221] Einstein, 1954, p. 24.

[222] Yourgrau, 2005, p. 56 and Collins, 1998, p. 727-728. Gödel even offered 'unsuspected cosmological solutions to the field equations of general relativity' and,

by arguing for the ideal nature of time based on the principles of relativity, he essentially worked out how a spaceship could possibly 'penetrate into any region of the past, present or future' (Yourgrau, 2005, p. 6). (See also Gödel, 1990, pp. 202-207.) Jaki notes further implications of Gödel's work:

> the more advanced is a physical theory the more mathematics it contains and the more advanced is the mathematics. From this the ground for connecting Gödel's theorem with physics readily follows. For insofar as Gödel's theorem states that no non-trivial system of arithmetic propositions can have its proof of consistency within itself, all systems of mathematics fall under this restriction, because all embody higher mathematics that ultimately rests on plain arithmetic. Then it follows that there can be no final physical theory which would be necessarily true at least in its mathematical part (Jaki, 2004, p. 4).

It is not that a final theory of fundamental particles is impossible to formulate, but that 'when it is on hand one cannot know rigorously that it is a final theory' (Jaki, 2004, pp. 8-9).

[223] Russell, 1968, p. 224.

[224] Russell, 1968, p. 224. See also Yourgrau, 2005, pp. 103-105. Gödel had also strongly criticized Russell's *Principia*, saying that it was 'a considerable step backwards as compared with Frege. What is missing, above all, is a precise statement of the syntax of the formalism' (Gödel quoted in Hylton in Bell and Cooper, 1990, p. 156).

[225] Yourgrau, 2005, p. 13.

[226] Dickson made this comment during my invited lecture in the Department of Physics, University of Liverpool, March 30, 2004. (Not in Works Cited.)

[227] Dummett, 1996, pp. 4-5.

[228] Dummett, 1996, p. 5. Black also writes that 'the analytic method adopted by the Viennese circle culminates in the judgment that there are no distinctive philosophical problems. Speculative philosophy must be transformed into a new methodology, the analysis of linguistic forms' (Black in Carnap, 1995, p. 13).

[229] Dummett, 1996, pp. 15-16.

[230] Hylton in Bell and Cooper, 1990, p. 164. Hylton continues:

> It ought to be a puzzle to us that so great a thinker as Russell can insist, in diametrical opposition to his earlier views, that logic has an essential characteristic which he cannot define, and cannot explain in a fashion which is at all illuminating. Without a clear understanding of the notion of tautology, how could he possibly have reason to believe that logic consists of tautologies? Part of the answer to this puzzle no doubt lies in the impact that Wittgenstein's personality had on Russell before the First World War, an impact that was evidently not dependent upon Russell's understanding of Wittgenstein's views (Hylton in Bell and Cooper, 1990, pp. 164-165).

Hylton also informs us that Russell had not read the *Tractatus* before making his claim about tautologies. Perhaps Collins may not be too far astray in claiming that

Wittgenstein seems to have been more interested in creating a personality cult centered on himself, using Russell and Moore and then treating them badly and usurping (if not plagiarizing) them (Collins, 1998, pp. 735-736).

[231] Henry Le Roy Finch writes that:
> no concept in Wittgenstein's later philosophy is more difficult to understand and has given rise to more differences in interpretation than the concept of forms of life. The expression occurs only five times in the *Philosophical Investigations*, and nowhere does Wittgenstein attempt to define it. Yet it is of fundamental importance and has to be understood along with the concept of language-games, with which it is closely associated. (Finch, 1977, p. 89.)

This point makes Baillie's attempt at defining analytic philosophy a bit awkward: 'it is probably the best policy to regard analytic philosophy as a "form of life," in the Wittgensteinian sense. Thus I will use the term "analytic philosophy" to mean philosophy in the tradition of such founders as Frege, Russell, and Wittgenstein, and which continues to constitute the dominant paradigm of philosophy practiced in the English-speaking world' (Baillie, 1997, p. x).

[232] Einstein, 1954, p. 25.

[233] Hacking, 1983, p. 50.

[234] Norris, 2000-B, p. 116.

[235] There are, however, signs of positive change. For example, see the Center for Consciousness Studies. But do not expect to see students of analytic philosophy practicing group mediation during class any time soon.

[236] Neurologist Richard Restak claims that 'if I become bored with the concert and wish to leave, there is a brain-bored-with-the-concert-intent-on-leaving...there is simply nothing to prove that anything exists other than the brain interacting with some aspect of external or internal reality' (Restak, 1984, p. 342). If he is correct, then he also cannot logically prove that there really are brains, or internal reality, or external reality. And what does he mean by 'internal' reality? On his account there can be no such thing. He also writes that 'the "I," or the mind, is a fiction that gives the illusion of permanence to my ever-changing perceptions' (Restak, 1984, p. 342). It would seem, therefore, that this illusion, this fiction, called the 'I' or 'mind' is so powerful that it can fool us into believing it is real. But how can something that does not exist cause me to believe anything? And who is it that is being fooled if 'I' do not even exist? In other words, how can 'I' be fooled if I do not actually exist? Restak has fallen into the trap of assuming that his implicit philosophical assumptions of materialism and empiricism are the *result* of his scientific data when, in fact, such assumptions form the *foundation* of his whole approach to interpreting scientific data in the first place. I suppose his confusion results from the fact that, according to his own materialist assumptions, he does not have a mind.

[237] Harris writes that 'Ayer's pronouncement that "if two states of affairs are distinct, a statement which refers to only one of them does not entail anything about the other" is totally belied. There can be no atomic facts and no atomic propositions' (Harris, 2000, p. 101).

[238] Bohm, 1980, p. 9.

[239] Sacks in Bell and Cooper, 1990, p. 173.
[240] Planck, 1931, pp. 25-26.
[241] Schrödinger, 1967, p. 30.
[242] Bohm, 1980, pp. xi-xii.
[243] Harris, 2000, p. 99. Even well known futurist Alvin Toffler notes that the same sort of integrated holistic approach in business is replacing the out-dated 'smokestack' economy based upon Cartesian mechanical assumptions. 'The new model of production that springs from the super-symbolic economy is dramatically different. Based on a systemic or integrative view, it sees production as increasingly simultaneous and synthesized. The parts of the process are not the whole, and they cannot be isolated from one another' (Toffler, 1990, p. 81).
[244] Hylton in Bell and Cooper, 1990, p. 141.
[245] It is interesting to note that although the doctrine of dependent co-arising is usually only assumed to be found in Buddhism, which is one of the connections people have made between quantum theory and Eastern philosophy, it is also part of the Platonic tradition. Rappe writes that 'Plotinus provisionally seems to suggest that we can understand the universe as a kind of mutual coming-to-be, or interdependent causal nexus: sentient beings all arise together as manifestations of a World Soul or universal form of life' (Rappe, 2000, p. 42). This Platonic view is in harmony with Buddhist teachings, but in the Platonic tradition, the One, the source and goal of all reality, is not dependent upon anything. However, I would argue that the Buddhist doctrine of emptiness is compatible with the Platonic notion of the One, an idea that I nevertheless will not pursue here.
[246] Pirsig, 1999, pp. 143-144.
[247] Bohm, 1980.
[248] Ayto, 1990.
[249] Liddell and Scott, 2004.
[250] Heisenberg paraphrasing Bohr in Heisenberg, 1958-A, p. 105. Bohr (1958, p. 9) makes the same basic point.
[251] Whitehead, 1948, p. 93. Despite Plato's copious philosophical writings, he also warns us that anyone who believes that 'anything in writing will be clear and certain, would be an utterly simple person' (Plato, 2001, p. 565, *Phaedrus*, 275c).
[252] Clark, 1986 p. ix.
[253] Dummett, 1996, pp. 4-5.
[254] Yourgrau, 2005, p. 165.
[255] Yourgrau, 2005, p. 168.
[256] Eddington, 1935, p. 208.
[257] Einstein, 1950-A, p. 59.
[258] See Rorty, 1991, p. 113.
[259] Eddington, 1929, p. 31.
[260] Russell, 1997, p. 175.
[261] Eddington, 1929, p. 38.
[262] K. Ward (1996) makes a related point.
[263] Sachs, 1988, p. 40 (original emphasis).
[264] Gibbins, 1989, p. 48.
[265] Stebbing, 1944, pp. 13-14.

266 Stebbing, 1944, pp. 196-197. See also Jeans, 1930. Eaton argues that Berkeley's immaterialism 'overcomes the bifurcation of nature and makes possible a science which fulfills the promise of the Christian doctrine of creation' (Eaton, 1987, p. 433). In a 1929 symposium on realism and modern physics, with presentations by J. Laird, C. E. M. Joad, and Stebbing, we can see the beginnings of much confusion over the interpretation of quantum theory (Laird, Joad, and Stebbing, 1929).

267 Stebbing, 1944, pp. 196-197.

268 Wilson writes that Jeans:
> judged his conclusions to be compatible with the idealistic systems of Berkeley and Hegel. Important here were the uniformity of nature, the constancy of existing things and the agreement among different minds. If phenomena were ultimately grounded in an entity independent of us—like Berkeley's God or Hegel's Absolute—uniformity, constancy and agreement could be explained (Wilson in van der Meer, 1996, p. 37).

Wilson has made clear how Berkeley was actually an extreme realist or ancient idealist. All things may be mind rather than matter, but this fact is only possible because God is the ultimate ground of all reality, the foundation that is independent of everything.

269 Schrödinger, 1967, p. 95 (original emphasis).

270 Schrödinger, 1967, p. 93.

271 See Schrödinger, 1967, p. 96. See also Huxley, 1946.

272 Polanyi, 1964, p. 62.

273 Harris, 2000, p. 66.

Notes for chapter four

274 More than a decade ago, a study in the UK showed that 'interdisciplinary research is pervasive throughout higher education. Around four-fifths of researchers report that they are engaged in at least some interdisciplinary work' (quoted in Sharp, Peters, and Howard, 2002, p. 9). (Originally from HEFCE, 2001, *Briefing Note 14* (not in Works Cited).) But there is still much work required to bring about the administrative and conceptual changes necessary to understand and properly utilize interdisciplinary research, especially between the arts and sciences (see Snow, 1998). However, recent developments in the UK (and perhaps elsewhere) indicate that some administrative officials may be forcing interdisciplinary research to conform to the 'official' ideology. In other words, the university may only recognize interdisciplinary research that focuses on a pre-set agenda, while ignoring or even punishing other areas. This sort of force-fed interdisciplinarity is not what I am defending. Academics should be free (and encouraged) to cross disciplinary boundaries in ways appropriate to their own research.

275 Mueller in Proclus, 1970, pp. xxvi-xxvii.

276 Proclus, 1970, p. 19, *A Commentary on the First book of Euclid's Elements*, Chapter VIII. 22.

277 Mueller in Proclus, 1970, pp. xxvi-xxvii.

[278] Mueller in Proclus, 1970, pp. xxvi-xxvii. Mueller's translation of *to kallos* as 'moral significance' could be a bit misleading without knowing that the literal meaning is 'beauty' (see Peters, 1967). To the degree that something is absolutely moral, it is also absolutely beautiful, and vice versa.

[279] Rowlands, 2003, p. 3.

[280] Barr, 2003, pp. 97-98.

[281] Brading and Castellani in Brading and Castellani, 2003, p. 1.

[282] Brading and Castellani in Brading and Castellani, 2003, p. 3.

[283] Brading and Castellani in Brading and Castellani, 2003, p. 6.

[284] Heisenberg, 1972, p. 133. He also acknowledges the Platonic roots of the notion that mathematical symmetry underpins 'the atomic structure of matter'. He continues by explaining that 'the existence of atoms or elementary particles as the expression of a mathematical structure was the new possibility that Planck opened up by his discovery, and here he is touching upon the basic problems of philosophy'. (Heisenberg, 1974, pp. 11-12.) And as Proclus writes, 'symmetry, therefore, is necessary to the union of the things that are mingled, and to an appropriate communion' (Proclus, 1816, p. 198, *The Theology of Plato*, BK III, Ch VI).

[285] Coldea et al., 2010, p. 177. They continue by saying, 'for example, the invariance of physical laws under spatial rotation ensures the conservation of angular momentum'. These physicists have recently conducted an experiment with results that 'demonstrate the power of symmetry to describe complex quantum behaviors', where 'just below the critical field, the spin dynamics shows a fine structure with two sharp modes at low energies, in a ratio that approaches the golden mean predicted for the first two meson particles of the E8 spectrum' (Coldea et al., 2010, p. 177).

[286] Alexandrakis in Wagner, 2002, p. 149 (original emphasis).

[287] Dirac quoted in Polanyi, 1964, p. 12. Ward also argues that the 'highest truth of all lies in the apprehension of an objective reality of supreme beauty and goodness' (K. Ward, 1996, p. 101).

[288] Trefil, 2007, p. 26. Rhodes, 2007, examines the role of beauty in physics and in the Neoplatonism of Pseudo-Dionysius.

[289] Jaki, 1978, p. 188.

[290] Jaki, 1978, p. 188.

[291] Trefil, 2007, p. 26.

[292] See Bronowski, 1961, p. 27. Coleridge also had 'an enthusiasm for the Neoplatonists, whose writings were being translated by Thomas Taylor' (Beer, 2007).

[293] Pauli in Jung and Pauli, 1955, pp. 205-206. Pauli writes that:

> we already find this contrast, for example, in antiquity in the two corresponding definitions of beauty: in the one it is proper agreement of the parts with each other and with the whole, in the other (going back to Plotinus) there is no reference to parts but beauty is the eternal radiance of the "One" shining through the material phenomena (Pauli in Jung and Pauli, 1955, pp. 205-206).

Heisenberg makes the same point (see Heisenberg, 1974, p. 167). Alexandrakis also notes that 'in contrast to Plato and the Pythagoreans, Plotinus *rejects* symmetry as a component of beauty. Beauty, for Plotinus, is not found in the symmetrical parts of the object: The One is Beauty (*kallos*), and the One has no parts. Plotinus' One stands for 'zero' (null), or what Scotus Eriugena calls 'Nothing'—that is, the negation of multiplicity in the nature of the One' (Alexandrakis in Wagner, 2002, p. 152). However, the One is above beauty, so cannot itself be beautiful. The 'eternal radiance', as Pauli puts it, is not actually reducible to the One, but I will set aside this discussion here.

[294] Pauli in Jung and Pauli, 1955, p. 206.

[295] Clark, 1995.

[296] Collins, 1998, p. 668.

[297] Collins, 1998, p. 695.

[298] For example, see Chapter 12, 'Quantum Theory and Atomic Structure' in *Principles of Modern Chemistry*, Oxtoby and Nachtrieb, 1986, p. 401.

[299] For example, see Bunge, 1982, p. 209-210. Bunge argues that '(a) chemists deal with things, namely chemical systems, that physicists are not normally interested in, yet (b) they do so with the help of physics, so chemistry is dependent upon the latter, and still (c) physics is not enough to do chemistry, for it must be enriched with peculiarly chemical concepts and hypotheses if it is to solve any chemical problems' (Bunge, 1982, p. 209-210).

[300] Heisenberg, 1974, p. 209.

[301] Schrödinger, 1952, p. 7.

[302] See also Baggott, 2004, p. 97. Baggott writes that 'quantum theory directly challenges our understanding of the nature of matter and radiation, and the process of measurement at their most elementary levels, and we cannot go forward unless we adopt some kind of interpretation, some way of trying to make sense of it all. As we will see, this interpretation has to be based on some philosophical position' (Baggott, 2004, p. 97).

[303] Psillos, 1999, p. 31.

[304] Heisenberg, 1958-B, p. 63. Grimes has suggested that it would be of great benefit to do an in-depth study showing exactly how classical training shaped Planck's thought, but such an important undertaking shall have to fall to someone else.

[305] Heisenberg, 1958-B, p. 53.

[306] Heisenberg, 1958-B, p. 64.

[307] Lowry, 1980, p. 25.

[308] Whittaker argues that ancient Greek philosophy in general was scientific in spirit (Whittaker, 1928, p. 8).

[309] Erikson in Waite, 1977, p. xiv (original emphasis).

[310] The quotes in the following two paragraphs come from a conversation between Heisenberg, Pauli, and Bohr, as recalled by Heisenberg (see Heisenberg, 1972, pp. 208-210).

[311] Heisenberg, 1972, p. 208, and also see p. 210.

[312] Heisenberg, 1972, p. 209 (emphasis added).

[313] Heisenberg, 1972, p. 209.

[314] Heisenberg, 1972, p. 209 (emphasis added).

[315] Schrödinger, 1964, p. 4 (original emphasis). He also notes that 'if we cut out all metaphysics it will be found to be vastly more difficult, indeed probably quite

impossible, to give any intelligible account of even the most circumscribed area of specialization within any specialized science you please' (Schrödinger, 1964, p. 3).

[316] Blackmore (1983) uses an interesting dialogue form to provide a brief defence of the claim that science and metaphysics are inseparable but still distinguishable.

[317] Jaki, 1978, p. 182.

[318] Einstein, 1954, p. 342.

[319] Jaki, 1978, p. 195. Recall also Einstein's review of Russell's book mentioned in Chapter Three.

[320] Jaki, 1978, p. 184.

[321] Jaki, 1978, p. 188.

[322] Einstein quoted in Jaki, 1978, p. 185.

[323] Frank, 1947, p. 215. See also Jaki, 1978, p. 185.

[324] Einstein quoted in Holton, 1968, p. 660 (original emphasis).

[325] Planck, 1932, p. 218 ('Faith' is written with the original capitalization).

[326] Corrigan, 2005, p. 34. See also Rappe, 2000, for an analysis of non-discursive thought in Neoplatonism.

[327] Trusted (1991) also outlines a similar tripartite division, but I have maintained my original distinctions because they are more fundamental and broader in scope.

[328] Jaki notes the views of astronomer William Herschel (1738-1822) who argued against Hume's empiricism:

> '[Herschel writes that] 'half a dozen experiments made with judgement by a person who reasons well, are worth a thousand random observations of insignificant matters of fact.' Clearly, to do science was to make rational judgements about facts, that is, to do metaphysics. But metaphysics has an even more important role than to make science possible. Those of us, Herschel continued, who love wisdom, 'by metaphysics…are enabled to prove the existence of a first cause, the infinite author of all dependent beings' (Jaki, 1978, pp. 110-111).

[329] Burtt writes that:

> even the attempt to escape metaphysics is no sooner put in the form of a proposition than it is seen to involve highly significant metaphysical postulates. For this reason there is an exceedingly subtle and insidious danger in positivism. If you cannot avoid metaphysics, what kind of metaphysics are you likely to cherish when you sturdily suppose yourself to be free from the abomination? (Burtt, 1925, p. 225).

[330] Frank writes that:

> the misinterpretation of scientific principles, as will be shown, can be avoided if, in every statement found in books on physics or chemistry, one is careful to distinguish an experimentally testable assertion about observable facts from a proposal to represent the facts in a certain way by word or diagram. If this distinction is sharply drawn, there will no longer be any room for an interpretation of physics in favour of a spiritualistic or a materialistic metaphysics (Frank, 1941, pp. 4-5).

[331] Trusted makes a similar point (see Trusted, 1991, pp. x-xi).
[332] Dummett, 1996, p. 6.
[333] Heisenberg, 1958-A, pp. 201-202.
[334] Heisenberg, 1958-A, p. 168.
[335] Hawking in Penrose, 1997, p. 169.
[336] Stenger, 1995.
[337] Cartwright in Penrose, 1997, p. 161.
[338] Balaguer, 1998.
[339] Fine, 1996.
[340] Einstein quoted in Yourgrau, 2005, pp. 107-108.
[341] Smolin, 2006-A.
[342] Gibbins, 1989, p. 1.
[343] Einstein, 1954, p. 9 (original capitalization).
[344] Einstein, 1954, p. 9.

[345] Einstein, 1954, p. 9. Adams and Spencer (2007) make the relatively easy argument that Einstein was much more of a genuine philosopher than Rorty, who desired tenure above the pursuit of truth.

[346] Penrose, 1997, p. 1 (original capitalization).

[347] More formally put: the uncertainty principle 'stated that the momentum and position values for a particle in any given direction could not be determined at the same time to an accuracy greater than $\Delta p_x \, \Delta x \geq h$. Or, more precisely, $\Delta p_x \, \Delta x \geq \hbar / 2$.' '$h$' = Planck's constant, which is about 6.62×10^{-34} joule-seconds; '$\hbar = h / 2\pi$'; 'Δp_x' = the uncertainty in a particle's momentum; and 'Δx' is the uncertainty in its position' (Rowlands, 1992, p. 221).

[348] See Barbour, 1966, pp. 289-315, which I draw from in this section.

[349] Einstein, 1954, p. 334.

[350] Baggott, 2004, p. 37. Brown discusses the important role of thought experiments in the history of science, saying that they 'are telescopes into the abstract realm' (Brown, 2004, p. 1131).

[351] Barbour, 1966, pp. 298-299.

[352] Baggott, 2004, p. 262. Rosen (quoted by Chase in Adams and Spencer, 2007, p. 10) and Marinoff (2002, p. 41) make a similar point.

[353] Atkins in Baggott, 2004, Foreword.

Notes for chapter five

[354] James, 2009, p. 95.
[355] Beck, 1959, p. 38.
[356] Lindsey, 1948, p. 398.
[357] Baggott, 2004, p. 115. See also Danforth, 1936.

[358] There are many possible sources dealing with religious faith. As one example, see Cobb (1994), who considers various aspects of religious faith, specifically Christian faith, but also touches upon Buddhism.

[359] For a detailed analysis of the notion of belief, see Price, 1969. Bradley, who aims to clarify the nature of faith in general, as well as religious faith, and how philosophy and faith are connected, makes an interesting observation. He writes that

'philosophy in my judgment cannot verify its principle in detail and throughout. If it could do this, faith would be removed, and, so far as it does this, faith ceases. But, so far as philosophy is condemned to act on an unverified principle, it continues to rest upon faith'. Nevertheless, he ends his essay by admitting that 'whether we are to assert or to deny that philosophy in the end rests on faith, is to my mind of no consequence' (Bradley, 1911, page 171). If such a question is really of no consequence, it is unclear why he would bother writing his paper in the first place.

[360] Lindley, 1993, p. 6.

[361] Some theorists argue that only the theistic hypothesis can ultimately make sense of the assumptions of scientific faith (for example, see Banner, 1990 and Clark, 1984).

[362] Polanyi, 1964, p. 28.

[363] Polanyi, 1964, p. 28.

[364] Polanyi, 1966, pp. 67-68.

[365] Pauli writes that 'many physicists have recently emphasized anew the fact that intuition and the direction of attention play a considerable role in the development of the concepts and ideas, generally far transcending mere experience, that are necessary for the erection of a system of natural laws (that is, a scientific theory)' (Pauli in Jung and Pauli, 1955, pp. 151-152). Clark also cites Planck as saying that innovative science depends upon 'the imaginative vision and faith in the ultimate success' (Clark, 1990, p. 37). And Medawar writes that:

> ever since Plato spoke of the divine rapture or divine fury of creativity, the act of poetic invention has been held in awe by those who exercise it, just because it seems to embody an infringement of divine copyright....If the generative act in science is imaginative in character, only a failure of the imagination—a total inability to conceive what the solution of a problem *might* be—could bring scientific inquiry to a standstill' (Medawar, 1985, pp. 84-85).

[366] Midgley, 1992, p. 57.

[367] Heisenberg, 1958-B, p. 65.

[368] Many scientists do, in fact, fabricate or otherwise manipulate and spin their data in order to get published and attract funding. As David Rowe, director of the Center for Regenerative Medicine and Skeletal Development at the University of Connecticut Health Center, admitted: 'the fact is that reviewing agencies want success. Therefore you spin your data in the most favorable way. That's where the dangers begin to come - that you spin it a little more than you can justify and then one thing leads to another. It's a very mushy, very fuzzy line' (quoted in Khan, 2012).

[369] Midgley, 1992, p. 57. Helm also argues that 'faith is not simply belief, it is also trust' (Helm, 2000, p. xv), while Clark uses the terms 'faith' and 'ungrounded intuition' interchangeably (Clark, 1984, p. 33). It is also interesting to note Alvin Toffler's observation more than twenty years ago that our economy is coming to be based upon the intangible and super-symbolic, which involves an important implicit faith (Toffler, 1990, p. 68).

[370] Helm, 2000, p. 48.

[371] Eddington, 1935, p. 192.

372 Einstein, 1954, p. 50.
373 Einstein quoted in Fine, 1996, p. 109.
374 Gibbins, 1989, p. 52.
375 Lowry, 1980, ix.
376 However, Laughlin and Pines argue that 'the central task of theoretical physics in our time is no longer to write down the ultimate equations but rather to catalogue and understand emergent behavior in its many guises, including potentially life itself. We call this physics of the next century the study of complex adaptive matter' (Laughlin and Pines, 2000, p. 30). They are unwisely calling for a return to a purely Aristotelian approach, which was precisely the approach that had to be superseded by Platonic/Pythagorean metaphysics in order for the birth of modern science to occur.
377 Bergmann, 1980, p. 2. In the sciences, we cannot be content with formal validity, for we also need to check our premises against physical reality, and so we are also aiming for sound arguments and theories. Formally, only deductive arguments can be sound because a sound argument must first be valid and also have true premises, and only deductive arguments can be valid. Moreover, an inductive argument only makes a conclusion probable, not certain, which gives us an interesting problem if we have a valid argument but cannot establish conclusively whether or not the premises are true. In such a case, even if the premises are acceptable, relevant, and adequate (Hughes and Lavery, 2008), unless we can say conclusively that they are true, we cannot say that the argument is formally sound. And, in any case, we can always logically question or doubt a premise that someone considers acceptable, relevant, or adequate. We also cannot say the argument is logically strong, because this notion applies only to inductive arguments; if the argument is valid, it cannot be inductive, but must be considered deductive. Thus, if an argument is valid and we cannot say for certain that the premises are true, but we can say that they meet the criteria of acceptability, relevancy, and adequacy, we should say that the argument is *informally sound*.
378 Eddington, 1929, p. 45.
379 Russell, 1997, pg. 27. Bronowski writes that 'there are, oddly, no technical rules for success in science. There are no rules even for using test tubes which the brilliant experimenter does not flout; and alas, there are no rules at all for making successful general inductions' (Bronowski, 1961, p. 71).
380 Agassi in Sachs, 1988, p. xviii.
381 Yourgrau, 2005, p. 13.
382 For example, see Rowlands, 1992.
383 Kepler quoted in Hummel, 1986, p. 63.
384 Clark, 1998, p. 26.
385 Clark, 1984, pp 93-94.
386 de Broglie, 1955, pp. 208-209. Titius makes a similar point:
> This life-faith is true faith; for we speak of faith where we lay hold on the unseen, where no external facts guarantee the truth, where we surrender ourselves to an inner conviction. Not that it is impossible for faith to experience also external corroboration—it is experienced by every life which asserts itself victoriously in the

struggle—but he only can receive corroboration who has acted in faith. Without such faith *no science* is possible; for how would science be possible without the naïve faith that our senses show us reality, and without the self-evident assumption that, when we think logically and correctly, we grasp reality with our thoughts (Titius, 1931, p. 25, original emphasis).

387 James, 2009, p. 51.

388 de Broglie, 1955, pp. 208-209.

389 Burtt, 1925, p. 2.

390 Gibbins, 1989, p. 5. Even well informed scholars often fail to understand the importance of the metaphysical beliefs of pioneering scientists. Lindberg, for example, dismisses Galen's theological and Platonist beliefs as being of no scientific consequence: 'Galen certainly believed that behind the admirable design found in living things could be discerned a designer; but this belief had no major influence on his analysis of disease or on his diagnostic and therapeutic procedures' (Lindberg, 1992, p. 131). On the contrary, it was precisely this metaphysical belief that informed Galen's entire outlook on and approach to medicine, health, and healing. Similarly, Heisenberg's mystical experience allowed him to understand the *Timaeus*, an experience that influenced his conception and development of quantum theory.

391 Rowlands, 1992, p. 18.

392 Brooke, 1991, p. 19. The mathematician George D. Birkhoff writes that 'all in all, it is a faith in the uniformity of nature which remains the guiding star of the physicist just as for the mathematician it is a faith in the self-consistency of all mathematical abstractions, although these faiths are more sophisticated than ever before. The minds of both are tinged with an unwavering belief in the supreme importance of their own fields' (Birkhoff, 1938, p. 607).

393 Brooke, 1991, p. 19.

394 See Clark, 1990, p. 37. Rowlands also shows how the 'development of science depends on individual creativity' (Rowlands 1992, p. 47).

395 Planck, 1932, p. 216

396 Planck, 1931, p. 61. See also Medawar, who writes that William Whewell used the phrase 'happy guesses', 'until, as if recollecting that he was Master of Trinity, he wrote "felicitous strokes of inventive talent"' (Medawar, 1985, p. 51).

397 John Stuart Mill makes the following relevant observations:
> Truths are known to us in two ways: some are known directly, and of themselves; some through the medium of other truths. The former are the subject of Intuitions, or Consciousness; the latter, of Inference. The Truths known by intuition are the original premises from which all others are inferred. Our assent to the conclusion being founded on the truth of the premises, we never could arrive at any knowledge by reasoning, unless something could be known antecedently to all reasoning (Mill, 2005, p. 3).

398 Clark, 1990, p. 87.

399 Einstein, 1954, p. 11. The question of personal existence after bodily death is an extraordinarily difficult one, and it is not nearly as simple as Einstein makes it

seem. Einstein apparently would have been in disagreement with his close friend Gödel on this point, although not necessarily.

[400] Einstein, 1954, p. 46.

[401] Bohm, 1980, p. 13.

[402] Bohm, 1980, p. 13. See also Vassilopoulou (2009, pp. 133-134) for a relevant discussion.

[403] See Knoblich and Oellinger (2006) for a discussion of the 'eureka moment'.

[404] See Rappe, 2000.

[405] See Medawar, 1985, p. 51.

[406] Heisenberg's relevant quotations here and my paraphrasing of his discussion are from Heisenberg, 1972, pp. 7-12.

[407] Eddington, 1935, p. 330 (see also pp. 304-339). It is interesting to note that the department of physics building at the University of Liverpool is named after the eminent physicist Oliver J. Lodge, who believed in spiritualism and became a devotee of psychic research (Asimov, 1967, p. 115).

[408] Schrödinger, 1964, p. 94.

[409] For example, see Austin, 1999.

[410] Polanyi, 1964, p. 9.

[411] Whitehead, 1953, p. 5.

[412] Heisenberg, 1958-A, p. 84. Clark writes: 'that the Humean analysis of causation is inadequate is an argument too vast to be attempted here. It is, at least, inadequate to the needs of practising scientists, who seek some intelligible description of events that will rule out magic' (Clark, 1984, p. 139-140). Elsewhere he states that 'David Hume's analysis of human knowledge eliminates most scientific theory in the process of eliminating scientific realism' (Clark, 1998, p. 27).

[413] Schäfer, 1997, p. 88. By 'non-rational' he does not mean 'irrational'; he means 'trans-rational'.

[414] Jaki, 1978, p. 104.

[415] Unfortunately, Hume ended up concluding that 'we have, therefore, no choice left but betwixt a false reason and none at all' (Hume, 1978, p. 268). But if there really is no reason at all, or only a false reason, then Hume's own arguments are either false or they are not reasonable.

[416] Barr, 2003, p. 266.

[417] Edward J. Larson & Larry Witham reported that amongst 'top' natural scientists, disbelief in God 'is greater than ever – almost total' (Larson and Witham, 1998). However, Elaine Howard Ecklund has more recently reported that her research shows that nearly fifty percent of the scientists she surveyed are religious. (See Ecklund, 2010.) A far more extensive study was the 1969 Carnegie Commission survey of the religious beliefs of more than 60,000 American college professors. Which group of professors do you believe was most likely to be religious? The lowest percentage of religious believers out of any group of professors was 29%, while the highest was 60%. Most people would probably assume that the scientists would be the group with the lowest percentage of religious believers. In fact, however, 60% of the mathematicians said that they were religious, followed by 55% of the physical and life scientists, 51% of the political scientists, 50% of the economists, 49% of the sociologists, 33% of the psychologists, and 29% of the

anthropologists. (See Stark et al, 1996, p. 436.) Those who study human behavior were most likely not to be religious, while those who study mathematics were most likely to be religious. The deeper and more abstract the science, the more likely it seems that we will find religiously inclined professors. Even Larson and Witham found that the mathematicians were more likely to be religious than any other group of scientists that they surveyed.

It is wise never to place too much emphasis on any such studies, especially since there is always ambiguity surrounding key concepts, such as what it even means to be 'religious' in this particular case. Would Einstein have said that he was religious? On the one hand, he was, and on the other hand he was not. Despite the likely shortcomings of this study, these percentages still reveal the interesting phenomenon that mathematicians are most likely to be religious. This statistic is noteworthy because the Platonists place mathematics as the intermediary between the divine realm and the sensible universe. (One may object that such statistics would not be relevant in the UK, for example, apparently because there are a higher percentage of atheists there than in the US. However, all that would matter for our purposes would be the degree to which the ratios between the different groups were similar. For example, if 80% of the anthropologists were religious while only 5% of the mathematicians claimed to be religious, that would be a fact of great interest. In any case, according to the Home Office statistics for the England and Wales 2001 census, 77.3% of the population is affiliated with a 'faith group', whereby 73% are Christians (O'Beirne, 2004, p. 6). Clearly, more than a few people in the UK still remain faithful to religion in some capacity.)

It seems that the further away from the abstract sciences we move into the messiness of human interaction, the more likely we are to encounter resistance to the religious impulse. I will set aside any conjectures as to why this may be the case, except to note that whereas the mathematical physicist sees order and symmetry, which imply intelligibility in the cosmos, the anthropologist tends to see disharmony between cultural and individual beliefs. Nevertheless, there are reasons underlying our behavior, and uncovering these reasons and understanding them is essential for lasting transformation, whether individually, socially, or globally. Such rational understanding points us toward the underlying unity in our differences, allowing for a flourishing and continually expanding variety of ways of life. There is rationality underlying the messiness of human interaction, but we must bring it to the surface. (See McIntyre (1996) for a defence of the potential for discovering laws in the social sciences.)

[418] Lindley, 1993, p. 156.

[419] Simon Blackburn states approvingly that 'Dawkins is an atheist, and indeed a strenuous, militant atheist' (Blackburn, no date).

[420] Bohr quoted in Fine, 1996, p. 22.

[421] Einstein quoted in Fine (1996, p. 109).

[422] Heisenberg, 1972, p. 73.

[423] Schrödinger as quoted in Heisenberg, 1972, p. 75. In a more sober, academic moment, Schrödinger writes: 'on grounds upon which we cannot enter here, we have to assume that a small system can by its very nature possess only certain discrete amounts of energy, called its peculiar energy levels. The transition from one

state to another is a rather mysterious event, which is usually called a 'quantum jump" (Schrödinger, 1967, p. 52).

[424] Hilgevoord, 2006.

[425] Baggott (2004, p. 39).

[426] See Grof, 2000.

[427] Directed panspermia may be a genuine scientific possibility but, as Clark reflects about Francis Crick's speculation, we must ask ourselves 'why it is easier to believe, without a shred of evidence, in such galactic engineers, than in the infinite intellect of God, who knows?' (Clark, 2000, p. 33).

[428] Plato, 1956, p. 490, *Phaedo* 85d.

[429] Zeh, 1993.

Notes for chapter six

[430] See Norris, 2002-A, pp. 1-2.

[431] Barbour, 1966, pp. 171-172 (original emphasis).

[432] Vision, 1988, p. 3.

[433] The following list could contain even further examples of different types of realism: naïve realism, external realism, internal realism, global realism, local realism (i.e. particular instances), local realism (i.e. Einsteinian limit of c for meaningful signal transmission), metaphysical realism, practical realism, scientific realism (in general), scientific realism (metaphysical stance), scientific realism (semantic stance), scientific realism (epistemic stance), scientific realism (for entities), scientific realism (for theories), physical realism, structural realism, direct realism, critical realism, dogmatic realism, empirical realism, agnostic realism, ontological realism, logical realism, conceptual realism, moral realism, rational realism, bare realism, realism-in-general, realism-in-particular, mechanical realism, political realism, thin realism, robust realism, and transcendental realism. Some of these terms are from specific authors while various authors use others more generally. The following references are samples of places to find these different kinds of realism: Einstein (1954, p. 20), Vision (1988, pp. 6, 8, 14, 74-77), Norris (2000-B, p. 54; 2002-A, pp. 3-5), Heisenberg (1958-A, pp. 81-83, 144-145), Hacking (1983, pp. 26-30), Psillos (1999, pp. xix, 146), Ladyman (1998, pp. 409-424), Turner (1925), Barbour (1966, pp. 171-172), Bhaskar (1978), Bhaskar and Norris (1999), Kant (1965, pp. 349-350), A. E. Taylor (1936, pp. 71-72), Yourgrau (2005, p. 102, 172), Shafer-Landau (2003), Banner (1990, pp. 34-35), Rogers (2005, pp. 13, 33, 106), Maddy (2005), and Goldstein and Pevehouse (2006).

[434] See especially Norris, 2002-A, p. 36.

[435] One could argue that atoms and subatomic particles maintain their geometric structures or forms, but this means that their form is abstract. To the degree that these entities are physical, they are in motion, even if only by hitching a ride with the earth as we orbit the sun. In other words, these particles are, at the very least, in different space-time positions at each moment of their physical manifestation.

[436] Yourgrau, 2005, p. 102.

[437] Plato, 1997, pp. 189-190; *Protagoras* 170b-171d.

[438] As Parmenides warned us so long ago, we cannot measure or see, or even say or think, what *is not*. See Parmenides' *Fragments* in Wheelwright, 1985, p. 95-100.

[439] One could make the same point more abstractly. If I am allowed to do X or Y in game G, then it may logically follow that I can also do Z, where Z could be a hitherto unknown corollary of accepting X and Y. Therefore, in some sense we could say that I have discovered Z when I finally realize that Z follows logically from accepting X and Y. At first glance, we have merely agreed with what many antirealists would also admit, because such an admission in no way necessitates correspondence with reality beyond the conventions of our arbitrary rules in G. I could have initially chosen A and B instead of X and Y as the rules for G, and then Z would never have been discovered because it would not have followed from A and B. But the antirealist cannot be let off so easily. Even within the rules of G, if Z is logically implied by X and Y, even though it was never formally stated as a rule and may or may not have originally been conceived to follow from X or Y, the fact that Z is a logical implication necessarily entails that something about *logical entailment itself* is not bound by the rules of G, or of any particular game. It would then follow that something about logic is belief- and game-independent, and so would require the same sort of ontological clarification as the verification-transcendent truths that realists endorse.

[440] Numerous branches of logic have been developed. For example, search 'logic' on the online Stanford Encyclopedia of Philosophy to find articles covering at least fifteen different types of logic. However, they are all subject to the sorts of metaphysical and internal logical difficulties I am briefly outlining here. The fact that there are different types of logic may lead one to falsely assume that there is no underlying unity to the various logical systems, which is to say that there is no fundamental logical law that underpins all logical systems. This assumption, however, would be mistaken. All logical symbols, operators, notations, etc. must be defined before we can begin using them in some logical system, LS_{-1}, but their definitions will necessarily entail some presupposed logical understanding. If such presupposed logical understanding is an outcome of, or dependent upon, LS_{-1}, then we are begging the question; if it is based on another logical system, LS_{-2}, then that system is necessarily more fundamental than LS_{-1}. We then need to inquire into the foundational axioms, assumptions, and rules of the more fundamental LS_{-2}, a process of logical inquiry that has no logically necessary ending point.

Furthermore, any statement of an axiom, A, in any logical system, LS_{-x}, must still follow syntactical rules of language. Each word in each definition of A in LS_{-x} is in need of unambiguous definition, but then each definition will also contain words which in turn require unambiguous definition, and so on without logical end. If there is any ambiguity in A of LS_{-x}, then there can never be any certainty in any conclusion arrived at in LS_{-x}. If there is no certainty but only plausibility, we will eventually run into the familiar problem of induction. In the end, however, we cannot escape presupposing some sort of foundational logical law.

There are very important, technical, abstract relationships between logic, arithmetic, geometry, and algebra, but we are not going to pursue this issue here except for a brief comment. After all, given that antirealists deny the objective reality of all the laws in these disciplines, it is first essential to prove that belief to be false, which is what we have been doing in this book.

Proclus writes that 'arithmetic is more precise than geometry, for its principles are simpler. A unit has no position, but a point has; and geometry includes among its principles the point with position, while arithmetic posits the unit'. Nevertheless, geometry and arithmetic 'have a certain community with one another, so that some theorems demonstrated are common to the two sciences, while others are peculiar to the one or the other' (Proclus, 1970, p. 48). Algebra is really an application of arithmetic that enable us to rely on variables to represent possible numbers, which then allows for a wide applicability when turned towards physical phenomena. But, still, algebra presupposes that prior reality of numbers, and relations between numbers, and all numbers are nothing more than a multiplicity of ones. For example, '2' is just 'two ones'.

Furthermore, if geometry is *like* arithmetic, then there is a likeness between them, which is to speak of an analogy, which is part of logic. Hence, these two branches of mathematics could not be like each other unless the possibility for analogies was metaphysically prior (a point made by Grimes in one of his many lectures). All we need to recognize here is that much more work is required to explore and clarify these extraordinarily difficulty issues, and so the work of Proclus is of paramount importance.

[441] Gibbins, 1989, p. 2.

[442] Einstein, 1954, p. 233. Similarly, Kepler wrote that the 'conclusions of mathematics are most certain and indubitable'. (Kepler quoted in Burtt, 1925, p. 57.)

[443] Popper, 1979, pp. 23-24. Given such a position, I have to wonder why he had such difficulty understanding Plato and instead grossly misinterpreted and misrepresented him.

[444] James, 2009, p. 12.

[445] Plato, 2001, p. 345, *Phaedo*, 100c.

[446] Gibbins makes a similar point: quantum mechanics' 'principal philosophical outcome (I suggest, humbly following Niels Bohr) lies in presenting us with the limits of theorizing in physics, the limits of our power to represent the physical world (Gibbins, 1989, p. 16).

[447] Baggott, 2004, p. 121. This is the general opinion, although there are some dissenters. See Torre *et al*, 2000.

[448] Baggott, 2004, pp. 121-122.

[449] Baggott, 2004, p. 287.

[450] For the Einstein-Bohr debate see Whitaker (1996) and Sachs (1988). I am also not going to discuss the much-debated Einstein-Podolsky-Rosen (EPR) paper that attacked the orthodox interpretation of quantum theory. Some sample literature discussing EPR includes Gibbins (1989) and Norris (2000-B). Fine has argued persuasively against the odd view that Einstein rejected (or did not understand) the new quantum theory because he had grown conservative or even senile. Rather, Fine claims that Einstein was actually 'more radical in his thinking' in hoping to replace the concepts of classical physics, which Bohr had wanted to keep. (Fine, 1996, p. 24.) Einstein was clearly a realist, but Fine has misunderstood this obvious and important point and tries to say that Einstein's realism has a deeply empiricist core

that makes it a realism 'more nominal than real' (Fine, 1996, p. 108). However, Nadeau and Kafatos write that the Einstein-Bohr debate 'eventually revolved around the issue of realism, and it is this issue that Einstein felt would determine the correctness of quantum theory', and they (mistakenly) claim that 'Bohr was the winner on all counts' (Nadeau and Kafatos, 1999, p. 65).

[451] (Gibbins, 1989, p. ix). Heisenberg also conflates what he calls 'metaphysical realism' and 'dogmatic realism' with materialism and strict determinism, respectively, both of which he believes quantum theory has abandoned (Heisenberg, 1958-A, pp. 81-83 and 1958-B, p. 39). He says that 'practical realism', which is similar to, but really a subset of, what I below call broad realism, has always been and always will be an essential part of science (Heisenberg, 1958-A, pp. 81-83), but he attacks Einstein's apparent dogmatic realism, equating it with 'the old materialistic ontology' (Heisenberg, 1958-A, p. 144). As Bhaskar rightly claims, one is only worried about quantum theory posing special problems for realists if 'one is wedded to normally implicit, atomistic presuppositions of empiricist ontology' (Bhaskar in Bhaskar and Norris, 1999).

[452] For example, see Gibbins (1989, pp. ix, 5-6, 89), Bohr (1963), Einstein (1954, p. 334), Nadeau and Kafatos (1999, p. 67), and Eddington (1935, p. 283).

[453] Plotinus generally considers matter to be evil, whereas Proclus defended the view that matter is good because it owes its existence to, and is ultimately produced by, the One. Experts may contest such interpretations, but I think there is a way to show how Plotinus and Proclus were both right. There is also some ambiguity about Plotinus' views with respect to qualities and matter, but I am not going to pursue these more arcane topics here. A sample of relevant readings includes Opsomer (2001), Katz (1954), Rist (1977), and Lee in Wagner (2002).

[454] Searle in Bunnin and Tsui-James, 2003, p. 12.

[455] Ward, 1996, p. 101.

[456] I doubt that my arguments would convince Dawkins or Marvin Minsky to consider that there just may be aspects of reality that are not physical. For example, Minsky claims that 'there is not the slightest reason to doubt that brains are anything other than machines with enormous numbers of parts that work together in perfect accord with physical laws' (Minsky, 1986, p. 288). Philosophers (or any thinking person), however, can find reasons to doubt pretty much anything. The point of greatest interest is that Minsky clearly believes in the reality of the laws of physics, which the entire physical world, including our own brains, must obey. Like most 'common-sense' materialists, he does not seem to know how to deal with the ontological and epistemological difficulties concerning the status of such laws. If they are real, and if all the parts of the brain must work in perfect accord with them, then he needs to be able to tell us what these laws are in reality, and how they interact with the brain. The usual tactic, however, is simply to ignore the problem, leaving us in ignorance.

[457] Plato, 1996, p. 373, *Sophist* 246b.

[458] Plato, 1996, p. 55, *Theaetetus* 155e-156a.

[459] When Plotinus was offering his philosophical/spiritual teaching more than seventeen centuries ago, he apparently did not even bother engaging with materialists, for doing so seemed to him to be pointless. He writes that those who

believe that the world is merely the result of chance and material causes are far from the divine, or the One. 'It is not such men as these that we address but such that admit the existence of a world other than the corporeal and at least acknowledge the existence of soul' (Plotinus, 1964, p. 79, *The Good or the One* VI.9.5). In a similar way, scientists publishing a research paper do not generally bother addressing antirealists who believe that the whole scientific enterprise is an elaborate fiction or lie.

[460] Banner's notion of 'bare realism' appears similar to what I have called 'broad realism', but in fact they are quite different—broad realism is the most fundamental realism in the widest sense. Banner's idea of bare realism concerns science, and it seems to be what others have called 'naïve realism' (see Banner, 1990, pp. 34-35).

[461] Clark puts the point succinctly: 'if I am *wrong* to be a realist, there is at least one 'fact of the matter' that is more than a social norm, and anti-realism is therefore false (and I am *not* wrong to be a realist.) So if I am wrong, I am not wrong; therefore I am not wrong'. (Clark, 1990, p. 87.)

[462] See Norris, 2000-B, pp. 82-83.

[463] Polanyi, 1964, p. 44.

[464] Clark, 1990, p. 12.

[465] Bhaskar in Bhaskar and Norris, 1999. See also Bhaskar, 1978.

[466] See Heisenberg, 1974, pp. 11, 22, 24, 26, 27, 105, 110, 116, 117, 118, 140, 171, 172, 173, 174, and 181.

[467] Heisenberg, 1974, pp. 73-74.

[468] Heisenberg, 1974, pp. 86-87. Consider Eddington's relevant observation: 'it is remarkable that just as Einstein found ready prepared by the mathematicians the Tensor Calculus which he needed for developing his great theory of gravitation, so the quantum physicists found ready for them an extensive action-theory of dynamics without which they could not have made headway' (Eddington, 1935, p. 180). Conversely, physicists can also develop the mathematics that they need, which is then made available to the mathematicians.

[469] Heisenberg, 1974, pp.185-186.

[470] Heisenberg, 1974, p. 186.

[471] Heisenberg, 1974, pp. 73-74.

[472] Heisenberg, 1974, p. 13 (emphasis added). At this time, there are considered to be at least 16 'universal' constants and more than 300 fundamental physical constants (see The National Institute of Standards and Technology). However, astronomer Michael Murphy believes that he has evidence that the fine structure constant might have 'changed by about one part in two-hundred-thousand during the last 10 billion years' (PhysOrg.com, 11 April, 2005). As I argue against Smolin in Chapter Eight, even if such constants and laws are changing, they are either changing randomly, which would make science impossible, or they change according to some higher order, constant, or law, which itself cannot change.

[473] Bohr quoted by Heisenberg, 1972, p. 138 (emphasis added).

[474] Bohr quoted by Heisenberg, 1972, p. 92.

[475] Schrödinger, 1967, p. 153. Plato was not the first, but he was one of the most prolific proponents of this view in the ancient world.

[476] Schrödinger, 1967, p 154.

477 Norris, 2002-A, p. 53. Norris also writes that the realist 'will say that there exists a vast number of objective truths about mathematics, physics, chemistry, biology, history and other 'areas of discourse' which we don't presently know – and may indeed have no possible means of finding out – but which nonetheless determine the truth-value of any statements we make concerning them' (Norris, 2002-A, p. 139). However, in some contexts Norris seems to be a materialist, as when he argues for Bohm's hidden-variables alternative, although Bohm himself was not a materialist. Indeed, Bohm's theory could accommodate nonlocality, which renders materialism false. But here Norris clearly is not a materialist for, along with Schrödinger, Einstein, Heisenberg, et al., he is claiming that there must exist (or be) nonphysical objective truths that we can '*discover*' (Norris, 2002-A, pp. 4-5, original emphasis). (See also Norris, 2002-A, p. 30.)

478 Frege quoted in Collins, 1998, p. 704.

479 Worrall quoted in Ladyman, 1998, p. 410.

480 Psillos, 1999, p. 146.

481 Eddington, 1935, p. 281.

482 For example, see Grimes in Grimes and Uliana, 1998, p. 218.

483 Siorvanes in Segonds and Steel, 2000, p. 54.

484 Heisenberg, 1974, p. 85.

485 Faye, 2008. A sample of further confusion: Hacking notes how Russell was a realist about theories but an antirealist about the entities posited by those theories, whereas Cartwright is an antirealist about theories and a realist about entities (Hacking, 1983, pp. 27-28, 37). Hacking claims to have become a realist about the material reality of electrons due to an experiment in which electrons are 'sprayed' (Hacking, 1983, p. 23), and he admits that he would not have been a realist about electrons prior to such an experiment. Fortunately, most physicists do not support Hacking's view. If we had really believed that electrons were not real, then we would not have done any experiments trying to understand them better, for only a fool would try to spray something that he was sure did not actually exist.

486 Concerning the debate amongst astronomers about whether or not Pluto should be classified as a planet, see BBC, 2006 and Cook, 2006.

487 See Nagel, 1974. Besides wondering what it is like to be a bat, he also offers a brief but relevant discussion about facts (p. 441).

488 Wheeler quoted in Norris, 2000-B, p. 207.

489 Wheeler quoted in Baggott, 2004, p. 18; also quoted in Norris, 2000-B, p. 254.

490 Bohr, 1963, p. 5.

491 Einstein quoted in Heisenberg, 1972, p. 63. Eddington makes a related point (Eddington, 1935, pp. 234-238). Bohm also makes a related point: 'when we look at the world through our theoretical insights, the factual knowledge that we obtain will evidently be shaped and formed by our theories' (Bohm, 1980, p. 5). Lindley notes that 'Herman Bondi, one of the founders of the steady-state theory, once declared that one should never believe an observation in cosmology until it is supported by theory' (Lindley, 1993, p. 183).

492 Heisenberg quoted in Norris, 2000-B, p. 97.

493 For example, Bohr claimed that 'quantum mechanics may be seen to fulfill all demands on rational explanation with respect to consistency and completeness'

(Bohr, 1963, p. 6). This view could entail that there is no going beyond quantum theory because it is already complete and consistent, or it could simply mean that within its own parameters of applicability it is complete and consistent, but that another theory could still supersede it.

[494] Heisenberg, 1974, p. 184. See also School of Mathematics and Statistics, University of St. Andrews. The same false belief seems to keep recurring: 'by around 1780, the belief had become widespread among leading mathematicians that mathematics had exhausted itself, that there was little left to discover. Unexpectedly, the following century was the most flamboyant in the history of the field, proliferating new areas and opening the realms of abstract higher mathematics' (Collins, 1998, p. 697).

[495] After noting that von Jolly tried to dissuade Planck from studying physics since it was supposed to have been essentially finished, Heisenberg states that 'nobody would wish to make any such false prediction nowadays' (Heisenberg, 1974, p. 184). Unfortunately, such predictions still happen.

[496] Baggott, 2004, p. 109

Notes for chapter seven

[497] Some examples of various incarnations of antirealism include: modern idealism, materialism, emotivism, behaviorism, phenomenalism, constructivism, operationalism, verificationism, instrumentalism, nominalism, and pragmatism (see Vision, 1988, p. 3). Vision has included materialism among this antirealist list, even though realists are often materialists. Perhaps he was correct, as materialists do deny the reality of the nonphysical, and so in that sense we could say that they are 'nonphysical antirealists'.

[498] Polanyi, 1964, p. 70.

[499] Baudrillard, 1994.

[500] Any further discussion of relativism or postmodernism is of no value for our goals here. For further criticisms of these positions, see Norris (1992, 1993), *The Philosopher's Magazine* (2002) and Sokal and Bricmont (2003).

[501] Taylor, 1936, p. 67.

[502] Baggott, 2004, p. 105. (Original emphasis.)

[503] Einstein, 1954, p. 334.

[504] Bohr in Baggott, 2004, p. 109.

[505] My example is similar to the well-known paradox of Schrödinger's cat. For those who are unfamiliar with this thought experiment, the very basic idea is that a cat is placed inside a box where it is supposed to be killed by poison, but the act of releasing the poison depends upon a quantum action. Consequently, the cat, according to the standard interpretation of quantum theory, remains literally half alive and half dead in a state of 'coherent superposition' that exists in 'transcendent potentia'. The cat becomes physically alive or dead once we make an observation (see Goswami, 1995, pp. 78-81). I am not going to discuss this thought experiment further, except to say that it excludes the conscious awareness of the cat. Surely the cat is aware, even if it is not aware that it is aware, but being aware of its own body by seeing, feeling, or inwardly sensing it should be enough to collapse its own wave

function and make it either alive or dead. If it is dead, and so not aware (unless having disembodied existence), then it cannot know that it is dead. Therefore, according to the standard interpretation that requires an observation or measurement for some event to become physically actual, the cat cannot be dead (or alive) unless someone else looks at it. But that implies that we too cannot be dead until someone else happens to look at us, and so, if nobody ever looks at me and I don't look at myself, then I should never die. (Maybe that is why God prefers not to be visible.)

[506] Although Goswami offers interesting insights regarding quantum mechanics, his goal was not to present a philosophically coherent or systematic understanding, and so we should not criticize him for failing to do so. However, it is a bit unsettling when he claims that quantum physics tells us that the moon does not exist when we are not looking at it (Goswami, 1995, pp. 59-60). I suppose that any creatures living on the moon would not be very happy about this, since their habitat could not exist unless some human happens to look at it.

[507] As Psillos reminds us, 'in scientific practice, an object is not supposed to have a property only when the test conditions S actually occur. For instance, bodies are taken to have masses, charges, temperatures and the like, even when these magnitudes are not being measured' (Psillos, 1999, p. 6).

[508] Bohm, 1980, p. 7. This notion of wholeness also resonates with Schrödinger's claim that consciousness is universal and singular.

[509] Bohm, 1980, p. 4

[510] Bohm, 1980, p. 17.

[511] For example, see Blau, 1995. The degree of Krishnamurti's influence on Bohm is debatable, but what Bohm himself made clear is that his thought deeply resonated with Krishnamurti's notion of the unity of the observer and the observed, which is what Bohr also believed. And, of course, this idea is also Platonic, which is not surprising given Krishnamurti's (usually unacknowledged) background in Neoplatonism.

[512] For example, Plotinus writes:

> For the soul keeps quiet then, and seeks nothing because it is filled, and the contemplation which is there in a state like this rests within because it is confident of possession. And, in proportion as the confidence is clearer, the contemplation is quieter, in that it unifies more, and what knows, in so far as it knows – we must be serious now – comes into unity with what is known. For if they are two, the knower will be one thing and the known another, so that there is a sort of juxtaposition, and contemplation has not yet made this pair akin to each other, as when rational principles present in the soul do nothing (Plotinus, 1999, p. 379, *Ennead* III.8.10-25).

[513] Heisenberg, 1958-A, p. 58. He also writes that 'the atomic physicist has had to resign himself to the fact that his science is but a link in the infinite chain of man's argument with nature, *and that it cannot simply speak of nature 'in itself'*....Thus even in science *the object of research is no longer nature itself, but man's investigation of nature*' (Heisenberg, 1958-B, pp. 15, 24, original emphasis). Gibbins puts it this way: 'our

best description of the world does not lead to propositions which are true or false and whose truth or falsity is independent of the means we have for determining them' (Gibbins, 1989, p. 56. See pp. 54-56).

[514] Heisenberg, 1958-B, pp. 15, 24.

[515] Some classicists may disagree with the claim that these geometric forms discussed in Plato's *Timaeus* are really what Plato meant by the 'Ideas', as Heisenberg believed, but his point about the reality of such nonphysical geometric forms at the foundation of all physical reality is still valid.

[516] For example, Gibbins notes Heisenberg's claim that 'if no experiment is available to measure a physical magnitude then it is meaningless to assert that it has a value at all' (Gibbins, 1989, p. 51). Heisenberg also states that 'quite generally there is no way of describing what happens between the two observations' of an electron. According to classical physics, he continues, we would say that 'the electron must have described some kind of path or orbit even if it may be impossible to know which path'. However, 'in quantum theory it would be a misuse of language, which, as we shall see later, cannot be justified' (Heisenberg, 1958-A, p. 48). In other words, when we make a measurement of an electron at time T_1, then again at time T_2, we cannot know what path the electron has taken in between T_1 and T_2; therefore, it is meaningless to say anything at all about the electron except when measuring it at T_1 or T_2. Of course, Heisenberg is not philosophically clear on this point.

[517] Consider the following familiar example. If a tree falls in the forest and no sentient creature with auditory apparatuses perceives the sound when it hits the ground, then is there any sound at all? If by 'sound' we mean that perceiving ears must hear it or else the sound would be nonexistent, then there would be no sound, given our definition from the outset. But there must be a sound, as all scientists must believe if they are to be consistent with their other fundamental scientific beliefs. When a tree of a certain mass falls at a certain speed and strikes the earth, a certain amount of energy will be released. If any auditory receiving apparatus, whether of a machine or sentient creature, is nearby to detect this energy release, then it will be experienced as a sound wave, although the experience will be somewhat differently perceived or interpreted depending upon the receiving machine or auditory apparatus. The paradox can only appear to exist when we define 'sound' in such a way that, by definition, it cannot exist unless something hears or receives the energetic release when the tree hits the ground. In reality, however, there is no paradox.

[518] Eddington, 1929, p. 55.

[519] Lindley, 1993, p. 63.

[520] Lindley, 1993, p. 6.

[521] Ralkowski (2009) shows Heidegger's debt to Platonism, which would not likely appeal to Rogers.

[522] Compare with Mach's misplaced denial of atoms. For example, Blackmore shows how Mach 'almost certainly did oppose, not just the reality of atoms and Boltzmann's gas theory which he had long done, but part or all of Einstein's theory of relativity, and particularly 'The Church of Physics"' (Blackmore, 1989, p. 537). See Spencer (2007) for an extended analysis of Rogers' position.

[523] Rogers, 2005, p. 172

524 Smith, et al, 2005, p. 67.
525 Smith, et al, 2005, p. 591.
526 Smith, et al, 2005, p. 439.
527 Lazcano and Miller (1994) argue that 'in spite of the many uncertainties involved in the estimates of time for life to arise and evolve to cyanobacteria, we see no compelling reason to assume that this process, from the beginning of the primitive soup to cyanobacteria, took more than 10 million years'.
528 Nadeau and Kafatos write the following:
> If electrons did not obey the [Pauli exclusion] principle, all elements would exist at the ground state and there would be no chemical affinity between them. Structures like crystals and DNA would not exist, and the only structures that would exist would be spheres held together by gravity. The principle allows for chemical bonds, which, in turn, result in the hierarchy of structures from atoms, molecules, cells, plants, and animals. (Nadeau and Kafatos, 1999, p. 34.)

529 Rogers, 2005, p. 172

530 Buddhism is not anti-intellectual, nor is it an escape from rationality, as it may sometimes appear to be when it is misinterpreted through the antirealist lens. For example, consider the words of the Buddha in the *Dhammapada*: 'clear thinking leads to Nirvana, a confused mind is a place of death. Clear thinkers do not die, the confused ones have never lived' (Buddha, 1967, p. 45). The sense of realism could not be stronger: follow the teachings of the Buddha or continue to suffer (or continue not even to be alive).

531 Buddhists also believe that it matters what we do here in conventional reality because of the effects of karma and so forth, but why would it matter what we do in conventional reality if it is all an illusion, and even more so if we, too, are illusions?

532 Wallace, 1996, p. 122. See also Sriraman, 2004.

533 For example, Garfield has clarified how Nagarjuna's dialectic method leads to the radical claim that 'emptiness itself is empty' (Garfield in Nagarjuna, 1995, p. 316). The motivation behind Nagarjuna's claim that Emptiness is empty of itself is that otherwise it would be 'a substantial entity, an independent existent, a nonconventional phenomena' (Garfield in Nagarjuna, 1995, p. 315). In other words, Buddhism would then have a fundamental metaphysical similarity to Platonism, as well as many other religions, from Christianity to Hinduism.

Although I will not argue for this claim here, I do, in fact, think that these traditions share foundational metaphysical similarities. We should, however, pause briefly on the nature of 'self', 'I', and 'soul' from Buddhist and Platonist perspectives, since I argue for the reality of soul in the final chapter, and I think that the common Buddhist *interpretation*, which rejects the reality of soul, self, and I, is mistaken. Where many Buddhists see incompatibility, I see unity.

Wallace claims that 'there is no intrinsic, personal self, nobody that stands apart from the constant fluctuation of mental and physical events' (Wallace, 1993, p. 124). Tulku Thondup Rinpoche writes that '*self is an illusion*, because everything in the experience of samsara is transitory, changing, and dying. Our ordinary mind thinks of self as something that truly exists as an independent entity. But in the Buddhist view, *self does not truly exist*' (Thondup, 1996, p. 18, emphasis added). Transpersonal psychologist Brian Lancaster argues 'that a view of self along the lines of the Buddhist concepts of no-self and the conditioned nature of "I" introduces a more parsimonious perspective on the neuropsychological data' (Lancaster, 1993).

At first reading, such views may not seem to make any sense; after all, to deny that we exist seems pretty crazy. But there is a very sophisticated way to understand how this Buddhist view is true, a way that is not at all antirealist. I am here only offering a very condensed and simplified clarification, which would require much more space to unpack in appropriate detail, but the basic point can still be made. There are two fundamental ways that we can view the nature of 'I' or 'self': the conditional self, and the essential self.

The *conditional self* is constructed from a variety of factors, including our genetics, environment, culture, and personal beliefs and thoughts. This conditional self does not have independent reality, which is to say that it vanishes with the death of the body and can be affected by altering brain chemistry or functionality. For example, I used to believe that I was a certain sort of person, but later in life this sense of self changed, and I now no longer view myself as the same self as I was in the past.

The *essential self*, however, is the background from which all experience is possible, the fundamental nature that I am, that which will not perish with the body. When the essential self forgets its true nature it then identities with the conditional self, and then we suffer because we are clinging to impermanence. We must awaken to our underlying nature, which is eternal. However, if we leave this physical world while clinging to conditional self, it can affect the fate of the essential self with respect to its next physical actualization.

This very simplified view can be metaphysically compatible with both Buddhist and Platonic notions of the self. If Buddhists, however, reject the reality of the essential self, which many of them seem to do, then there is, in fact, no coherent way to make sense of our experience or even our very existence. There would certainly be enormous legal ramifications, since, on this Buddhist interpretation of self, nobody is ever the same person from moment to moment so we could not say that Joe committed a crime in the *past* because 'Joe' would *now* be a different person, a different self. In fact, 'Joe' would not really exist at all, and neither would 'Joe's' body.

Giles (1993) argues for the similarities between this Buddhist notion of 'no-self' and Hume's related theory of personal identity (or lack of personal identity). However, the similarities are only superficial, for it would be rather difficult to argue that Hume could accommodate the Buddhist notion of a 'stream of awareness' (which functions very much like the Platonic notion of soul, a point that appears to upset many Buddhists caught in the existential trap of believing that they do not really exist, even though they tend to believe in some version of reincarnation). After reaffirming that 'there is no notion of a self-sufficient, self-supporting "I", self, soul, or ego' (Dalai Lama in Houshmand, Livingston, and Wallace, 1999, p. 113), the Dalai Lama then asserts that 'the continuum of awareness that conjoins with the foetus does not depend upon the brain...[and] it must arise fundamentally not from a physical base but from a proceeding continuum of awareness' (Dalai Lama in Houshmand, Livingston, and Wallace, 1999, p. 114). This stream or continuum of awareness is what accounts for the supposed transmigration from one life to another (see, for example, Dalai Lama in Houshmand, Livingston, and Wallace, 1999, p. 70). It is certainly puzzling to deny the reality of soul and yet assert that a stream of awareness, which is also nonphysical, could perform the same function as soul.

Consider the Buddha's words in the *Dhammapada*: 'value your self, look after your self…You are your own refuge…One's self is the lord of oneself; there is no other lord…You cannot save another, you can only save yourself' (Buddha in P. Lal, 1967, pp. 93-94).

On one of the rare occasions that the Buddha would make an explicit metaphysical statement, we find him saying that

> verily, there is an unborn, unoriginated, uncreated, unformed. If there were not this unborn, unoriginated, uncreated, unformed, then an escape from the world of the born, the originated, the created, and the formed would not be possible. (The Buddha in Hanh, 1995, pp. 138-139, from Udana viii, 3).

Clearly, that which is prior to creation is more real than that which has been created, and to escape suffering in the created world (the realm of becoming), we must attain to the uncreated (the realm of being, or the realm beyond being, depending on further clarification). Moreover, there is someone, a self, that is currently suffering and which can also escape from the realm of suffering. At the highest levels of unification with the uncreated, there is a sense in which even the essential self would cease to be what it is and become completely one with the origin of all reality. But any attempt within any tradition to articulate such notions inevitably encounters great difficulties with expression. Plotinus, for example, offers an interesting analogy of mystical union with the One:

> the man who obtains the vision [of the One] becomes, as it were, another being. *He ceases to be himself, retains nothing of himself.* Absorbed in the beyond he is one with it, like a center coincident with another center. While the centers coincide, they are one. They become two only when they separate. (Plotinus, 1964, p. 87; VI. 9. 10, emphasis added.)

Thich Nhat Hanh offers a balanced historical perspective by showing how the Buddha reacted against the corruption within the Vedic priesthood by attacking the 'notion of *Atman*, Self, which was at the center of Vedic beliefs, [and] was the cause of much of the social injustice of the day' (Hanh, 1995, p. 54). Hanh continues: 'but when Buddhists began worshipping the idea of emptiness, he said, "It is worse if you get caught in the non-self of a flower than if you believe in the self of a flower."' (Hanh, 1995, p. 54).

Buddhists make a very important argument against our naïve understanding of personal identity, but the antirealist spin put on this powerful ancient tradition by Wallace and others is not the answer.

[534] Wallace, 1996, p. 74.
[535] Wallace, 1996, p. 76.
[536] Wallace, 1996, p. 76.
[537] Wallace, 1996, p. 122.

[538] Planck, 1931, pp. 8-9. Schäfer makes a related point:
"in the same way, viewed from the outside, it may not be possible to prove the existence of an **objective outer reality**. However, whenever that body which is associated with this (in some sense) independent self-conscious mind takes part in a process of that outer reality, **experience by the mind of this interaction** leaves no doubt that an objective outer reality exists' (Schäfer, 1997, p. 89, original emphasis).

[539] Vision, 1988, pp, xiv.

[540] Norris has argued strongly against the antirealist views of both Rorty and van Fraassen, and while there will be some overlap with my criticisms here, I am generally taking a different, albeit compatible, approach. For example, see Norris, 1997. In addition to my arguments against Rorty's views, see also Gary Hatfield in Sorrell and Rogers (2005). Rorty's 'arguments' (or whatever he actually offered) in *The Mirror of Nature* depend upon his historical evaluation of previous philosophers, such as Plato, Descartes, Locke, and Kant. But as Hatfield makes clear, Rorty's rendering of these philosophers:

is at best an outdated caricature, at worst a 'just so' story fabricated to portray the 'authority' of past philosophy as resting on a rhetorical ploy that would fail in the sophisticated present. The moral of his tale is that philosophy today can make no direct contribution to intellectual discussion. Its role can only be to 'edify', by describing the results of one (non-philosophical) area of discourse to the participants of another (non-philosophical) area of discourse (Hatfield in Sorrel and Rogers, 2005, pp. 97-98).

Hatfield also points out a serious inconsistency in Rorty's historicism, where he claims that philosophical problems have both changed *and* remained essentially the same. Rorty 'has failed to see how it could be true both that philosophy has been concerned since the time of Plato with questions about the knower's relation to the known, and also that the theories and purposes of philosophers have changed from epoch to epoch, or even from writer to writer' (Hatfield in Sorrel and Rogers, 2005, p. 100). Finally, Hatfield also notes the irony that 'Rorty's image of the philosophy of the past is remarkably similar to the actual practice of the detached and imperious analytic philosophers of the 1960's, the very time when he framed his project' (Hatfield in Sorrel and Rogers, 2005, p. 101).

[541] The ancient Cynics, the 'dogs', indeed lived this way. See Hadot, 2002, pp. 108-111 and John Christian Laursen in Popkin, 2008, p. 149.

[542] Rorty, 1991, pp. 1, 8.

[543] Rorty, 1991, pp. 21-24, 31 (footnote), 40, 90.

[544] Antirealists reveal their realist faith every time they fly in an airplane or believe that their loved ones exist when they are not looking at them. I have many significant disagreements with Dawkins, but he makes a good rhetorical point on this matter: 'show me a cultural relativist at 30,000 feet, and I'll show you a hypocrite. Airplanes built according to scientific principles work' (Dawkins, 1996, pp. 31-32). Similarly, while criticizing 'gender feminists' who claim that logic and rationality are 'phallocentric' and who attack the 'rational standards and methods that have been

the hallmark of scientific progress' (Sommers, 1995, pp. 64-65), Sommers quotes Margarita Levin who 'dryly remarks': 'one still wants to know whether feminists' airplanes would stay airborne for feminist engineers' (Levin in Sommers, 1995, p. 73).

[545] Rorty, 1990-A, p. 1.

[546] Here is Rorty addressing the Faculty of Law at the University of Toronto: 'for we [anti-essentialists] cannot afford to sneer at *any* human project, any chosen form of human life' (Rorty, 1990-A, p. 17, original emphasis). Would Rorty really embrace '*any*' human project, even the most horrifying atrocities?

[547] Rorty, 1980, p. 727.

[548] Rorty, 1991, pp. 2-3. When using the number 17 as a 'model for reality', Rorty says of this number that 'by not bothering to ask whether these things really exist or whether he has invented them, he becomes both freer and happier' (Rorty, 1990-A, p. 29).

[549] Rorty, 1991, p. 7.

[550] Rorty, 1991, pp. 7, 99, 151.)

[551] Rorty, 1990-A, p. 13.

[552] Rorty, 1979, p. 279.

[553] Rorty, 1979, p. 264.

[554] Rorty, 1990-A, p. 24.

[555] Rorty, 1990-B, p. 6.

[556] As Rorty further writes:
> By dropping a representationalist account of knowledge, we pragmatists drop the appearance-reality distinction in favor of a distinction between beliefs which serve some purposes and beliefs which serve other purposes — for example, the purposes of one group and those of another group. We drop the notion of beliefs being made true by reality, as well as the distinction between intrinsic and accidental features of things (Rorty, 1990-B, p, 8).

[557] The role of instrumentalism in the history of philosophy has also been overplayed. See Lloyd, 1978.

[558] Rorty, 1991, p. 14.

[559] Clark, 1995. Harris adds that 'relativism is inextricably entangled with scepticism. The denial of objectivity is the denial of truth, and that cannot but infect the asseverations of the relativist and the skeptic themselves' (Harris, 2000, p. 4).

[560] As Polanyi writes: 'it is logically false to deny the existence of truth since the very statement asserting this is based on the assumption that truth can be established' (Polanyi, 1964, p. 78). Norris also emphasizes that realists allow their beliefs 'to be put to the test rather than hedged around with protective disclaimers' (Norris, 2002-A, p. 45).

[561] Bernard Williams also notes that since Rorty admits that it is 'overwhelmingly convenient' to say that physics describes a world that is already there, 'then what everyone should be saying is simply: physics describes a world that is already there. So why does Rorty go on telling us *not* to say that?' (Williams in Malachowski and Burrows, 1990, p. 32, original emphasis.)

[562] Rorty, 1990-A, p. 25.

563 Thomas A Russman, whose former PhD director at Princeton University was Rorty, offers some relevant observations. He shows how Rorty does indeed simply want to poke fun at those he disagrees with rather than seriously arguing with them, for arguing would entail that he too needs to defend his views (Russman, 1987, pp. 89-92). Russman continues:

> if Rorty were to say that from his viewpoint no arguments are allowed, and therefore, a priori, no arguments against the myth of the framework are allowed, then I shall be content to reply: Let all those who reject this move as insipidly self-serving line up with me. It is surely we who shall be able to carry on the conversation using the fullest range of conversational tools, including arguments (Russman, 1987, pp. 91-92).

In Plato's dialogue the *Meno*, Socrates says that he would fight to the end, in word and deed, for the belief that we should try our best to uncover what we presently do not know. In fact, we would 'be better and braver and less idle' to seek to discover the unknown 'than if we believed that what we do not know it is impossible to find out and that we need not even try' (Plato, 1956, p. 51, *Meno* 86c). After spinning around in the confusion and deception of Rorty's mind, Clark's bluntness is a welcome relief: 'those who say there is no Truth are liars; those who say we cannot find it out admit they have no reason for what they say' (Clark, 1998, p. 23).

564 See Stern, 1989, p. 210.

565 Clark, 1990, 21.

Notes for chapter eight

566 For example, Norris writes that 'the realist position – starting out with Aristotle and still very active in our day...' (Norris, 2001, p. 280). See also Norris, 2000-C, p. 40. Despite writing an informative paper on quantum mechanics, Santos also continues the confusion between realism and idealism by equating the views of Aristotle (and Einstein) with the former and those of Plato (and Bohr) with the latter (Santos in Ferrero and van der Merwe, 1995, p. 1). Even though abstractionism is often attributed to Aristotle, this common view has been questioned (see Cleary, 1995). But what matters more is not whether or not Aristotle was really an abstractionist, but whether or not abstractionism is really true. In a similar way, the common assumption that Aquinas was an Aristotelian (if not anti-Platonist) has been shown to be incorrect. See Quinn (1996), The Introduction to Pseudo-Dionysius (1987), and Hankey (2004).

We find similar admiration and condemnation of Aristotle as we do of Plato, and there is much disagreement about what Aristotle actually believed and the intelligibility of his views. For example, as Brumbaugh writes, incompatible views about Aristotle include believing that 'he is wrong at almost every point, that he has brilliant insights but develops them to anticlimaxes, and that he is using an idiom that is wholly unfamiliar' (Brumbaugh, 1955, pp. 379-380).

567 Mueller in Proclus, 1970, p. xxvi.

568 James Ross is an interesting example of a philosopher who is an antirealist abstractionist about the mathematical laws of physics. However, he makes two key

mistakes with respect to Platonism. First, he falsely assumes that modal logic, all possible worlds, and many-world theories are aspects of Platonism. For the Neoplatonists, as for Plato, there is only one universe, which was brought into order out of chaos by the demiurge, the working god:

> Have we been correct in speaking of *one* heaven, or would it have been more correct to say that there are many, in fact infinitely many? There is but one, if it is to have been crafted after its model....[T]he maker made neither two, nor yet an infinite number of worlds. On the contrary, our heaven came to be as the one and only thing of its kind, is so now, and will continue to be so in the future (Plato, 2000, p. 16, *Timaeus* 31a-b, original emphasis).

The demiurge is not the absolute supreme One, but is a lesser deity. Proclus writes:

> Those, then, who say that the first God is Demiurge or Father are not correct; for the Demiurge and the Father is a particular god.... Let the One then be termed simply God, as being the cause for all gods of their being gods, but not for some particular gods, as for instance demiurgic or paternal or any other particular type of godhead, which is a type of qualified divinity, not divinity in the simple sense' (Proclus, 1987, pp. 443-444, *Proclus' Commentary on Plato's* Parmenides, BK VI. 1096-1097).

This is a complicated issue that will need to be set aside here. See, for example, Robinson (1967), Gurtler in Wagner (2002), and Finamore (1988). What is important to note, however, is that although Platonists assert that there are different metaphysical *levels* of reality, there is only one physical universe, and the ultimate nature of all reality (both physical and nonphysical) is the unifying principle known as the One.

Ross' second key mistake occurs when he confuses universals, such as 'chair', whose actualization depends upon our mind's conception, with a Platonic Form or Idea, such as a law of physics or absolute Beauty, whose reality is not dependent upon us or the physical cosmos. According to Ross, Platonism is also circular, hostile to creation, and ontologically inert; he even claims that it is 'satanic' and 'bacterial' (presumably pathogenic, not probiotic). He dismisses anything, including various types of logic and developments in modern biology, that are a threat to his belief that his whole body, character, memory, and phenomenal person will, at the end of the world and after God's judgment (including purgatory, if necessary), be 'miraculously transformed into the renewed and everlasting creation and glorified body', which is a view that he believes fits within 'the foundations of physical science' (Ross, 2001. See also Ross, 1989 and 2008).

[569] See Ross (2008), Cleary (1995), and Brumbaugh (1955), from whom I have drawn while developing my story.

[570] Ross writes: 'the intelligible repeatable structures of things and processes are real. They are constitutive of particulars. There is such software everywhere in nature. We can discern, formulate, and use the structures; indeed a good part of science and technology does that.... The intelligible structures are like "little minds in things" determining what things are capable of and do' (Ross, 2008, pp. 129-130).

⁵⁷¹ Let us consider an example. A fertilized human egg, free of genetic manipulation, has the potential to grow into a human, but not an apple or a zebra. Thus, there must be some reason-principle or intelligence built into the very structure of a zygote. The abstractionist may want to say that the genetic information is a material reason-principle, for this information is what makes growth into a human possible. But we cannot accept this claim so easily, since we need an explanation for how the egg and sperm contain their own DNA that later combine in the zygote. What was the reason-principle that produced the egg in the ovaries and then sent it through the fallopian tubes in search of sperm? Since the eggs were originally produced while the woman herself was still a fetus, the original genetic information in the zygote that eventually developed into this woman must have contained the reason-principle to produce the eggs that would contain the genetic information for further propagation.

The complexity of this situation could quickly become overwhelming if we were to analyze this process in much further detail, which, fortunately, will not be necessary. The main point is that there must be some sort of intelligence in our genetic information in order for us to develop into the creatures that we are. Surely biologists do not want to say that human development is a completely random process; to do so would be to destroy the entire science of biology. If all biological processes were truly random, then a human zygote should be able to develop into a full-grown human in 3 seconds, or in 50 million years, or perhaps develop into a tree. These scenarios are ruled out because human development is not random, even though there are myriad relevant external factors that contribute to the dynamic flexibility of human development, within the parameters and potentiality of the particular human reason-principle interacting with all the other relevant reason-principles in its environment. But to say that our DNA is programmed for us to develop in a certain way does not explain how the DNA received its program, how the program is actually embedded, nor how such a program is imbued with its innate intelligence and power to grow on its own. It also does not tell us the nature of the quantum structure from which DNA is composed. Biologists have become technical wizards, but they still lack a deeper underlying understanding of how biological processes are even possible in the first place.

⁵⁷² Proclus, 1970, pp. 40-41, Book II, Chap I. par. 50-51, *A Commentary on the First Book of Euclid's* Elements.

⁵⁷³ See Plato, 2000.

⁵⁷⁴ See Zeyl's note in Plato 2000, p. 14.

⁵⁷⁵ See also Whittaker, 1969.

⁵⁷⁶ Dawkins, 1989, p. 218. Hume also writes, in a somewhat different, though not entirely unrelated, context that the apparent connection between our correspondent impressions and ideas shows that 'such a constant conjunction, in such an infinite number of instances, can never arise from chance' (Hume, 1978, p. 4). Morris writes the following:

> it appears, therefore, that we exist in a very improbable kind of universe, one that was fine-tuned to an accuracy of one part in 10^{15} at a time of one second after the big bang...Scientists distrust coincidences. When they find that a number is that close to a

critical value, they are generally unwilling to believe this could have happened by chance. They are not satisfied until they find a reason for why the fine-tuning should be that exact (Morris, 1990, pp. 53-54).

[577] Chaisson and McMillan, 1999, pp. 338-339.

[578] Kuhn, 1970, p. 111.

[579] For further reading, see Boyd (1973), Laudan and Leplin (1991), Douven (2000), and Hoefer and Rosenberg (1994).

[580] Although I briefly consider related speculations in the next chapter (under the section *Dreaming the future*), I am not here concerned with whether or not we are in a virtual reality controlled by aliens, or in a dream-world created by an evil genius, because in any such case there would still be objective constraints placed upon us. But assuming that we are in virtual reality, if we wish to succeed (or escape) we had better figure out the rules governing the virtual reality to the best of our ability, and Bostrom (2003) argues that we may currently be living in a computer simulation. Even if this were true, some creature or immaterial entity first had to design the program and build the computer, which has to, or had to, exist somewhere. Even if we were nothing but ones and zeros projected into cyberspace, or given any other such scenario, it would still be the case that all these possibilities presuppose the reality of mathematical laws, which in turn presupposes the one eternal mathematical law. Realism is true regardless of how believable the illusion of our simulation may be. There could be no simulation without an underlying reality.

[581] See Einstein, 1954, p. 322.

[582] Addey in Addey and Wyndham-Jones, 2003, p. 80 (original emphasis).

[583] Smolin, 2006-B, p. 33.

[584] Recall Bergethon, 1998.

[585] See Halacy, 1979. Also see Stein, 1987.

[586] Babbage, 1838, p. 40.

[587] Babbage, 1838, pp. 42-43.

[588] When asked how to play Calvinball, *Calvin and Hobbes* author Bill Watterson replies: 'it's pretty simple: you make up the rules as you go'. But there is still one rule, as Calvin reminds us: 'the only permanent rule in Calvinball is that you can't play it the same way twice' (Watterson, 1995, p. 129). This example is actually quite significant, for we cannot reject all rules unless we have an unchanging rule that says 'reject all rules'. Even Calvin, with all of his heroic powers and anarchistic disposition, cannot escape this consequence.

[589] Newton in Olson, 2004, p. 119.

[590] Newton in Olson, 2004, p. 119. Olson is actually quoting from Manuel who is quoting from Newton. See Manuel, F. (1974). *The Religion of Isaac Newton*. Oxford University Press. (Not listed in Works Cited.) Coudert also writes the following:

> Newton believed that polytheism went hand in hand with bad science, while monotheism was the source of good science dedicated to finding the simple, unifying cause of all natural phenomena. Just as there was only one all-powerful divine Creator and Father, there was only one cause for all the disparate phenomena in the created world. The law of simplicity was valid

for both the interpretation of nature and scripture. (Coudert in Force and Popkin, 1999, pp. 31.)

And Newton writes the following:

This most elegant system of the sun, planets, and comets could not have arisen without the design and dominion of an intelligent and powerful being. And if the fixed stars are the centers of similar systems, they will all be constructed according to a similar design and subject to the dominion of *One*, especially since the light of the fixed stars is of the same nature as the light of the sun, and all the systems send light into all the others (Newton, 1999, p. 940; *General Scholium*, original emphasis).

[591] Einstein, 1954, p. 27. Einstein also thought that Mach's idea of simplicity was much too subjective. 'In reality', Einstein told Heisenberg, 'the simplicity of the natural laws is an objective fact as well, and the correct conceptual scheme must balance the subjective side of this simplicity with the objective. But that is a very difficult task. Let us rather return to your lecture' (Einstein as quoted by Heisenberg, 1972, pp. 65-66). He was right to say that we need a correct conceptual scheme to balance the subjective and objective, which is one of the goals of this book, and he was also correct in saying that the task of clarifying this issue is very difficult.

[592] Heisenberg, 1974, p. 83.

Notes for chapter nine

[593] Clarke, 1998, p. 122.

[594] As a sample of Cartwright's work, see Cartwright 1983, 1994-A, 1994-B. Smith (2001) provides a fair critique of specific aspects of Cartwright's position.

[595] Proclus provides a fuller expression of Coleridge's point when he describes the essence of the dialectic: 'dialecticians must seek the dissimilarities in kindred things and the similarities in divergent things' (Proclus, 1987, p. 528, *Commentary on Plato's Parmenides*, BK VII, 1178). For what seems to be the most precise and detailed account of the dialectic, see Proclus, 1987, pp. 352, *Commentary on Plato's* Parmenides, BK V, 1002.

[596] Cartwright, 1994-B, p. 288.

[597] Cartwright, 1983, p. 13.

[598] See Bergethon, 1998.

[599] Oxtoby and Nachtrieb, 1986, p. 401.

[600] Higdon et al, 1978, p. v.

[601] See Epp, 2004.

[602] Becker et al., 1996, p. 841.

[603] Becker et al., 1996, p. 377.

[604] Clarke, 1998, p. 35.

[605] Hume, too, struggled with the notion of unity: 'but all my hopes vanish, when I come to explain the principles that unite our successive perceptions in thought or consciousness' (Hume, 1826, p. 55).

[606] Emerson writes that:

there is one mind common to all individual men. Every man is an inlet to the same and to all of the same. He that is once admitted to the right of reason is made a freeman of the whole estate. What Plato has thought, he may think; what a saint has felt, he may feel; what at any time has befallen any man, he can understand. Who hath access to this universal mind is a party to all that is or can be done, for this is the only and sovereign agent (Emerson, 1941, p. 1).

Heisenberg experienced such insight when he knew directly for himself what Plato had been pointing at in the *Timaeus*. Also recall Schrödinger's belief that consciousness is universal and singular.

[607] Gerson, 2005-B, p. 32 (original italics have been removed). The other five aspects of Platonism, as provided by Gerson, are:

3. The divine constitutes an irreducible explanatory category.
4. The psychological constitutes an irreducible explanatory category.
5. Persons belong to the systematic hierarchy and personal happiness consists in achieving a lost position within the hierarchy.
6. Moral and aesthetic valuation follows from the hierarchy.
7. The epistemological is included within the metaphysical order. (Gerson, 2005-B, pp. 33-34.)

[608] Proclus, 1963, p. 3, *The Elements of Theology*, Prop. 1. See Spencer in Cohen, 2006, for an introduction to this principle and Proclus' reasoning.

[609] Taylor, 2001, p. xi.

[610] Norton, 2000, p. 137.

[611] Harris, 2000, p. 242. For example, Einstein writes that an important, subtle motive for the desire to devise new theories 'is the striving toward unification and simplification of the premises of the theory as a whole (*i.e.*, Mach's principle of economy, interpreted as a logical principle)' (Einstein, 1950-B, original emphasis). As already noted, this positivist-sounding belief is actually an aspect of Platonic realism, but the positivists lack the metaphysical muscle to support it.

[612] Planck, 1931, p. 68.

[613] See Proclus, 1963, p. 59, prop. 60.

[614] See Proclus, 1963, p. 59, prop. 62.

[615] As Harris notes, 'if this testimony [from the great pioneering physicists] is accepted for the way in which scientific knowledge is progressively developed and supplemented, it carries with it significant implications for the much criticized doctrine of degrees of truth, another corollary of the coherence theory' (Harris, 2000, p. 244).

[616] Rowlands, 2007, pp. 61-62.

[617] Rowlands, 2007, p. 62. Heisenberg also stated that 'the great comprehensive formulation of natural laws, such as first became possible in Newtonian mechanics, are concerned with idealizations of reality, not with reality itself' (Heisenberg, 1974, p. 185). Heisenberg was not as philosophically precise as he should have been here, because these 'idealizations' are not abstractionist fictions. They are ideal,

nonphysical laws, which the physical universe must obey. As already noted, Heisenberg believed that mathematical laws remain valid within their applicable domains forever, and so in the above quote he simply should have said 'physical reality' instead of 'reality'. These errors may seem minor, but they can soon be misinterpreted and extrapolated into the philosophical and theoretical mess we find in quantum physics today.

618 Rowlands, 1992, p. 21. Whitehead makes a similar point: 'The history of seventeenth century science reads as though it were some vivid dream of Plato or Pythagoras...the paradox is now fully established that the utmost abstractions are the true weapons with which to control our thought of concrete fact' (Whitehead, 1953, p. 41). It should be noted that he does not mean 'abstractions' in the sense of 'abstractionism'.

619 Harris, 1983, p. 158.

620 Lee in Wagner, 2002, p. 38.

621 See Beauregard, 2012.

622 Malin, 2001, p. 230. Schrödinger had earlier used the phrase 'Subject of Cognizance' (Schrödinger, 1967, p. 127). Eddington offers a similar discussion, but he refers to the observer as 'Mr. X' (Eddington, 1935, pp. 252-263).

623 Harris, 2000, p. 97.

624 Schrödinger, 1967, p. 138.

625 Whatever one may think of the idea, the Dalai Lama has said that it is possible for a person (a stream of awareness) to reincarnate into a computer. He states that 'if all the external conditions and the karmic action were there, a stream of consciousness might actually enter into a computer' (The Dalai Lama in Hayward and Varela, 2001, p. 152). In such a case, the nonphysical stream of awareness could animate a computer, making genuine self-awareness possible.

626 Proclus, 1963, p. 19; *The Elements of Theology*, prop. 15.

627 Proclus, 1963, p. 77, *The Elements of Theology*, prop. 83. Siorvanes also notes that 'in Neo-Platonism, cause and effect, known and knower, are mutually implied' (Siorvanes in Segonds and Steel, 2000, p. 54).

628 Proclus, 1963, p. 163, *The Elements of Theology*, prop. 186.

629 Bohr, 1963, p. 4.

630 Bohr, 1963, p. 12.

631 See Harris, 2000, p. 35.

632 Harris notes two further points regarding the significance of the relation between theories and experiments. First, all quantitative measurements presuppose that a relevant question has been asked, which itself can only arise out of a pre-existing body of scientific knowledge, all of which depends on conscious beings (Harris, 2000, p. 88). Second, there can be an extraordinary number of potentially observable facts in most observable situations. Even if a tightly focused experiment is arranged, there are still innumerable relevant facts from which we must consciously select those that serve our purposes as best as possible, as various scientists from Darwin to Einstein have in principle recognized (Harris, 2000, pp. 88-89). But that does not reduce truth to whatever serves our purposes. On the contrary, the very fact that we can choose which facts to focus on necessarily

presupposes that there are objective facts in the first place. In any case, all these points inevitably inform us of the fundamental role of the observer.

[633] Rowlands, 2007, p. 83. He continues by providing a further important distinction: 'observation and theory, in physics, necessarily use incompatible types of mathematics because observation depends only on one member of the parameter group (space), while theory sets up the properties of the other members in opposition' (Rowlands, 2007, p. 83). Penrose makes a related point:
> the more deeply we probe the fundamentals of physical behaviour, the more we find that it is very precisely controlled by mathematics. Moreover, the mathematics that we find is not just of a direct calculational nature; it is of profoundly sophisticated character, where there is subtlety and beauty of a kind that is not to be seen in the mathematics that is relevant to physics at a less fundamental level (Penrose, 2004, p. 1026).

[634] Consider what Harris writes:
> contemporary physics, we have observed, has abolished the classical materialism and has made it impossible any longer to maintain the sharp separation between the physical object and the observer. It has established the internal character of the physical relations and the unity of the physical universe. Philosophically considered, the essential nature of the whole so revealed requires a dialectical structure that leads inevitably to religious conclusions (Harris, 2000, p. 270).

I think that Harris is correct on this point, except that I would use the word 'spiritual' rather than 'religious'. Biblically, the term 'religious' (*threskos*-θρησκος) meant to be 'careful of the externals of divine service', whereas 'spiritual' (*pneumatikos*-πνευματικος) was used variously, but a key meaning was to denote 'things that have their origin in God [but higher in the scale of being than man in his natural state], and which, therefore, are in harmony with His character' (Vine, 1952). Such a view, where there are grades of reality that have more being when closer to the One (or God) is, as I have been arguing throughout, an essential aspect of Platonism and physics.

[635] Heisenberg cites Bohr as stating that 'naturally, it still makes no difference whether the observer is a man, an animal or a piece of apparatus, but it is no longer possible to make predictions without reference to the observer or the means of observation' (Heisenberg, 1972, p. 88). Heisenberg claims that the 'transition from the 'possible' to the 'actual' takes places as soon as the interaction of the object with the measuring device, and thereby with the rest of the world, has come in to play; it is not connected with the act of registration of the result by the mind of the observer' (Heisenberg, 1958-A, pp. 54-55). Physicists, however, are never really clear about how exactly the wave in potentia interacts with a physically actualized macro object.

[636] Harris, 2000, p. 100.

[637] Schrödinger, 1967, p. 18.

⁶³⁸ Cromwell, 1997. As Eddington has also argued, we 'do not quite attain that thought of the unity of the whole which is essential to a complete theory' without understanding the relationship between the 'pointer readings' (or measurements), the physical laws, and the background behind such symbols with the isolated consciousness that also resides in that background. In other words, 'no complete view can be obtained so long as we separate our consciousness from the world of which it is a part' (Eddington, 1935, p 317).

⁶³⁹ Baggott, 2004, p. 83.

⁶⁴⁰ Baggott, 2004, p. 243.

⁶⁴¹ This conclusion is similar to what Goswami (1995) struggled to articulate, but he sometimes becomes ensnared in antirealist traps.

⁶⁴² Eddington, 1935, p. 278.

⁶⁴³ Eddington, 1929, p. 24.

⁶⁴⁴ Weber in Radin, 1997, p. 273. Burtt also writes that 'Berkeley, Hume, Kant, Fichte, Hegel, James, Bergson—are all united in one earnest attempt, the attempt to reinstate man with his high spiritual claims in a place of importance in the cosmic scheme' (Burtt, 1925, p. 11).

⁶⁴⁵ For example, Thomas Taylor, the nineteenth century philosopher and translator of Plato, Aristotle, and Neoplatonic philosophers, writes that 'the soul is the generator of mathematical forms, and the source of the productive principles with which the mathematical sciences are replete' (Taylor, 2006, p. 3). See The Prometheus Trust for more information about Taylor.

⁶⁴⁶ Proclus made a clear distinction between soul and mind (see Proclus, 1987, pp. 284-285, *Commentary on Plato's Parmenides*, 930). As Whittaker writes:
> Proclus saw quite clearly that Plato's theory of ideas, while it has psychological references, could not be understood as merely psychological. His own development has strikingly Kantian turns; and it may be said in his favour that, by his distinction between "soul" and "mind" (the associate of a particular body and the intellect in which it shares), he makes more clear than Kant did that it is not the merely individual intelligence that is conceived as "projecting" the forms of knowledge (Whittaker, 1928, pp. 257-258).

⁶⁴⁷ Gerson, 2005-B, p. 34. Proclus, for example, argues that 'every soul is indestructible and imperishable' (Proclus, 1963, p. 163, *The Elements of Theology*, prop. 187).

⁶⁴⁸ Plato, 1997, p. 589, *Alcibiades* 130c.

⁶⁴⁹ Malin, 2001, p. 219. As well, 'Plotinus says that we *are* soul' (Corrigan, 2005, p. 80). (As Corrigan notes, see I, 1, 13 (1-3).)

⁶⁵⁰ Heisenberg, 1972, p. 216.

⁶⁵¹ Gerson writes that 'for Plato, the falsity of materialism establishes the identity of the knowable as immaterial. Then, assuming that knowledge is at least possible, the way is open for an argument that it is only possible for a knower who is also immaterial' (Gerson, 2005-A, p. 268). Whittaker notes that the soul can only be described by means that are nonsensical to materialists. But 'within the soul, [Plotinus] finds all the metaphysical principles in some way represented. In it are

included the principles of unity, of pure intellect, of moving and vitalising power, and, in some sense, of matter itself' (Whittaker, 1928, pp. 42-43). Despite the soul's ability to turn inwardly and intuit (feel, recognize, discover, or remember) intellect and absolute unity, these metaphysical principles are metaphysically prior to the soul.

[652] Norris, 2002-A, p. 50.

[653] See Rosán, 1949, p. 164.

[654] Gödel quoted in Wang, 2001, p. 316. The complete list of Gödel's fourteen philosophical theses are as follows:
1. The world is rational.
2. Human reason can, in principle, be developed more highly (through certain techniques).
3. There are systematic methods for the solution of all problems (also art, etc.).
4. There are other worlds and rational beings of a different and higher kind.
5. The world in which we live is not the only one in which we shall live or have lived.
6. There is incomparably more knowledge a priori than is currently known.
7. The development of human thought since the Renaissance is thoroughly intelligible (*durchaus einsichtige*).
8. Reason in mankind will be developed in every direction.
9. Formal rights comprise a real science.
10. Materialism is false.
11. The higher beings are connected to the others by analogy, not by composition.
12. Concepts have an objective existence.
13. There is a scientific (exact) philosophy and theology, which deals with concepts of the highest abstractness; and this is also most highly fruitful for science.
14. Religions are, for the most part, bad—but religion is not.

[655] Wang, 2001, p. 319. We are fortunate that Wang has shared his experience with Gödel with the reader.

[656] Biologist W. H. Thorpe noted that if we had taken seriously Wittgenstein's belief that 'there may be no deep structure', then 'subatomic physics, to say the least, would have dwindled and died' (Thorpe, 1978, p. 2). Barnes adds that Wittgenstein's 'notion of elucidatory nonsense is one that only a very subtle mind in a very stupid moment could have conceived' (Barnes quoted in Harris, 2000, p. 61).

[657] Wang, 2001, p. 318.

[658] Gödel summarized by Wang in Wang, 2001, p. 316.

[659] Nonlocal correlations could possibly be explained by the power of the universal soul to be in communication with itself through its own self-knowledge as it animates the entire universe.

[660] Whittaker, 1928, pp. 225-226.

[661] Wedberg notes that 'Plato envisages the possibility that the theorems of the mathematical sciences may obtain a foundation in Dialectic [Republic 510c-511e, 533a-d]. Thus, it seems that, in its final stage, Dialectic is here a deductive science which is the logical basis for the entire field of rational knowledge and which derives all its conclusions from a first self-evident principle, expressing the supreme insight into the Idea of the Good' (Wedberg, 1955, p. 44).

662 See Whittaker, 1928, pp. 53-54, for a further explanation of certain aspects of Plotinus' metaphysics that are most relevant here. See also Forrester (1972) and Dodds (1928).

663 Proclus, 1963, p. 163, *The Elements of Theology*, prop. 187.

664 Proclus, 1963, p. 167, *The Elements of Theology*, prop. 191. There are also 'divine souls'; see props. 201-205.

665 Einstein, 1954, p. 11. Einstein rejected what could be referred to as a naïve understanding of 'a God who rewards and punishes his creatures, or has a will of the kind that we experience in ourselves', and he thought that only 'feeble souls' would cherish the idea of an individual surviving his bodily death. Platonism, however, holds a very sophisticated view of God and the soul, one that is implied by Einstein's other metaphysical beliefs.

666 Penrose, 2004, p. 1029. See also Clark: 'Being and Intellect and Beauty are the same' (Clark, 1998, p. 5), and also Rist, 1977.

667 Eddington, 1935, p. 324. Philosopher John Norris (1657-1711) also held a similar view. He followed Malebranche 'in asserting that humans have not only empirical knowledge of nature but also direct access to ideas in the divine mind' (Nartonis, 2005, p. 437).

668 As Gerard O'Daly states, for Plotinus 'the self is not a static *datum*, even if it exists potentially in its entirety: it is essentially a faculty of conscious self-determination, a *mid-point* which can be directed towards the higher or the lower' (O'Daly, 1973, p. 49, emphasis added on 'mid-point'). Proclus, too, argues that 'every soul is intermediate between the indivisible principles and those which are divided in association with bodies' (Proclus, 1963, p. 167, *The Elements of Theology*, prop. 190).

669 Galileo in Koyré, 1968-B, pp. 13, 42 and in Clark, 1990, p. 37.

670 'The basic character of the [Neoplatonic] metaphysical system was dynamic. Proclus saw things in continuous procession and return, driven by power' (Siorvanes, 1996, p. 42-43). The essence of this universal power is terribly misunderstood by most of us, but we will have to resist exploring this topic further here.

671 Plato calls 'divine mania' the greatest gift when sent from the gods (Plato, 2001, p. 465, *Phaedrus*, 244a).

WORKS CITED

Adams, G. P. and Montague, W. M. P. (Eds.) (1930). *Contemporary American Philosophy: Personal Statements, Vol. II.* London: George Allen & Unwin Ltd; NY: The MacMillan Company.

Adams, J. and Spencer, J. H. 'Conference Report: Philosophy as a Way of Life — Rediscovering the Past for a Better Future', *Practical Philosophy*, Aug 2005, Vol. 7 No. 2.

———. (Eds.) 'Editorial Introduction — The Philosophical Life: Deep Reflections and Practical Applications', *Practical Philosophy*, Special Edition, Vol. 8. No. 2 Winter 2007.

Addey, T. and Wyndham-Jones, G. (2003). *Beyond the Shadows: The Metaphysics of thePlatonic Tradition.* Frome, Somerset, UK: The Prometheus Trust.

Afshar, S. S., Flores, E., McDonald, K. F., and Knoesel, E. 'Paradox in Wave-Particle Duality', *Foundations of Physics*, Vol. 37, No. 2, February 2007.

Arieti, J. A. and Wilson, P. A. (2003). *The Scientific and the Divine: Conflict and Reconciliation from Ancient Greece to the Present.* Rowman & Littlefield Publishers, Inc.

Aristotle. (Trans. 1963). *The Philosophy of Aristotle* (Ed. and Intro. R. Bambrough) (Trans. J. L. Creed and A. E. Wardman). A Mentor Book from New American Library.

Asimov, I. (1967). *Science, Numbers, and I.* London: Rapp & Whiting.

Atmanspacher, H., et al. Hosted a conference on 'Wolfgang Pauli's Philosophical Ideas and Contemporary Science', Monte Verita, Ascona, Switzerland, May 20 - 25, 2007. Housed at: www.solid.ethz.ch/pauli-conference/index.htm [Last accessed January 27, 2012.]

Audi, R. 'Philosophy in American Life: The Profession, the Public, and the American Philosophical Association', *Proceedings and Addresses of the American Philosophical Association*, Vol. 72, No. 5. (May,1999), pp. 139-148.

———. (Ed.) (2001). *The Cambridge Dictionary of Philosophy* (2nd ed.). Cambridge University Press.

Austin, J. H. (1999). *Zen and the Brain: Toward an Understanding of Meditation and Consciousness.* Cambridge, Massachusetts: The MIT Press.

Ayer, A. J. (1952). *Language, Truth and Logic* (2nd ed.). Dover Publications.

Ayto, J. (1990). *Dictionary of Word Origins.* NY: Arcade Publishing.

Babbage, C. (1838). *The Ninth Bridgwater Treatise: a Fragment* (2nd ed.). London: John Murray, Albemarle Street.

Baggini, J. and Fosl, P. S. (2003). *The Philosopher's Toolkit: A Compendium of Philosophical Concepts and Methods*. Blackwell Publishing.

Baggott, J. (2004). *Beyond Measure: Modern Physics, Philosophy and the Meaning of Quantum Theory*. Oxford University Press.

Baillie, J. (1997). *Contemporary Analytic Philosophy*. Upper Saddle River, NJ: Prentice Hall.

Balaguer, M. (1998). *Platonism and Anti-Platonism in Mathematics*. Oxford University Press.

Banner, M. C. (1990). *The Justification of Science and the Rationality of Religious Belief*. Oxford. Clarendon Press.

Barbour, I. G. (1966). *Issues in Science and Religion*. SCM Press LTD.

———. (1997). *Religion and Science: Historical and Contemporary Issues* (revised). HarperSanFrancisco.

Barr, S. M. (2003). *Modern Physics and Ancient Faith*. University of Notre Dame.

———. 'The Devil's Chaplain Confounded', *First Things: The Journal of Religion and Public Life*, August/September 2004. http://www.firstthings.com/article/2007/09/001-the-devils-chaplain-18 [Last accessed January 27, 2012.]

Baudrillard, J. http://www.egs.edu/faculty/jean-baudrillard/articles/radical-thought/#text9 [Last accessed January 27, 2012.] [Original source given as: 'Radical Thought' (Trans. F. Debrix) in *Collection Morsure* (Eds. Sens and Tonka). Paris, 1994.]

BBC. 'Pluto fate to be decided by vote', 24 August 2006. [No author listed.] http://news.bbc.co.uk/2/hi/science/nature/5281468.stm [Last accessed January 27, 2012.]

Beauregard, M. (2012). *Brain Wars: The Scientific Battle Over the Existence of the Mind and the Proof that Will Change the Way We Live Our Lives*. HarperOne.

Beck, S. (1959). *The Simplicity of Science*. Penguin Books Ltd.

Becker, W. M., Reece, J. B., and Poenie, M. F. (plus other contributors) (1996). *The World of the Cell* (3rd ed.). The Benjamin/Cummings Publishing Company.

Beer, J. 'Coleridge, Samuel Taylor (1772—1834)', *Oxford Dictionary of National Biography*, Oxford University Press, Sept 2004; online edition, May 2007. http://www.oxforddnb.com/view/article/5888 [Last accessed January 27, 2012.]

Bell, D. and Cooper, N. (1990). *The Analytic Tradition: Meaning, Thought and Knowledge*, 'Philosophical Quarterly Monographs, Volume 1'. Basil Blackwell.

Benn, A. W. 'The Relation of Greek Philosophy to Modern Thought', *Mind*, Vol. 7, No. 25 (Jan., 1882), pp. 65-88.

Bergethon, P. R. (1998). *The Physical Basis of Biochemistry: The Foundations of Molecular Biophysics*. NY: Springer.

Bergman, M. et al. (1980). *The Logic Book*. Random House.

Bergvall, A. 'Reason in Luther, Calvin, and Sidney', *The Sixteenth Century Journal*, Vol. 23, No. 1 (Spring, 1992), pp. 115-127.

Berkeley, G. (1992). *Philosophical Works: including the works on vision* (Intro and Notes: M. R. Ayers). London: J. M Dent & Sons Ltd; Rutland, Vermont: Charles E. Turtle Co., Inc.

Berlin, I. (1957). *The Great Ages of Western Philosophy IV: The Age of Enlightenment—The Eighteenth Century Philosophers* (Ed. I Berlin.). NY: Georger Braziller, Inc.

Bernal, M. (1987). *Black Athena: the Afroasiatic Roots of Classical Civilization* (Volume 1: The Fabrication of Ancient Greece 1785-1985). Rutgers University Press.

Bhaskar, R. (1978). *A Realist Theory of Science*. Sussex, UK: The Harvester Press.

Bhaskar, R. and Norris, C. Bhaskar Interview in 'Realism Under Attack: Five interviews with some of the world's leading thinkers on realism', *The Philosopher's Magazine Online*, Issue 8 Autumn 1999. Housed at the Website for Critical Realism: http://www.raggedclaws.com/criticalrealism/archive/rbhaskar_rbi.html [Last accessed January 27, 2012.]

Bible. (1992). *The New Student Bible, New International Version* (Notes. P. Yancey and T. Stafford). Grand Rapids, Michigan: Zondervan Publishing House.

Birkhoff, G. D. 'Intuition, Reason and Faith in Science', *Science*, New Series, Vol. 88, No. 2296 (Dec. 30, 1938), pp. 601-609.

Bissig, K., Stefan, F., Wieland, S. F., Phu Tran, P., Isogawa, M., Le, T. T., Chisari, F. V., and Verma, I. M. 'Human Liver Chimeric Mice Provide a model for Hepatitis B and C Virus Infection and Treatment', *J Clin Invest*, February 22, 2010.
http://www.jci.org/articles/view/40094?search[article_text]=human+liver+mouse&search[authors_text]= [Last accessed January 27, 2012.]

Blackburn, S. (no date given).
http://www.phil.cam.ac.uk/~swb24/reviews/Dawkins.htm [Last accessed January 27, 2012.]

Blackmore, J. 'Ernst Mach Leaves 'The Church of Physics'', *The British Journal for the Philosophy of Science*, Vol. 40, No. 4 (Dec., 1989), pp. 519-540.

Blackmore, J. T. 'Should we abolish the distinction between science and metaphysics?', *Philosophia*, Volume 12, Numbers 3-4, 393-400. 1983.

Blau, E. (1995). *Krishnamurti: 100 Years*. Stewart, Tabori, & Chang.

Blumenthal, H. and Markus, R. A. (Eds.) (1981). *Neoplatonism and Christian Thought: Essays in Honour of A. H. Armstrong*. London: Variorum Publications.

Bohm, D. (1980). *Wholeness and the Implicate Order*. Routledge.

Bohr, N. (1958). *Atomic Physics and Human Knowledge*. NY: John Wiley and Sons, Inc.

———. (1963). *Essays 1958-1962 on Atomic Physics and Human Knowledge*. Suffolk, UK. Richard Clay and Company, Ltd.

Bos, E. P. and Meijer, P. A. (Eds.) (1992). *On Proclus and his Influence in Medieval Philosophy*. E. J. Brill.

Bostrom, N. 'Are We Living in a Computer Simulation?', *The Philosophical Quarterly*, Vol. 53, No. 211, April 2003.

Boyd, R. N. 'Realism, Underdetermination, and a Causal Theory of Evidence', *Noûs*, Vol. 7, No. 1. (Mar., 1973), pp. 1-12.

Brading, K. and Castellani, E. (Eds.) (2003). *Symmetries in Physics: Philosophical Reflections*. Cambridge University Press.

Bradley, F. H. 'Faith', *The Philosophical Review*, Vol. 20, No. 2 (Mar., 1911), pp. 165-171.

Brittan, G. G. Jr. (1978). *Kant's theory of Science*. Princeton University Press.

Bronowski, J. (1961). *Science and Human Values*. London. Hutchinson & Co. (P*ublishers*) LTD.

Brooke, J. H. (1991). *Science and Religion: Some Historical Perspectives*. Cambridge University Press.

Brown, J. R. 'Peeking into Plato's Heaven', *Philosophy of Science*, Vol. 71, No. 5, Proceedings of the 2002 Biennial Meeting of the Philosophy of Science Association. Part II: Symposia Papers (Dec., 2004), pp. 1126-1138.

Brumbaugh, R. 'Aristotle as a Mathematician', *The Review of Metaphysics*, Vol. 8, No. 3 (Mar., 1955), pp. 379-393.

Buddha. (Trans. 1967). *The Dhammapada* (Trans. P. Lal). NY: Farrar, Straus & Giroux.

Bunge, M. 'Is Chemistry a Branch of Physics?', *Zeitschrift für allgemeine Wissenschaftstheorie / Journal for General Philosophy of Science*, Vol. 13, No. 2 (1982), pp. 209-223.

Bunnin, N. and Tsui-James, E. P. (Eds.) (2003). *The Blackwell Companion to Philosophy* (2nd ed.). Blackwell Publishing.

Burtt, E. A. (1925). *The Metaphysical Foundations of Modern Science: A Historical and Critical Essay*. London. Kegan Paul, Trench, Trubner & Co., LTD.

Capra, F. (1991). *Tao of Physics: an Exploration of the Parallels Between Modern Physics and Eastern Mysticism* (3rd ed.). Boston: Shambhala Publications.

Carnap, R. (1995). *The Unity of Science* (Trans. M. Black). Bristol. Thoemmes Press.

Cartwright, N. (1983). *How the Laws of Physics Lie*. New York: Oxford University Press.

———. 'The Metaphysics of the Disunified World', *PSA: Proceedings of the Biennial Meeting of the Philosophy of Science Association*, Vol. 1994, Volume Two: Symposia and Invited Papers (1994-A), pp. 357-364.

———. 'Fundamentalism vs. the Patchwork of Laws', *Proceedings of the Aristotelian Society, New Series*, Vol. 94 (1994-B), pp. 279-292.

Center for Consciousness Studies, The University of Arizona. http://www.consciousness.arizona.edu/ [Last accessed January 27, 2012.]

Chaisson, E. and McMillan, S. (1999). *Astronomy Today* (3rd ed.). NJ: Prentice Hall.

Clark, S. R. L. (1984). *From Athens to Jerusalem: the Love of Wisdom and the Love of God*. Oxford University Press.

———. (1986). *The Mysteries of Religion*. Blackwell.

———. (1990). *A Parliament of Souls*. Oxford: Clarendon Press.

———. 'The Possible Truth of Metaphor', *International Journal of Philosophical Studies*. March 1994; 2(1): 19-30.

———. (1995). 'Objective Values, Final Causes: Stoics, Epicureans, and Platonists', *The Electronic Journal of Analytic Philosophy*. http://ejap.louisiana.edu/EJAP/1995.spring/clark.abs.html [Last accessed January 27, 2012.]

———. 'A Plotinian Account of Intellect', *American-Catholic-Philosophical-Quarterly*. Sum 1997; 71(3): 421-432.

———. (1998). *God, Religion and Reality*. London: Society for Promoting Christian Knowledge.

———. (2000-A). *Biology and Christian Ethics*. Cambridge, UK: The Press Syndicate of the University of Cambridge.

———. 'Nothing Without Mind', *Consciousness Evolving* (Advances in Consciousness Research Vol. 34), ed. J. H. Fetzer, pp. 139-160, 2002. John Benjamins Publishing Company.

Clarke, S. (1998). *Metaphysics and the Disunity of Scientific Knowledge*. Ashgate.

Cleary, J. J. (1995). *Aristotle and Mathematics: Aporetic Method in Cosmology and Metaphysics*. E. J. Brill.

———. (Ed.) (1999). *Traditions of Platonism: Essays in Honour of John Dillon*. Ashgate.

Cobb, J. B. Jr. 'Faith', *Buddhist-Christian Studies*, Vol. 14 (1994), pp. 35-41.

Cohen, M. (Ed.) (2006). *The Essentials of Philosophy and Ethics*. London: Hodder Education.

Cohn, D. 'Does Socrates Speak for Plato? Reflections on an Open Question', *New Literary History*, The University of Virginia, Vol. 32.3 (2001), pp. 485-500.

Coldea, R., Tennant, D. A., Wheeler, E. M., Wawrzynska, E., Prabhakaran, D., Telling, M., Habicht, K., Smeibidl, P., and Kiefer, K. 'Quantum Criticality in an Ising Chain: Experimental Evidence for Emergent E8 Symmetry', *Science*, 8 January 2010: Vol. 327. No. 5962, pp. 177 - 180.

Collins, R. (1998). *The Sociology of Philosophies: A Global Theory of Intellectual Change*. The Belknap Press of Harvard University Press.

Computational Physics, The University of Groningen, the Netherlands. http://rugth30.phys.rug.nl/quantummechanics/intro.htm [Last accessed February 5, 2012.]

Cook, N. 'Crunch time for planet Pluto', 20 June 2006. http://news.bbc.co.uk/2/hi/science/nature/5099292.stm [Last accessed January 27, 2012.]

Corrigan, K. (2005). *Reading Plotinus: A Practical Introduction to Neoplatonism*. West Lafayette, Indiana: Purdue University Press.

Crease, R. P. 'This is your philosophy', *Critical Point*, Physicsworld.com., Apr 4, 2002. http://physicsworld.com/cws/article/print/5279 [Last accessed January 27, 2012.]

Cromwell, D. Zen and the art of Theories of Everything. *Science Tribune*, August 1997. Housed at: http://www.tribunes.com/tribune/art97/crom.htm [Last accessed January 27, 2012.]

Danforth, W. E. 'Einstein Stresses Faith as the Basis of Science', *The Science News-Letter, Vol. 29*, No. 781 (Mar. 28, 1936), pp. 203-204.

Daniels, Mark. 'What's New in…Ancient Philosophy', *Philosophy Now*, Issue 20, Spring 1998. Housed at: http://philosophynow.org/issues/20/Whats_New_in_Ancient_Philosophy [Last accessed January 27, 2012.]

Daniels, Michael. (2005). *Shadow, Self, Spirit*. Exeter, UK: Imprint Academic.

Dawkins, R. (1989). *The Selfish Gene*. Oxford University Press.

———. (1996). *River Out Of Eden: A Darwinian View Of Life* (Science Masters Series). Basic Books.

de Broglie, L. (Trans. 1955). *Physics and Microphysics* (Trans. M. Davidson). NY: Grosset & Dunlap.

Dennett, D. C. (1991). *Consciousness Explained*. USA: Little, Brown & Company Limited.

Dillon, J. M. (1977). *The Middle Platonists: a study of Platonism, 80 B.C. to A.D. 220*. London: Duckworth.

Dodds, E. R. 'The Parmenides of Plato and the Origin of the Neoplatonic One': *The Classical Quarterly*, Vol. 22, No. 3/4 (Jul.-Oct., 1928), pp. 129-142.

Douven, I. 'The Anti-Realist Argument for Underdetermination', *The Philosophical Quarterly*, Vol. 50, No. 200. (Jul., 2000), pp. 371-375.

Dummett, M. (1996). *Origins of Analytical Philosophy*. Harvard University Press.

Eaton, J. C. 'Berkeley's Immaterialism and the Scientific Promise of the Christian Doctrine of Creation', *The Harvard Theological Review*, Vol. 80, No. 4 (Oct., 1987), pp. 431-447.

Ecklund, E. H. (2010). *Science vs. Religion: What Scientists Really Think*. OUP.

Eddington, A. (1929). *Science and the Unseen World*. London. George Allen & Unwin LTD.

———. (1935). *The Nature of the Physical World*. London: J. M. Dent & Sons LTD.

Einstein, A. (1950-A). *Out of My Later Years*. Norwich: Jarrold and Sons Limited.

———. (1950-B). 'On the Generalized Theory of Gravitation: An account of the newly published extension of the general theory of relativity against its historical and philosophical background', *Scientific American*, April, 1950, Vol. 182, No. 4, p. 13.

———. (Trans. 1954). *Ideas and Opinions* (Ed. C. Seelig) (Trans. S. Bargmann). London: Alvin Redman Limited.

Emerson, R. W. (1941). *Essays of Ralph Waldo Emerson: Including Essays, First and Second Series, English Traits, Nature* and *Considerations by the Way*. NY: A.S. Barnes & Company.

———. (2000). *The Essential Writings of Ralph Waldo Emerson*. NY: The Modern Library.

Epp. S. S. (2004). *Discrete Mathematics with Applications* (3rd ed.). Thompson: Brooks/Cole.

Faye, J. 'Copenhagen Interpretation of Quantum Mechanics', *Stanford Encyclopedia of Philosophy*, First published Fri May 3, 2002; substantive revision Thu Jan 24, 2008. http://plato.stanford.edu/entries/qm-copenhagen/ [Last accessed January 27, 2012.]

Ferenstein, G. 'Steve Jobs Biographer: Apple Founder Was Driven By Simplicity, Mystical Thinking, And Occasional LSD Use', *Fast Company*, October 26, 2011. http://www.fastcompany.com/1790791/steve-jobs-biography-walter-isaacson [Last accessed April 06, 2012.]

Ferrero, M. and Van der Merwe, A. (Eds.) (1995). *Fundamental Problems in Quantum Physics*. Kluwer Academic Publishers.

Feynman, R. P. (1985). *QED: The Strange Theory of Light and Matter*. Princeton University Press.

Finamore, J. F. 'Qeoi Qewn: An Iamblichean Doctrine in Julian's Against the Galilaeans', *Transactions of the American Philological Association* (1974), Vol. 118 (1988), pp. 393-401.

Finch, H. (1977). *Wittgenstein—the Later Philosophy: an Exposition of the Philosophical Investigations*. Atlantic Highlands, NJ: Humanities Press.

WORKS CITED

Fine, A. (1996). *The Shaky Game: Einstein Realism and the Quantum Theory* (2nd ed.). The University of Chicago Press.

Force, J. E. and Popkin, R. H. (Eds.) (1999). *Newton and Religion: Context, Nature, and Influence*. Springer.

Forrester, J. W. M. 'Plato's Parmenides: The Structure of the First Hypothesis', *The Journal of the History of Philosophy*, Vol X, 1972.

Forsyth, T. M. 'The New Cosmology in Its Historical Aspect: Plato, Newton, Whitehead', *Philosophy*, Vol. 7, No. 25 (Jan., 1932), pp. 54-61.

Frank, P. (1941). *Between Physics and Philosophy*. Harvard University Press.

———. (Trans. 1947). *Einstein — His Life And Times* (Trans. G. Rosen). Alfred A. Knopf, Inc.

Frattaroli, E. (2001). *Healing the Soul in the Age of the Brain: Why Medication Isn't Enough*. Penguin Books.

Freud, S. (Trans. 1960). *Group Psychology and the Analysis of the Ego* (Trans. J. Strachey). Bantam Books.

Gersh, S. (1986). *Middle Platonism and Neoplatonism. The Latin Tradition. Publications in Medieval Studies*, 23. 2 vols. University of Notre Dame Press.

Gerson, L. P. 'What is Platonism?', *Journal of the History of Philosophy* 43 (2005-A), 253-276.

———. (2005-B) *Aristotle and other Platonists*. Cornell University Press.

Gibbins, P. (1989). *Particles and Paradoxes: The Limits of Quantum Logic*. Cambridge University Press.

Giles, J. 'The No-Self Theory: Hume, Buddhism, and Personal Identity', *Philosophy East and West*, Vol. 43, No. 2 (Apr., 1993), pp. 175-200.

Gödel, K. (1990). *Kurt Gödel: Collected Works, Volume II, Publications 1938-1974*. (Eds. S. Feferman et al.).

Goldstein, J. S. and Pevehouse, J. C. (2006). *International Relations* (7th Ed.). Pearson Longman.

Goswami, A. (1995). *The Self-Aware Universe: How Consciousness Creates the Material World*. NY: Jeremy P. Tarcher / Putnam.

Greene, W. C. 'Platonism and Its Critics', *Harvard Studies in Classical Philology*, Vol. 61 (1953), pp. 39-71.

Grimes, P. and Uliana, R. L. (1998). *Philosophical Midwifery: A New Paradigm for Understanding Human Problems; With its Validation*. Costa Mesa, CA: Hyparxis Press.

Grof, S. (2000). *Psychology of the Future: Lessons from Modern Consciousness Research* (Suny Series in Transpersonal and Humanistic Psychology). NY: State University of New York.

Hacking, I. (1983). *Representing and Intervening: Introductory topics in the philosophy of science*. Cambridge University Press.

Hadot, P. (1995). *Philosophy as a Way of Life*. Malden, MA: Blackwell Publishing company.

———. (2002). *What is Ancient Philosophy?* Cambridge, MA: The Belknap Press of Harvard University.

Halacy, D. (1979). *Charles Babbage: Father of the Computer*. Crowell-Collier Press.

Hammond, L. M. 'Plato on Scientific Measurement and the Social Sciences', *The Philosophical Review*, Vol. 44, No. 5 (Sep., 1935), pp. 435-447.

Hanh, T. N. (1995). *Living Buddha, Living Christ*. NY: Riverhead Books.

Hankey, W. *Participatio divini luminis*, Aquinas' doctrine of the Agent Intellect: Our Capacity for Contemplation, *Dionysius*, Vol. XXII, Dec. 2004.

Harris, E. (1983). *The Foundations of Metaphysics in Science*. University Press of America, Inc.

———. (2000). *The Restitution of Metaphysics*. NY. Humanity Books an imprint of Prometheus Books.

Hawking, S. (1988). *A Brief History of Time: From the Big Bang to Black Holes*. NY: Bantam Books.

Hawking, S. and Mlodinow, L. (2010). The Grand Design. NY: Bantam Books.

Hayward, J. W. and Varela, F. J. (Eds.) (2001). *Gentle Bridges: Conversations with the Dalai Lama on the Sciences of Mind*. Boston and London: Shambhala.

Helm, P. (2000). *Faith with Reason*. Oxford: Clarendon Press.

Heisenberg, W. (1958-A). *Physics and Philosophy: The Revolution in Modern Science*, NY: Harper and Row.

———. (1958-B). *The Physicists Conception of Nature* (Trans. A. J. Pomerans). Hutchinson & Co. (*Publishers*) LTD.

———. (Trans. 1972). *Physics and Beyond: Encounters and Conversations* (Trans. A. J. Pomerans). Harper Torchbooks.

———. (Trans. 1974). *Across the Frontiers* (Trans. P. Heath). Harper & Row, Publishers.

Hewitt-Horsman, C. and Vedral, V. 'Entanglement without nonlocality', *Physical Review A*, Volume 76, Issue 6, 2007. http://pra.aps.org/abstract/PRA/v76/i6/e062319. [Last accessed June 20, 2012.]

Higdon, A., Ohlsen, E. H., Stiles, W. B., Weese, J. A., and Riley, W. F. (1978). *Mechanics of Materials* (3rd ed.). John Wiley and Sons.

Hilgevoord, J. 'The Uncertainty Principle', *Stanford Encyclopedia of Philosophy*, First published Mon Oct 8, 2001; substantive revision Mon Jul 3, 2006. http://plato.stanford.edu/entries/qt-uncertainty/ [Last accessed January 27, 2012.]

Hoefer, C. and Rosenberg, A. 'Empirical Equivalence, Underdetermination, and Systems of the World', *Philosophy of Science*, Vol. 61, No. 4. (Dec., 1994), pp. 592-607.

Holton, G. 'Mach, Einstein, and the Search for Reality', *Daedalus*, Vol. 97, No. 2, Historical Population Studies (Spring, 1968) (pp. 636-673).

Horsten, L. 'Philosophy of Mathematics', *Stanford Encyclopedia of Philosophy*, Sep 25, 2007. http://plato.stanford.edu/entries/philosophy-mathematics/ [Last accessed January 27, 2012.]

Horstman, J. (2009). The Scientific American Day in the Life of Your Brain. San Francisco: Jossey-Bass (A Wiley Imprint).

Houshmand, Z., Livingston, R. B., and Wallace, B. A. (Eds.) (1999). *Consciousness at the Crossroads: Conversations with the Dalai Lama on Brain Science and Buddhism*. Ithaca, NY: Snow Lion Publications.

Huby, P. and Neal, G. (1989). *The Criterion of Truth: Essays in Honour of George Kerferd together with a text and translation (with annotations) of Ptolemy's* On the Criterion and Hegemonikon. Liverpool University Press.

Hughes, W. and Lavery, J. (2008). *Critical Thinking: an Introduction to the Basic Skills* (5th ed.). Broadview Press.

Hume, D. (1826). *The Philosophical Works of David Hume, Vol. 11*. London: Adam Black and William Tait.

———. (Edited and Revised 1978). *A Treatise of Human Nature* (Ed. L. A. Selby-Bigge, Revised P. H. Nidditch). Oxford at the Clarendon Press.

Hummel, C. E. (1986). *The Galileo Connection: Resolving Conflicts Between Science and theBible*. Intervarsity Christian Fellowship of the United States of America.

Huxley, A. (1946). *The Perennial Philosophy*. London: Chatto & Windus.

Hyland, G. J. and Rowlands, P. (Eds.) (2006). *Herbert Fröhlich, FRS: a Physicist Ahead of his Time*. The University of Liverpool.

Jaki, S. (1978). *The Road of Science and the Ways to God*. Scottish Academic Press.

———. 'A Late Awakening to Gödel in Physics.' *Sensus communis* 5 (2004), pp. 153-162. Housed at: http://www.sljaki.com/online.html [Last accessed February 4, 2012.]

James, W. (2009). *The Will to Believe: And Other Essays in Popular Philosophy*. BiblioBazaar.

Janiak, A. 'Newton's Philosophy', *Stanford Encyclopedia of Philosophy*, First published Fri Oct 13, 2006. http://plato.stanford.edu/entries/newton-philosophy/#ActDis. [Last accessed June 20, 2012.]

Jayne, S. 'Ficino and the Platonism of the English Renaissance', *Comparative Literature*, Vol. 4, No. 3 (Summer, 1952), pp. 214-238.

Jeans, J. (1930). *The Mysterious Universe*. Cambridge at the University Press.

Jewish Encyclopedia. (1906). Housed at http://www.jewishencyclopedia.com/articles/12116-philo-judaeus [Last accessed April 07, 2012.]

Jung, C. G. and Pauli, W. (Trans. 1955). *The Interpretation of Nature and the Psyche* [Jung: 'Synchronicity: An Acausal Connecting Principle' (Trans. R. F. C. Hull); Pauli: 'The Influence of Archetypal Ideas on the Scientific Theories of Kepler' (Trans. P. Silz).] London: Routledge & Kegan Paul.

Kant, I. (Trans. 1965). *Critique of Pure Reason* (Unabridged) (Trans. N. K. Smith). NY: St. Martin's Press.

Kahney, L. (Expanded Edition) (2009). *Inside Steve's Brain*. Portfolio (Penguin Group).

Katz, J. 'Plotinus and the Gnostics', *Journal of the History of Ideas*, Vol. 15, No. 2 (Apr., 1954), pp. 289-298.

Kepler, J. (Trans. 1997). *The Harmony of the World* (Trans. E. J. Aiton, A. M. Duncan and J. V. Field). American Philosophical Society.

Khan, A. 'Resveratrol researcher faked data, report says; what drives academic fraud?', *Los Angeles Times* (Jan. 2012)

Knobe, J. and Nichols, S. (Eds.) (2008). *Experimental Philosophy*. Oxford University Press.

Knoblich, G. and Oellinger, M. 'The Eureka Moment', *Scientific American*, October 04, 2006. Housed at: http://www.scientificamerican.com/article.cfm?id=the-eureka-moment&SC=I100322 [Last accessed January 27, 2012.]

Koyré, A. (1968-A). *From the Closed World to the infinite Universe*. Baltimore & London: The John Hopkins Press.

———. (1968-B). *Metaphysics and Measurement: Essays in Scientific Revolution*. London: Chapman & Hall.

Krauss, L. (2012). *A Universe from Nothing: Why There Is Something Rather than Nothing*. Free Press.

Krishna, G. (1993). *Living with Kundalini: The Autobiography of Gopi Krishna* (Ed. L. Shepard.). Boston, Massachusetts: Shambhala Publications, Inc.

Kuhn, T. S. (1970). *The Structure of Scientific Revolutions* (2nd ed.). The University of Chicago Press.

Ladyman, J. 'What is Structural Realism?', *Studies in History and Philosophy of Science*, 29, pp. 409-424, 1998.

———. 'Review: Discussion: Empiricism versus Metaphysics', (Reviewed work(s): The Empirical Stance by Bas C. van Fraassen), *Philosophical Studies: An International Journal for Philosophy in the Analytic Tradition*, Vol. 121, No. 2 (Nov., 2004), pp. 133-145.

Laird, J., Joad, C. E. M., and Stebbing, L. S. 'Symposium: Realism and Modern Physics', *Proceedings of the Aristotelian Society*, Supplementary Volumes, Vol. 9, Knowledge, Experience and Realism (1929), pp. 112-161.

Lancaster, B. 'Self or No-self? Converging Perspectives from Neuropsychology and Mysticism', *Zygon*, vol. 28, no. 4, December 1993.

Larson, E. J. and Witham, L. 'Leading scientists still reject God', *Nature* **394**, 313 (23 July 1998).

Laszlo, E. (2004-A). *Science and the Akashic Field: An Integral Theory of Everything*. Rochester, Vermont: Inner Traditions.

———. 'Why I Believe in Science and Believe in God: A Credo', *Journal of Religion and Science*, Vol. 39, No. 3, September 2004-B.

Laudan, L. 'Review Article of Koyré's 'Metaphysics and Measurement: Essays in the Scientific Revolution", *The British Journal for the Philosophy of Science*, 1969, pp. 180-181. The British Society for the Philosophy of Science, Oxford University Press.

Laudan, L. and Leplin, J. 'Empirical Equivalence and Underdetermination' *The Journal of Philosophy*, Vol. 88, No. 9. (Sep., 1991), pp. 449-472.

Laughlin, R. B. and Pines, D. 'The Theory of Everything', *Proceedings of the National Academy of Sciences of the United States of America*, Vol. 97, No. 1 (Jan. 4, 2000), pp. 28-31.

Lazcano, A. and Miller, S. L. 'How long did it take for life to begin and evolve to cyanobacteria?', *Journal of Molecular Evolution*, 1994, Volume 39, Number 6, 546-554.

Levinson, R. (1953). *In Defense of Plato*. Harvard University Press.

Liddell and Scott. (2004). *Greek-English Lexicon* (Abridged). Oxford at the Claredon Press.

Lindberg, D. C. (1992). *The Beginnings of Western Science: The European Scientific Tradition in Philosophical, Religious, and Institutional Context, 600 B.C. to A.D. 1450*. The University of Chicago Press.

Lindley, D. (1993). *The End of Physics: The Myth of a Unified Theory*. BasicBooks.

Lindsey, A. W. 'The Faith of Science', *The Scientific Monthly*, Vol. 66, No. 5 (May., 1948), pp. 395-398.

Lloyd, A. C. (1982). 'Procession and Division in Proclus' in *Soul and the Structure of Being in Late Neoplatonism: Syrianus, Proclus and Simplicius* (Ed. H. J. Blumenthal and A. C. Lloyd). Liverpool University Press.

Lloyd, G. E. R. 'Plato as a Natural Scientist', *The Journal of Hellenic Studies*, Vol. 88 (1968), pp. 78-92.

———. 'Saving the Appearances', *The Classical Quarterly*, New Series, Vol. 28, No. 1 (1978), pp. 202-222.

Lowry, J. M. P. (1980). *The Logical Principles of Proclus' ΣΤΟΙΧΕΙΩΣΙΣ ΞΕΟΛΟΓΙΚΗ as Systematic Ground of the Cosmos*. Amsterdam: Rodopi.

Maddy, P. 'Mathematical Existence', *The Bulletin of Symbolic Logic*, Vol. 11, No. 3, Sept. 2005. Housed at: http://www.lps.uci.edu/lps_bios/pjmaddy [Last accessed January 27, 2012.]

Malachowski, A. R. and Burrows, J. (Eds.) (1990). *Reading Rorty: Critical Responses to* Philosophy and the Mirror of Nature *(and Beyond)*. Basil Blackwell.

Malin, S. (2001). *Nature Loves to Hide: Quantum Physics and the Nature of Reality, a Western Perspective*. Oxford University Press.

Mandt, A. J. 'The Triumph of Philosophical Pluralism? Notes on the Transformation of Academic Philosophy', *Proceedings and Addresses of the American Philosophical Association*, Vol. 60, No. 2. (Nov.,1986), pp. 265-277.

Marinoff, L. (2002). *Philosophical Practice*. Academic Press.

Mattick, P. 'Marxism and the New Physics', *Philosophy of Science*, Vol. 29, No. 4 (Oct., 1962), pp. 350-364.

McFarlane, T. J. (Ed.) (2002). *Einstein and Buddha: The Parallel Sayings*. Berkeley, CA: Seastone.

McGinn, B. (1991). *The Foundations of Mysticism: Origins to the Fifth Century*. NY: Crossroad.

McGrath, A. 'God is not Dead yet', *The Times Higher Education Supplement*, 22 October 2004.

McGuire, J. E. and Rattansi, P. M. 'Newton and the 'Pipes of Pan''. *Notes and Records of the Royal Society of London 21* (1966):108-42.

McIntyre, L. C. (1996). *Laws and Explanation in the Social Sciences: Defending a Science of Human Behavior*. WestviewPress (A Division of HarperCollins*Publishers*).

Medawar, P. (1985). *The Limits of Science*. Oxford University Press.

Mermin, D. 'Is the moon there when nobody looks? Reality and the quantum theory', *Physics Today*, April 1985, 38-47.

Midgley, M. 'Gene-juggling', *Philosophy*, 54 (1979), pp. 439-458.

———. (1992). *Science as Salvation: A Modern Myth and its Meaning*. Routledge.

Mill, J. S. (2005). *A System of Logic: Ratiocinative and Inductive*. Elibron Classics.

Millar, A. 'Understanding Theism', *Religious Studies*, Vol. 17, No. 3 (Sep., 1981), pp. 311-321.

Minsky, M. (1986). *Society of Mind*. NY: Simon and Schuster.

Mohler, C. and Monton, B. 'Constructive Empiricism', *Stanford Encyclopedia of Philosophy*, First published Wed Oct 1, 2008. Housed at: http://plato.stanford.edu/entries/constructive-empiricism/ [Last accessed January 27, 2012.]

Mortensen, C. 'Plato's Pharmacy and Derrida's Drugstore' *Language and Communication* 20 (2000), pp. 329-346.

Morris, R. (1990). *The Edges of Science: Crossing the Boundary from Physics to Metaphysics*. NY: Prentice Hall Press.

Murphy, C. (2007). *Are We Rome? The Fall of an Empire and the Fate of America*. Boston and NY: A Mariner Book (Houghton Mifflin Company).

Nadeau, R. and Kafatos, M. (1999). *The Non-Local Universe: The New Physics and Matters of the Mind*. Oxford University Press.

Nagel, T. 'What Is It Like to Be a Bat?', *The Philosophical Review*, Vol. 83, No. 4 (Oct., 1974), pp. 435-450.

Nartonis, D. 'Louis Agassiz and the Platonist Story of Creation at Harvard, 1795-1846', *Journal of the History of Ideas*, Vol. 66, No. 3, July 2005. pp. 437-449.

National Institute of Standards and Technology. http://physics.nist.gov/cuu/Constants/index.html [Last accessed January 27, 2012.]

Newton, I. (Trans. 1999). *Isaac Newton: The Principia: Mathematical Principles of Natural Philosophy* (Trans. I. B. Cohen and A. Whitman). University of California Press.

Newman, M. 'Philosophers call for Liverpool v-c to reconsider department closure', *TimesHigher Education*, 30 March, 2009. www.timeshighereducation.co.uk/story.asp?storycode=405983 [Last accessed January 27, 2012.]

Nichols, R. (2006). 'Why is the History of Philosophy Worth Our Study?', *Metaphilosophy*, 37 (1), pp. 34-52.

Norris, C. (1992). *Uncritical Theory: Postmodernism, Intellectuals And The Gulf War*. Lawrence & Wishart Ltd.

———. (1993). *The Truth About Postmodernism*. Blackwell.

———. 'Ontology According to Van Fraassen: Some Problems with Constructive Empiricism', *Metaphilosophy*, Vol. 28, No. 3, July 1997.

———. 'Should Philosophers Take Lessons from Quantum Theory?', *Inquiry*, 42, 311-42, 1999.

———. (2000-A). *Quantum Theory and the Flight from Realism: Philosophical Responses to Quantum Mechanics*. Routledge.

———. 'Structure and Genesis in Scientific Theory: Husserl, Bachelard, Derrida', *British Journal for the History of Philosophy* 8(1) 2000-B.

———. 'Quantum Nonlocality and the Challenge to Scientific Realism', *Foundations of Science* 5: 3-45, 2000-C.

———. "Courage Not Under Fire': Realism, Anti-realism, and the Epistemological Virtues', *Inquiry*, 44, 269-90, 2001.

———. (2002-A). *Truth Matters; Realism, anti-realism and response-dependence*. Edinburgh University Press.

———. 'Ambiguities of the Third Way: Realism, Anti-Realism, and Response-dependence', *The Philosophical Forum*, Vol. XXXIII, No. 1, Spring 2002-B.

———. 'The Perceiver's Share: Realism, Scepticism, and Response Dependence', *Metaphilosophy*, Vol. 34, No. 4, July 2003.

Norton, J. D. 'Nature is the Realisation of the Simplest Conceivable Mathematical Ideas: Einstein and theCanon of Mathematical Simplicity, Stud. Hist. Phil. Mod. Phys., Vol. 31, No. 2, pp. 135-170, 2000.

O'Beirne, M. 'Home Office Research Study 274 — Religion in England and Wales: findings from the 2001 Home Office Citizenship Survey', *Home Office Research, Development and Statistics Directorate*, March 2004. Housed at: http://www.homeoffice.gov.uk/rds/pdfs04/hors274.pdf [Last accessed January 27, 2012.]

O'Brien, L. 'The GSP Workout: The Evolution of the Ultimate Fighter', *MensHealth*, 2012. Housed at: http://www.menshealth.com/celebrity-fitness/georges-st-pierre [Last accessed April 02, 2012.]

O'Daly, G. (1973). *Plotinus' Philosophy of Self*. Irish University Press.

Olson, R. G. (2004). *From Copernicus to Darwin: Science and Religion 1450 — 1900*. The John Hopkins University Press.

Opsomer, J. 'Proclus vs Plotinus on Matter ("De mal. subs." 30-7)', *Phronesis*, Vol. 46, No. 2 (May, 2001), pp. 154-188.

Oxford Dictionary of English. Edited by Angus Stevenson. Oxford University Press, 2010. *Oxford Reference Online*. Oxford University Press. University of British Columbia. 12 April 2012 <http://www.oxfordreference.com.ezproxy.library.ubc.ca/views/ENTRY.html?subview=Main&entry=t140.e0690160>

Oxtoby, D. W. and Nachtrieb, N. H. (1986) *Principles of Modern Chemistry*. NY & Philadelphia, PA: CBS College Publishing.

Penrose, R. (1997). *The Large, the Small, and the Human Mind*. Cambridge University Press.

———. (2004). *The Road to Reality: A Complete Guide to the Laws of the Universe*. London: Jonathan Cape.

Peters, F. P. (1967). *Greek Philosophical Terms: A Historical Lexicon*. New York UniversityPress.

PhysOrg.com. 'Sacred Constant Might be Changing', 11 April, 2005. http://www.physorg.com/news3665.html [Last accessed January 27, 2012.]

Philo. (Trans. 1941). *Philo IX* (Tans. F. H. Colson). Loeb Classical Library, Harvard University Press; London: William Heinemann LTD.

Pirsig, R. M. (1999). *Zen and the Art of Motorcycle Maintenance*. Vintage.

Planck, M. (1931). *The Universe in the Light of Modern Physics*. London: George Allen &Unwin LTD.

———. (1932). *Where is Science Going?*. NY: W. W. Norton * Company, INC. Publishers.

Plato. (Trans. 1956). *The Great Dialogues of Plato* (Trans. W. H. D. Rouse). NY: Signet Classic reprint (1999).

———. (Trans. 1996). *Plato: Theaetetus, Sophist* (Rev.) (Trans. H. N. Fowler). London: Loeb Classical Library, Harvard University Press.

———. (Trans.1997). *Plato: Complete Works* (Ed. J. M Cooper and Assoc Ed. D. S. Hutchinson). Indianapolis/Cambridge: Hackett Publishing Company.

———. (Trans. 2000). *Timaeus* (Trans. D. J. Zeyl). Cambridge: Hackett Publishing Company.

———. (Trans. 2001). *Plato: Euthyphro, Apology, Crito, Phaedo, Phaedrus* (Trans. H. N. Fowler). London: Loeb Classical Library, Harvard University Press.

Plotinus (Trans. 1964). *The Essential Plotinus* (Trans. E. O'Brien). Indianapolis, Indiana: Hackett Publishing Company, Inc.

———. (Trans. 1999). *Ennead III* (Trans. A. H Armstrong.) Loeb Classical Library, Harvard University Press.

———. (Trans. 1991). *Plotinus: The Enneads* (Trans. S. MacKenna) (Abridgement, Intro., and Notes J. Dillon). Penguin Books.

Polanyi, M. (1964). *Science, Faith and Society*. The University of Chicago Press.

———. (1966). *The Tacit Dimension*. London. Routledge & Kegan Paul Ltd.

Poli, R. 'The Basic Problem of the Theory of Levels of Reality,' *Axiomathes 12*, Nos. 3-4, pp. 261-283, 2001.

Pope Benedict XVI. 'Saint Augustine of Hippo (3)', *General Audience*, Paul VI Audience Hall Wednesday, 30 January 2008. http://www.vatican.va/holy_father/benedict_xvi/audiences/2008/documents/hf_ben-xvi_aud_20080130_en.html [Last accessed January 27, 2012.]

Pope Paul VI. 'Declaration on Religious Freedom (*Dignitatis Humanae*): On the Right of the Person and of Communities to Social and Civil Freedom in Matters Religious', *Vatican*, December 7, 1965. http://www.vatican.va/archive/hist_councils/ii_vatican_council/documents/vat-ii_decl_19651207_dignitatis-humanae_en.html [Last accessed January 27, 2012.]

Popkin, J. D. (Ed.). (2008). *The Legacies of Richard Popkin*. Springer.

Popper, K. R. (1979). *The Growth of Scientific Knowledge*. Vittorio Klostermann Frankfurt am Main.

Powell, A. 'A line on string theory', *physorg.com*, November 12, 2009. http://www.physorg.com/news177262216.html [Last accessed January 27, 2012.]

Price, H. H. (1969). *Belief*. London. George Allen & Unwin LTD.

Proclus. (Trans. 1816). *The Theology of Plato* (Trans. T. Taylor). Somerset, UK: Prometheus Trust edition, 1995.

———. (Trans. 1963). *The Elements of Theology* (2nd ed.) (Trans. E. R. Dodds). NY: Oxford University Press.

———. (Trans. 1970). *Proclus: A Commentary on the First Book of Euclid's* Elements (Trans. G. R. Morrow). Princeton University Press.

———. (Trans. 1987). Proclus' Commentary on Plato's Parmenides (Trans. G. R. Morrow and J. M. Dillon). Princeton, NJ: Princeton University Press.

Pseudo-Dionysius (Trans. 1987). *Pseudo-Dionysius: The Complete Works: The Classics of Western Spirituality* (Trans. C. Luibheid). Mahwah, NJ: Paulist Press.

Psillos, S. (1999). *Scientific Realism: How Science Tracks Truth*. Routledge.

Putnam, H. 'Sense, Nonsense, and the Senses: An Inquiry into the Powers of the Human Mind', *The Journal of Philosophy*, Vol. 91, No. 9. (Sep., 1994), pp. 445-517.

Quinn, P. (1996). *Aquinas, Platonism, and the Knowledge of God*. Avebury.

Radin, D. (1997). *The Conscious Universe: The Scientific Truth of Psychic Phenomena*. HarperEdge.

Ralkowski, M. A. (2009). *Heidegger's Platonism*. Continuum.

Rasmussen, W. 'Whose Platonism?', *International Journal of Hindu Studies*, Vol. 9, No. 1/3 (Jan., 2005), pp. 131-152.

Rappe, S. (2000). *Reading Neoplatonism: Non-discursive Thinking in the texts of Plotinus, Proclus, and Damascius*. Cambridge University Press.

Restak, R. M. (1984). *The Brain*. NY: Bantam Books.

Restivo, S. P. 'Parallels and Paradoxes in Modern Physics and Eastern Mysticism: I - A Critical Reconnaissance', *Social Studies of Science*, Vol. 8, No. 2 (May, 1978), pp. 143-181.

Rhodes, M. C. 'The Sense of the Beautiful and Apophatic Thought: Empirical Being as Ikon', *Zygon*, Vol. 42, No. 2 (June., 2007).

Rist, J. M. (1977). *Plotinus: The Road to Reality*. Cambridge University Press.

Robinson, T. M. 'Demiurge and World Soul in Plato's Politicus', *The American Journal of Philology*, Vol. 88, No. 1 (Jan., 1967), pp. 57-66.

Rogers, G. A. J. 'Locke, Newton, and the Cambridge Platonists on Innate Ideas', *Journal of the History of Ideas*, Vol. 40, No. 2 (Apr. - Jun., 1979), pp. 191-205.

Rogers, K (2005). *On the Metaphysics of Experimental Physics*. Palgrave Macmillan.

Rorty, R. (1979). *Philosophy and the Mirror of Nature*. Princeton University Press.

———. 'Pragmatism, Relativism, and Irrationalism', *Proceedings and Addresses of the American Philosophical Association*, Vol. 53, No. 6, (Aug., 1980), pp. 717-738.

———. 'Anti-essentialism in General: The Number Seventeen as a Model for Reality', *Legal Theory Workshop Series*, Faculty of Law, University of Toronto, January 12, 1990-A.

———. 'Feminism and Pragmatism', *The Tanner Lectures on Human Values*, delivered atUniversity of Michigan, December 7, 1990-B. Housed at: http://www.tannerlectures.utah.edu/lectures/atoz.html#r [Last accessed January 27, 2012.]

———. (1991). *Objectivity, Relativism and Truth*. Cambridge University Press.

———. 'Analytic Philosophy and Transformative Philosophy', Nov 10, 1999. Housed at: http://evans-experientialism.freewebspace.com/rorty02.htm [Last accessed January 27, 2012.]

Rosán, L. (1949). *The Philosophy of Proclus: The Final Phase of Ancient Thought*. NY: Cosmos.

Rosen, E. 'Was Copernicus A Neoplatonist?' *Journal of the History of Ideas*, Vol. 44, No. 4, Oct. - Dec., 1983), pp. 667-669.

Rosenthal, A. L. 'What Ayer Saw When He Was Dead', *Philosophy*, Vol. 79, No. 310 (Oct., 2004), pp. 507-531.

Ross, J. 'The Crash of Modal Metaphysics', *The Review of Metaphysics*, Vol. 43, No. 2 (Dec., 1989), pp. 251-279.

———. 'Together With the Body I Love' [abstract only], *Presidential Address for American Catholic Philosophical Association*, November 2001, Albany, New York. Housed online at: http://www.sas.upenn.edu/~jross/Essays1.htm [Last accessed January 27, 2012.]

———. (2008). *Thought and World: The Hidden Necessities*. University of Notre Dame Press.

Rowan University. 'Physicists Modify Double-Slit Experiment to Confirm Einstein's Belief', *physorg.com*, March 12, 2007. Housed at: www.physorg.com/news92937814.html[Last accessed January 27, 2012.]

Rowlands, P. (1992). *Waves versus Corpuscles: The Revolution that Never Was*. Liverpool, UK: PD Publications.

———. 'Why Does Physics Work?', *Hevelius*, 1, 76-103, 2003.

———. (2007). *Zero to Infinity: The Foundations of Physics*. World Scientific Publishing Company.

Russman, T. A. (1987). A Prospectus for the Triumph of Realism. Macon, GA: Mercer University Press.

Russell, B. (1903). *A Free Man's Worship*. Housed at Just Response: http://www.justresponse.net/Russell3.html [Last accessed January 27, 2012.]

———. (1968). *The Autobiography of Bertrand Russell 1914-1944 (Vol. II)*. London: George Allen and Unwin LTD.

———. (1997). *Religion and Science*. Oxford University Press.

Sachs, M. (1988). *Einstein versus Bohr: The Continuing Controversies in Physics*. La Salle, Illinois: Open Court.

Santillana, de. G. (1957). *The Great Ages of Western Philosophy II: The Age of Adventure — The Renaissance Philosophers*. NY: George Braziller.

Schäfer, L. (1997). *In Search of Divine Reality: Science as a Source of Inspiration*. The University of Arkansas Press.

Schiermeier, Q. 'Quantum Physics: The Philosopher of Photons': *Nature* **434**, 1066 (28 April 2005). Housed at: www.nature.com/nature/journal/v434/n7037/full/4341066a.html[Last accessed January 27, 2012.]

Schnädelbach, H. 'Transformations of the Concept of Reason', *Ethical Theory and Moral Practice*, Vol. 1, No. 1 (Mar., 1998), pp. 3-14.

School of Mathematics and Statistics, University of St. Andrews. http://turnbull.mcs.st-and.ac.uk/~history/ [Last accessed January 27, 2012.]

Schrödinger, E. (1952). *Science and Humanism: Physics in Our Time*. Cambridge at the University Press.

———. (1964). *My View of the World*. Cambridge University Press.

———. (1967). *What is Life? & Mind and Matter*. Cambridge at the University Press.

Schuré, E. (1923). *Pythagoras and the Delphic Mysteries* (revised edition). London: William Rider & Son, Limited.

Segonds, A. Ph. And Steel, C. (Eds.) (2000). Proclus et la Théologie Platonicienne.

Selleri, F. (Ed) (1988). *Quantum Mechanics versus Local Realism: The Einstein-Podolsky-Rosen Paradox*. NY and London: Plenum Press.

Shafer-Landau, R. (2003). *Moral Realism: A Defence*. Oxford Scholarship Online (via Dalhousie University library).

Sharp, J. A., Peters, J., and Howard, K. (2002). *The Management of a Student Research Project* (3rd ed.). Gower House, UK: Gower Publishing Limited.

Sheppard, A. 'Proclus' Attitude to Theurgy', *Classical Quarterly*, New Series 32 (i) 212-224, (1982).

Shorey, P. 'Platonism and the History of Science', Proceedings of the American Philosophical Society, 1927, pp. 159-181.

Shrimplin-Evangelidis, V. 'Sun Symbolism and Cosmology in Michelangelo's Last Judgment', The Sixteenth Century Journal, Vol. 21, No. 4 (Winter, 1990), pp. 607-644.

Sibelius, P. 'A Major Failure within Modern Analytic Philosophy', *Philosophy of Science*, Vol. 60, No. 4. (Dec., 1993), pp. 558-567.

Siorvanes, L. (1996). *Proclus: Neo-Platonic Philosophy and Science*. Edinburgh University Press.

Sloss, R. R. (2000). *Lives in the Shadow with J. Krishnamurti*. iUniverse.com, Inc. (an Authors Guild Backinprint.com edition).

Smith, C., Marks, A. D., Lieberman, M. A., and Marks, D. B. (2005). *Marks' Basic Medical Biochemistry: A Clinical Approach* (2nd ed.). Philadelphia, PA: Lippincott Williams and Wilkins.

Smith, S. R. 'Models and the Unity of Classical Physics: Nancy Cartwright's Dappled World', *Philosophy of Science*, Vol. 68, No. 4 (Dec., 2001), pp. 456-475.

Smolin, L. 'A Crisis in Fundamental Physics', *Academy Publications: The New York Academy of Sciences*, Jan/Feb 2006-A. http://www.nyas.org/publications/UpdateUnbound.asp?UpdateID=41#top [Accessed before January 2008, but last attempt on February 4, 2012 was unsuccessful.]

———. 'Never Say Always: Do the Laws of Nature Last Forever?', *New Scientist*, 23 September 2006-B.

Snobelen, S. D. '"God of Gods, and Lord of Lords": The Theology of Isaac Newton's General Scholium to the Principia', *Osiris*, 2nd Series, Vol. 16, Science in Theistic Contexts: Cognitive Dimensions (2001), pp. 169-208.

———. (2006). http://www.isaac-newton.org/ [Last accessed January 27, 2012.] [There are several relevant articles on this website.]

Snow, C. P. (1998). *The Two Cultures*. Cambridge University Press.

Snyder, L. J. 'William Whewell', *Stanford Encyclopedia of Philosophy*, First published Sat Dec 23, 2000; substantive revision Wed Oct 11, 2006. http://plato.stanford.edu/entries/whewell/ [Last accessed January 27, 2012.]

Sokal, A. and Bricmont, J. (2003). *Intellectual Impostures*. London: Profile Books

Solso, R. (2003) *The Psychology of Art and the Evolution of the Conscious Brain*. Massachusetts Institute of Technology.

Sommers, C. H. (1995). *Who Stole Feminism? How Women Have betrayed Women*. A Touchstone Book.

Sorrell, T. and Rogers, G. A. J. (Eds.) (2005). *Analytic Philosophy and History of Philosophy*. Oxford University Press.

Spencer, J. H. 'Defending Realism: Reflections on Karl Rogers's Metaphysics of Experimental Physics', *Journal of Critical Realism*, London: Equinox Publishing Ltd., Volume 6.1, 2007.

Sriraman, B. 'The Influence of Platonism on Mathematics Research and Theological Beliefs', *Theology and Science*, Vol. 2, No. 1, 2004.

Stark, R., Iannaccone, L. R., and Finke, R. 'Religion, Science, and Rationality': *The American Economic Review*, Vol. 86, No. 2, 1996, Papers and Proceedings of the Hundredth and Eighth Annual Meeting of the American Economic Association.

Stebbing, L. S. (1944). *Philosophy and the Physicists*. Penguin Books.

Stein, D. (1987). *Ada: A Life And A Legacy*. The MIT Press.

Steinkraus, W. E. (Ed.). (1966). *New Studies in Berkeley's Philosophy*. Holt, Rinehart and Winston, Inc.

Stenger, V. J. (1995). *The Unconscious Quantum: Metaphysics in Modern Physics and Cosmology*. NY: Prometheus Books.

Stern, P. 'Antifoundationalism and Plato's "Phaedo"', *The Review of Politics*, Vol. 51, No. 2 (Spring., 1989), pp. 190-217.

Stoljar, D. 'Physicalism', *Stanford Encyclopedia of Philosophy*, First published Tue Feb 13, 2001; substantive revision Wed Sep 9, 2009. http://plato.stanford.edu/entries/physicalism/ [Last accessed January 27, 2012.]

Stokes, M. (2010). *Christian Encounters Series: Isaac Newton*. Nashville, Tennessee: Thomas Nelson

Tait, W. W., 'Plato's Second Best Method', *The Review of Metaphysics*, Vol. 39, No. 3 (Mar., 1986), pp. 455-482.

Tanner, N. P. (Ed.). (1990). *Decrees of the Ecumenical Councils*: Volume Two *Trent to Vatican II*. Sheed & Ward and Georgetown University Press.

Taylor, A. E. (1936). *Elements of Metaphysics* (10th ed.). London. Methuen & Co. LTD.

Taylor, J. C. (2001). *Hidden Unity in Nature's Laws*. Cambridge University Press.

Taylor, T. (2006). *The Theoretic Arithmetic of the Pythagoreans*. Sturminster Newton, England: The Prometheus Trust.

The American Institute of Physics. http://www.aip.org/history/heisenberg/p08.htm [Last accessed January 27, 2012.]

The Philosopher's Magazine. 'Postmodernism R.I.P.', Issue 20, Autumn, 2002.

The Prometheus Trust. http://www.prometheustrust.co.uk [Last accessed January 27, 2012.]

Thondup, T., (1996). *The Healing Power of Mind: Simple Meditation Exercises for Health, Well-being, and Enlightenment.* Boston, Massachusetts: Shambhala Publications, Inc.

Thorpe, W. H. (1978). *Purpose in a World of Chance.* Oxford University Press.

Tilby, A. (1993). *Soul: God, Self and the New Cosmology.* Doubleday.

Tipler, F. J. (1994). *The Physics of Immortality: Modern Cosmology, God and the Resurrection of the Dead.* NY: Anchor Books.

Titius, A. 'Natural Science and the Christian Faith', *The Journal of Religion*, Vol. 11, No. 1 (Jan., 1931), pp. 20-29.

Toffler, A. (1990). *Power Shift: Knowledge, Wealth, and Violence at the Edge of the 21st Century.* Bantam Books.

Torre, Daleo, and García-Mata, 'The photon-box Bohr-Einstein debate demythologized,' *European Journal of Physics Online*, 21, 253-260 (May, 2000). http://www.iop.org/EJ/abstract/0143-0807/21/3/308/ [Last accessed January 27, 2012.]

Trefil, J. 'Relativity's Infinite Beauty', *Astronomy: Collector's Edition Cosmos*, Jan., 2007.

Trusted, J. (1991) *Physics and Metaphysics: Theories of Space and Time.* Routledge.

Turner, J. E. (1925). *A Theory of Direct Realism: and the Relation of Realism to Idealism.* London: George Allen & Unwin LTD.

Twist, J. 'Universe 'too queer' to grasp, BBC News, 12 July, 2005. http://news.bbc.co.uk/1/hi/sci/tech/4676751.stm [Last accessed January 27, 2012.]

UCD News. 'US Philosopher awarded UCD Ulysses Medal', 05 March 2007.http://www.ucd.ie/news/mar07/030507_Putnam_Award.htm [Last accessed January 27, 2012]

van der Meer, J. M. (Ed.) (1996). *Facets of Faith & Science. Volume I: Historiography and Modes of Interaction.* Lanham, Maryland. University Press of America.

van Fraassen, B. (1989). *Laws and Symmetry.* Oxford University Press.

———. (1991). *Quantum Mechanics: An Empiricist View.* Oxford University Press.

———. 'Constructive Empiricism Now', *Philosophical Studies: An International Journal for Philosophy in the Analytic Tradition*, Vol. 106, No. 1/2, Selected Papers Presented in 2000 at the 74th Annual Meeting of the Pacific Division of the American Philosophical Association (Nov., 2001), pp. 151-170.

———. (2002). *The Empirical Stance*. Yale University Press.

———. *The Scientific Image*. Oxford Scholarship Online, November 2003.

———. 'Commenting on Professor Haldane's Aquinas Lecture' [No actual title is given], *New Blackfriars*, Volume 80, Issue 938, 177-181. Published Online: 27 Jul 2007.

Vassilopoulou, P. 'Book Review of James Wilberding's *Plotinus' Cosmology: A Study of Ennead II.1 (40). Text, Translation, and Commentary*', *Journal of the History of Philosophy*, 47:1 January 2009, pp. 133-134.

Vatican. 'The Successor of Peter Teaches Infallibly', *General Audience*, March 17, 1993.http://www.vatican.va/holy_father/john_paul_ii/audiences/alpha/data/aud19930317en.html [Last accessed January 27, 2012.]

Vine, W. E. (1952). *An Expository Dictionary of New Testament Words with their Precise Meanings for English Readers*. London/Edinburgh: Oliphants LTD.

Vision, G. (1988). *Modern Anti-Realism and Manufactured Truth*. Routledge.

von Staden, H. 'Anatomy as Rhetoric: Galen on Dissection and Persuasion', *Journal of the History of Medicine and Allied Sciences*, Oxford Journals, 1995, 50: 47-66.

Wagner, M. F. (Ed.) (2002). *Neoplatonism and Nature: Studies in Plotinus Enneads*. State University of New York Press.

Waite, R. G. L. (1977). *The Psychopathic God: Adolf Hitler*. NY: Basic Book, Inc., Publishers.

Wallace, B. A. (1993). *Tibetan Buddhism from the Ground Up: A Practical Approach for Modern Life*. Somerville, Massachusetts: Wisdom Publications.

———. (1996). *Choosing Reality: A Buddhist View of Physics and the Mind*. Ithaca, NY: Snow Lion Publications.

Walsh, W. H. (1963). *Metaphysics*. London: Hutchinson.

Wang, Hao. 2001. *A Logical Journey: From Gödel to Philosophy*. The MIT Press.

Ward, J. Review of 'Kant's Platonic Revolution in Moral and Political Philosophy' by T. K. Seung, *The Journal of Politics*, Vol. 58, No. 1. (Feb., 1996), pp. 280-282.

Ward, K. (1996). *God, Chance & Necessity*. Oxford: Oneworld Publications.

Watterson, B. (1995). *The Calvin and Hobbes Tenth Anniversary Book*. Kansas City: Andrewsand McMeel (A Universal Press Syndicate Company).

Wedberg, A. (1955). *Plato's Philosophy of Mathematics*. Stockholm: Almqvist & Wiksell.

Wertheimer, L. K. 'Finding my Religion', *The Boston Globe*, July 30, 2006. http://www.boston.com/news/globe/magazine/articles/2006/07/30/finding_my_religion/?page=3 [Last accessed January 27, 2012.]

Westfall, R. S. 'The Foundations of Newton's Philosophy of Nature', *The British Journal for the History of Science*, Vol. 1, No. 2 (Dec., 1962), pp. 171-182.

Wheelwright, P. (1985). *The Presocratics* (Ed. P. Wheelwright). NY: Macmillan Publishing Company.

Whitehead, A. N. (1948). *Essays in Science and Philosophy*. London. Rider and Company.

———. (1953). *Science and the Modern World*. Cambridge University Press.

Whitaker, A. (1996). *Einstein, Bohr and the Quantum Dilemma*. Cambridge University Press.

Whittaker, J. '"Timaeus" 27D 5 ff', *Phoenix*, Vol. 23, No. 2 (Summer, 1969), pp. 181-185.

Whittaker, T. (1928). *The Neo-Platonists: A Study in the History of Hellenism* (2nd ed.). Cambridge at the University Press.

Wigner, E. 'The Unreasonable Effectiveness of Mathematics in the Natural Sciences', *Communications in Pure and Applied Mathematics*, Vol. 13, No. I (February 1960). Also see: http://www.dartmouth.edu/~matc/MathDrama/reading/Wigner.html [Last accessed January 27, 2012.]

Wilber, K. (1984). *Quantum Questions: Mystical Writings of the World's Great Physicists*.Boulder, Colorado: Shambhala Publications, Inc.

Wilson, R. 'What Happens When Entire Departments Get the Ax', The Chronicle of Higher Education, Jul 18, 2009. Housed at http://chronicle.com/article/When-an-Entire-Department-Gets/44519/ [Last accessed February 4, 2012.]

Wittgenstein, L. (1918). *Tractatus Logico-Philosophicus*. Hypertext of the Ogden bilingual edition. Housed at: http://www.kfs.org/~jonathan/witt/tlph.html [Last accessed January 27, 2012.]

Yourgrau, P. (2005). *A World Without Time: The Forgotten Legacy of Godel and Einstein*. Allen Lane *an imprint of* Penguin Books.

Zeh, H. D. 'There are no Quantum Jumps, nor are there Particles!', *Physics Letters*, A172, 189* (1993). Housed at: www.rzuser.uni-heidelberg.de/~as3/no-quantum-jumps.pdf [Last accessed January 27, 2012.]

Zhmud, L. 'Plato as "Architect of Science"', *Phronesis* XLIII/3, 1998, pp. 211-244.

Zukav, G. (1980). *The Dancing Wu Li Masters: An Overview of the New Physics*. Bantam.

INDEX

't Hooft, Gerard, 94
Abstractionism, 163, 164, 165, 166, 168, 169, 171, 202
Agassi, Joseph, 107
Anti-essentialism, 158, 159, 161
Antirealism, 237, 243
  Antirealism & abstractionism, 165, 168
  Antirealism & quantum theory, 52, 131, 140, 142, 143, 147, 151
  Creation of reality, 1, 55, 140, 151, 197, 232
  Extreme antirealism, 152, 154, 158
  Qualified antirealism, 22, 134, 232
  Rejection of objective reality, 2, 90, 140, 143, 151, 155, 232
  Skepticism of ontology, 130, 223
Anti-representationalism, 158, 160
Archimedes, 38, 133
Aristotle, 24, 25, 42, 45, 48, 49, 50, 68, 163, 188, 200, 202, 235, 240, 241, 245, 277, 285
Atheism
  Atheism & faith, 15, 89, 116, 226
  Atheism & materialism, 38
  Atheistic fundamentalism, 170
Atkins, Peter, 97
Audi, Robert, 66
Babbage, Charles, 177, 178
Baggott, Jim, 35, 36, 96, 97, 99, 118, 130, 139, 144, 196, 197, 207, 247, 255, 285
Baillie, James, 66, 68, 251
Balaguer, Mark, 239
Barbour, Ian, 75, 95, 120, 154, 207, 208, 209, 213, 263
Barr, Stephen, 83, 116, 238
Baudrillard, Jean, 141
Beauty, 254, 287
  Absolute beauty, 3, 5, 31, 45, 129, 133, 170, 202, 203, 278
  Beauty & morality, 84
  Beauty & symmetry, 83

Beauty in physics, 14, 49, 85, 203
Beauty, hierarchical place of, 31, 203
Berkeley, George, 80, 252, 253, 285
Bhaskar, Roy, 132, 263, 266
Bohm, David, 75, 76, 77, 90, 112, 114, 115, 123, 147, 150, 195, 197, 207, 208, 238, 268, 270
Bohr, Niels, 36, 39, 40, 41, 59, 64, 75, 76, 79, 87, 118, 131, 132, 133, 135, 137, 138, 139, 142, 144, 145, 151, 155, 195, 196, 207, 208, 210, 215, 216, 237, 238, 242, 265, 268, 270, 277, 284
Brahe, Tycho, 111
Buddhism, 75, 154, 155, 194, 212, 252, 272, 273
Burtt, Edwin, 49, 54, 241, 256, 285
Capra, Fritjof, 24
Carnap, Rudolph, 63, 67, 71, 246
Cartwright, Nancy, 22, 93, 185, 186, 187, 190, 192, 268, 281
Catholicism, 99, 156, 219, 220, 226, 227
Chaos, 3, 10, 217
Clark, Stephen R. L., 78, 85, 109, 111, 132, 160, 162, 185, 225, 242, 245, 258, 261, 263, 267, 276, 277, 287
Clarke, Steve, 185, 187, 190, 192
Coleridge, Samuel Taylor, 84, 254, 281
Collins, Randall, 62, 63, 66, 247, 248, 250
Complementarity, principle of, 40, 41, 42, 76
Comte, Auguste, 54, 65
Consciousness, 244, 251, 270, 281, 282
  Consciousness in quantum theory, 212
  Immaterial consciousness, 74, 197
  Role of the observer, 196, 197
  Universal/pure consciousness, 43, 80, 123, 202

Copenhagen interpretation, the, 35, 36, 40, 42, 65, 74, 99, 114, 119, 134, 144, 146, 152, 154, 196, 197, 237
Copernicus, Nicolaus, 15, 48, 49, 70, 109, 241
Corrigan, Kevin, 44, 90, 242, 285
Cottingham, John, 68
Creativity, 4, 12, 13, 85, 112, 174, 175, 258
Darwin, Charles, 283
Dawkins, Richard, 15, 36, 38, 89, 90, 117, 131, 170, 238, 262, 266, 275
De Broglie, Louis, 39, 109, 110, 207
Deductive reasoning, 105, 106, 107, 211, 259, 286
Dennett, Daniel, 32
Descartes, René, 91, 235, 275
Determinism, 32, 144, 208, 210, 211, 212, 213, 216
    Ideal determinism, 215, 216
    Strict determinism, 213, 215

Dickson, Dominic, 250
Dirac, Paul, 83
Direct perception, 84, 89, 90, 107, 111, 112, 113, 114, 123, 129, 166, 205
Dummett, Michael, 34, 63, 65, 69, 72, 78, 93, 159, 256
Eddington, Arthur, 55, 78, 79, 80, 103, 107, 114, 134, 151, 193, 197, 204, 244, 267, 283, 285, 287
Einstein, Albert, 238, 247, 256, 257, 260, 262, 263, 265, 266, 267, 268, 277, 281, 282, 283, 287
    Einstein & metaphysics, 17, 83, 88, 89, 111, 112, 170, 179, 207
    Einstein & mysticism, 9, 111, 203
    Einstein & Platonism, 47, 48, 62, 94, 134, 190
    Einstein & quantum theory, 39, 40, 64, 78, 118, 144, 207

Einstein-Bohr debate, the, 130, 265
Emerson, Ralph Waldo, 44

Empiricism, 20, 47, 48, 65, 92, 104, 221, 222, 225
    Constructive empiricism, 219, 222, 226, 228
Epistemology, 18, 35, 54, 60, 62, 90, 130, 140
Erikson, Erik, 87
Eternal law, the, 3, 6, 10, 18, 25, 26, 27, 31, 48, 88, 124, 169, 170, 174, 175, 176, 178, 180, 182, 184, 185, 189, 190, 191, 198, 202, 203, 204, 280
Faith, 15, 17, 19, 20, 89, 98, 99, 100, 102, 106, 109, 111, 112, 116, 117, 118, 161, 170, 187, 205, 211, 219, 220, 223, 224, 225, 226, 227, 257, 258, 259
    Blind faith, 15, 100
    Scientific faith, 98, 99, 101, 102, 109, 111, 115, 117, 118, 119, 226

Feynman, Richard, 36
Fine, Arthur, 93, 265
Frank, Philipp, 26, 64, 66, 88, 89, 92, 256
Frattaroli, Elio, 16
Free will, 27, 211
Frege, Gottlob, 61, 62, 63, 65, 68, 72, 93, 114, 134, 250, 251
Freud, Sigmund, 11
Fröhlich, Herbert, 56, 84
Fundamentalism, 2, 15, 242
Galileo, 15, 36, 48, 49, 109, 204, 287
Gerson, Lloyd, 42, 44, 45, 46, 51, 52, 188, 189, 199, 240, 282, 285
Gibbins, Peter, 28, 37, 39, 40, 42, 79, 94, 104, 128, 130, 209, 265, 270, 271
Gödel, Kurt, 71, 72, 78, 109, 123, 200, 201, 211, 249, 250, 260, 286
Goethe, Johann Wolfgang von, 84, 208
Good, the, 14, 51, 85, 94, 95, 202
Hacking, Ian, 73, 247, 248, 263, 268
Harris, Errol, 54, 55, 75, 81, 191, 192, 193, 195, 197, 216, 217, 251, 276, 282, 283, 284, 286

INDEX

Hawking, Stephen, 3, 50, 60, 61, 81, 93, 130, 245
Heidegger, Martin, 152, 271
Heisenberg, Werner, 18, 24, 31, 36, 39, 40, 59, 64, 66, 81, 83, 85, 86, 87, 93, 102, 104, 105, 113, 115, 118, 131, 133, 138, 142, 147, 148, 149, 150, 151, 155, 179, 185, 190, 196, 199, 202, 209, 210, 238, 247, 254, 255, 256, 260, 261, 263, 266, 268, 269, 270, 271, 281, 282, 283, 284, 285
   Heisenberg matrix mechanics, 39, 181
   Heisenberg uncertainty principle, 40, 82, 95, 113, 150, 207, 213, 214, 216, 257

Helm, Paul, 258
Hidden-variables interpretation, the, 40, 41, 74, 208, 238, 239, 268
Hierarchies, 31, 141
Hume, David, 115, 116, 117, 127, 237, 256, 261, 273, 279, 281, 285
Huxley, Aldous, 66, 80
Hylton, Peter, 246, 250
Idealism, 31, 32, 51, 141, 243
   Ancient idealism, 51, 141, 242
   Modern idealism, 51, 141, 242

Immaterialism, 33, 130, 131
Indeterminism, 130, 209, 210, 212, 214, 216
Induction, 259
Inductive reasoning, 106, 107, 108, 165, 177, 211
Instrumentalism, 160
Interdisciplinarity, 27, 66, 76, 82, 85, 86, 253
Intuition, 17, 54, 55, 57, 63, 77, 102, 107, 111, 112, 113, 114, 205, 258
   Intuitive type, 84

Isaacson, Walter, 29
Jaki, Stanley, 64, 83, 88, 116, 256
James, William, 98, 109, 129, 159
Jeans, James, 80, 101, 253

Jobs, Steve, 29
Jung, Carl, 15, 198, 203, 208
Justice, 45, 135, 212
Kant, Immanuel, 50, 68, 224, 237, 245, 263, 275, 285
Kepler, Johannes, 15, 45, 48, 49, 50, 109, 111, 117, 265
Koyré, Alexandre, 50, 287
Krauss, Lawrence, 245, 246
Krishnamurti, J., 13, 66, 147, 270
Kuhn, Thomas, 173, 248
Law of physics, 7, 30, 34, 43, 51, 101, 127, 163, 169, 175, 176, 178, 192
Lee, Bruce, 13
Lindley, David, 100, 116, 152
Linguistic turn, the, 59, 72, 73
Logic, 264
Logos, 17, 18, 19, 166, 203, 204, 217, 236
Mach, Ernst, 64, 65, 66, 88, 89, 104, 247, 271, 281
Malin, Shimon, 193, 197, 199, 236, 283, 285
Many-worlds interpretation, the, 23, 40
Materialism, 32, 74, 135, 191, 211, 236, 237, 243
   Materialism & abstractionism, 169
   Materialism & classical physics, 37
   Materialism & quantum theory, 32, 39, 79, 130
   Materialism vs realism, 33, 140
   Rejection of materialism, 64, 143, 191

Mcginn, Bernard, 55
Medawar, Peter, 113, 258, 260
Metaphysics, 8, 53, 90, 220, 221, 256, 259
   Applied metaphysics, 91, 94, 95, 117
   Metaphysical reasoning, 19, 22, 33, 40, 53, 79, 88, 95, 96, 125, 132, 143, 149, 170, 212, 214
   Metaphysics & physics, 23, 51, 86, 96

Platonic metaphysics, 31, 52, 98, 190, 192
Presupposed metaphysics, 91, 92, 228
Pure metaphysics, 91, 92, 94, 95, 201, 202, 214, 227
Rejection of metaphysics, 71, 87, 88, 220, 224

Midgley, Mary, 102, 238
Mind, 285
  Mind & reality, 124, 125
  Philosophy of mind, 131
  Role in logical reasoning, 107
  Universal/pure mind, 123, 202, 204

Mlodinow, Leonard, 60
Moore, G. E., 63, 250
Mueller, Ian, 82, 253
Mysticism, 6, 7, 9, 22, 24, 44, 52, 55, 56, 57, 63, 77, 107, 115, 198, 236
  Mystical experience, 22, 23, 30, 56, 65, 113, 205, 225, 260

Neoplatonism, 44, 45, 50, 188, 239, 254, 256, 270, 278
New Age, 6
Newton, Sir Isaac, 15, 22, 37, 48, 49, 50, 84, 109, 110, 112, 134, 179, 208, 238, 241, 280, 281
  Newton & Platonism, 25, 241

Newton. Sir Isaac
  Newton & Platonism, 241

Nichols, Ryan, 69
Nonlocality, 40, 214, 238, 268
Norris, Christopher, 33, 66, 73, 74, 121, 124, 134, 199, 200, 208, 210, 219, 237, 238, 248, 263, 265, 267, 268, 269, 275, 276, 277, 286, 287
Occam's Razor, 171
One, the, 3, 31, 46, 50, 52, 76, 124, 155, 169, 170, 185, 189, 190, 191, 202, 203, 204, 205, 252, 254, 266, 274, 278, 284
Ontology, 35, 54, 90, 121, 140, 227

Order, 3, 10, 17, 22, 38, 52, 82, 83, 84, 96, 101, 109, 110, 111, 113, 141, 166, 169, 170, 171, 178, 179, 180, 187, 192, 199, 200, 215, 216, 217, 262
  Cosmic order, 17, 18, 19, 25, 37, 114, 133, 170, 180, 188, 199, 203, 227

Parmenides, 78, 188, 263, 285
Pauli, Wolfgang, 17, 25, 36, 40, 72, 84, 87, 118, 133, 142, 170, 180, 207, 208, 254, 255
  Pauli exclusion principle, 40

Penrose, Roger, 3, 14, 93, 94, 95, 130, 187, 203, 217, 218, 245, 284, 287
Pirsig, Robert, 13, 75
Planck, Max, 15, 16, 39, 66, 75, 86, 88, 89, 90, 93, 100, 101, 103, 109, 111, 112, 115, 117, 123, 133, 139, 156, 191, 207, 212, 254, 255, 257, 258, 269
Platonism, 42, 43, 155, 188
  Criticism of Platonism, 45, 155
  Mathematical Platonism, 46, 47, 49, 50, 62, 65, 104, 231
  Metaphysical foundation of Platonism, 46, 113, 179, 188
  Platonism & contemporary philosophy, 3, 62, 65, 73, 80
  Platonism & modern philosophy, 76
  Platonism & modern science, 3, 24, 33, 48, 49, 58, 190
  Platonism & unity, 43, 46, 147, 188, 190, 226
  Scientific Platonism, 47, 49, 104, 111, 135

Plotinus, 16, 42, 44, 45, 48, 83, 84, 90, 188, 193, 199, 242, 252, 254, 266, 270, 274, 285, 287
Polanyi, Michael, 81, 101, 115, 132, 141, 276
Popper, Karl, 129, 265

INDEX

Positivism, 8, 35, 54, 62, 63, 64, 65, 73, 79, 80, 81, 85, 87, 88, 89, 92, 93, 96, 104, 164, 200, 230, 239, 246, 256
  Logical positivism, 62, 63
Postmodernism, 140, 141
Power, 205, 206, 287
Pragmatism, 105, 159
Proclus, 36, 44, 50, 55, 82, 85, 86, 91, 135, 167, 188, 189, 191, 194, 200, 202, 203, 254, 264, 265, 266, 278, 281, 283, 285, 287
Psillos, Stathis, 86, 134, 263, 270
Putnam, Hilary, 69, 70, 249
Pythagoreanism, 42, 49, 83, 134, 188, 241, 254
Quantum theory, 40, 211, 255
  New quantum theory, 39, 265
  Old quantum theory, 39
  Quantum jump, 39, 262
  Quantum theory & metaphysics, 23, 87
  Quantum theory & Platonism, 24, 34, 45, 69
  Quantum theory & positivism, 8
Quine, Willard, 62, 67, 69, 70, 71
Rationalism, 47, 48
Realism, 1, 21, 31, 32, 33, 124, 132, 236, 237, 242, 243, 263, 272
  Abstract realism, 121, 127, 133, 135
  Broad realism, 121, 131, 132, 133, 267
  Factual realism, 121, 127, 136, 137
  Logical realism, 123
  Naïve realism, 34, 36, 104, 120, 263, 267
  Platonic realism, 33, 34, 36, 47, 48, 51, 52, 55, 65, 75, 81, 117, 119, 120, 121, 131, 141, 142, 143, 148, 154, 160, 163, 171, 178, 180, 189, 190, 224, 226, 230, 240, 242, 282
  Realism & faith, 98, 99, 102, 111
  Realism & idealism, 32

Realism & physics, 33, 41, 89, 96, 129, 130, 131, 133, 145, 150, 151, 180, 197
Scientific realism, 221
Structural realism, 134, 135, 136, 263
Reality, 121
  Objective reality, 1, 2, 3, 120, 131, 137, 201, 206
  Ultimate reality, 3, 5, 8, 25, 43, 52, 56, 82, 91, 147, 155, 185, 205, 227
Reason, 17, 18, 84, 111, 114, 170, 203, 217, 227
Reason-principle, 19, 166, 167, 169, 171, 203, 204, 279
Reification, 155
Relativism, 140, 141, 158, 243, 275, 276
Relativity, theory of, 23, 48, 55, 64, 70, 75, 83, 84, 95, 104, 195, 218, 249, 271
Religious beliefs, primary, 25
Religious beliefs, secondary, 25
Rogers, Karl, 152, 153, 154, 155, 227, 263, 271, 275
Rorty, Richard, 44, 45, 59, 67, 69, 70, 72, 156, 157, 158, 159, 160, 161, 219, 257, 275, 276, 277
Rowlands, Peter, 37, 39, 83, 110, 187, 191, 192, 195, 208, 260, 283, 284
Russell, Bertrand, 61, 62, 63, 65, 66, 68, 71, 72, 75, 79, 84, 107, 123, 166, 200, 246, 248, 250, 251, 256, 268
Sachs, Mendel, 79, 133
Santillana, Giorgio de, 38, 49, 241
Schäfer, Lothar, 25, 52, 115, 236, 243, 261, 274
Schlick, Moritz, 88, 89, 246
Schrödinger, Erwin, 15, 16, 18, 40, 45, 64, 75, 80, 81, 85, 86, 88, 114, 118, 119, 134, 193, 194, 197, 198, 203, 208, 255, 262, 268, 270, 282, 283, 284
  Schrödinger's equations, 40

Schrödinger's wave mechanics, 40, 181
Scientia, 26
Searle, John, 131
Self-awareness, 16, 194, 197, 283
Shorey, Paul, 48
Simplicity, 33, 39, 49, 171, 172, 179, 189, 190
Siorvanes, Lucas, 46, 47, 50, 135, 283, 287
Skepticism, 79, 115, 116
Smolin, Lee, 22, 93, 176, 177, 178, 227, 267
Socrates, 9, 42, 123, 129, 161, 188, 199, 206, 235, 247, 277
Sorrell, Tom, 59, 67, 275
Soul, 272, 273, 285, 286, 287
   Reality of soul, 15, 185
   Soul & quantum theory, 193
   Soul & self-awareness, 194
   Soul in Platonism, 48, 198, 199, 202, 203, 204, 205
Spirituality, 15, 16, 25
Stebbing, Susan, 80, 253
Stenger, Victor, 58
Symmetry, 31, 45, 49, 82, 83, 84, 133, 170, 172, 191, 203, 254, 262
Tautology, 72, 105, 250
Taylor, A. E., 54, 141, 142, 189, 190, 254, 263, 285
Technoculture, 28
Theurgy, 47, 240
Tipler, Frank, 26
Trefil, James, 83
Trusted, Jennifer, 53, 54, 243, 256
Truth
   Belief-independent truth, 34
   Coherence theory of truth, the, 103
   Correspondence theory of truth, the, 103, 159

Objective truth, 1, 2, 11, 66, 120, 127, 129, 131, 141, 151, 158, 160, 205, 206
Verification-transcendent truth, 33, 34, 111, 121, 125, 264
Underdetermination, 173, 184
Unity, 272, 281, 286
   Unity in physics, 49, 83, 84, 104, 116, 172, 173, 176, 189
   Unity of ultimate reality, 29, 75, 180
   Unity underlying complexity, 12, 44
Van Fraassen, Bas, 156, 219, 220, 221, 222, 223, 224, 225, 226, 227, 228, 275
Vienna Circle, the, 63, 64, 66, 78, 87, 248
Virtual reality, 137, 280
Vision, Gerald, 121, 156, 263, 269
Von Jolly, Philipp, 139, 269
Von Neumann, John, 40, 197, 238
Wallace, B. Alan, 154, 155, 156, 167, 272, 273, 274
Wang, Hao, 200, 201, 286
Ward, Keith, 32, 131
Wave/particle duality, 40, 42, 80, 104, 138
Weber, Renee, 285
Wheeler, John, 137
Whewell, William, 26, 260
Whitehead, Alfred North, 49, 55, 77, 115, 134, 236, 283
Whittaker, Thomas, 51, 202, 235, 255, 285, 286, 287
Wholeness, 75, 76, 84, 90, 147, 150, 185, 194, 195, 270
Wigner, Eugene, 51, 197
Wilber, Ken, 240, 244
Wittgenstein, Ludwig, 62, 66, 68, 71, 72, 159, 201, 245, 250, 251, 286
Worrall, John, 134
Yourgrau, Palle, 71, 78, 216, 247, 263